Probability and
Random Variables

A Beginner's Guide

This is a simple and concise introduction to probability theory. Self-contained and readily accessible, it is written in an informal tutorial style with a humorous undertone. Concepts and techniques are defined and developed as necessary. After an elementary discussion of chance, the central and crucial rules and ideas of probability, including independence and conditioning, are set out. Counting, combinatorics, and the ideas of probability distributions and densities are then introduced. Later chapters present random variables and examine independence, conditioning, covariance, and functions of random variables, both discrete and continuous. The final chapter considers generating functions and applies this concept to practical problems including branching processes, random walks, and the central limit theorem. Examples, demonstrations, and exercises are used throughout to explore the ways in which probability is motivated by, and applied to, real-life problems in science, medicine, gaming, and other subjects of interest. Essential proofs of important results are included.

Since it assumes minimal prior technical knowledge on the part of the reader, this book is suitable for students taking introductory courses in probability and will provide a solid foundation for more advanced courses in probability and statistics. It would also be a valuable reference for those needing a working knowledge of probability theory and will appeal to anyone interested in this endlessly fascinating and entertaining subject.

Probability and Random Variables

A Beginner's Guide

David Stirzaker

University of Oxford

CAMBRIDGE
UNIVERSITY PRESS

PUBLISHED BY THE PRESS SYNDICATE OF THE UNIVERSITY OF CAMBRIDGE
The Pitt Building, Trumpington Street, Cambridge, United Kingdom

CAMBRIDGE UNIVERSITY PRESS
The Edinburgh Building, Cambridge CB2 2RU, UK www.cup.cam.ac.uk
40 West 20th Street, New York, NY 10011-4211, USA www.cup.org
10 Stamford Road, Oakleigh, Melbourne 3166, Australia
Ruiz de Alarcón 13, 28014 Madrid, Spain

First published 1999

Printed in the United Kingdom at the University Press, Cambridge

Typeset in Times 10/12.5pt, in 3B2 [KT]

A catalogue record for this book is available from the British Library

Library of Congress Cataloguing in Publication data
Stirzaker, David.
Probability and random variables : a beginner's guide / David
Stirzaker.
p.   cm.
ISBN 0 521 64297 3 (hb)
ISBN 0 521 64445 3 (pb)
1. Probabilities.   2. Random variables.   I. Title.
QA273.S75343   1999
519.2–dc21   98-29586 CIP

ISBN 0 521 64297 3 hardback
ISBN 0 521 64445 3 paperback

Contents

Synopsis

This is a simple and concise introduction to probability and the theory of probability. It considers some of the ways in which probability is motivated by, and applied to, real-life problems in science, medicine, gaming, and other subjects of interest. Probability is inescapably mathematical in character but, as befits a first course, the book assumes minimal prior technical knowledge on the part of the reader. Concepts and techniques are defined and developed as necessary, making the book as accessible and self-contained as possible.

The text adopts an informal tutorial style, with emphasis on examples, demonstrations, and exercises. Nevertheless, to ensure that the book is appropriate for use as a textbook, essential proofs of important results are included. It is therefore well suited to accompany the usual introductory lecture courses in probability. It is intended to be useful to those who need a working knowledge of the subject in any one of the many fields of application. In addition it will provide a solid foundation for those who continue on to more advanced courses in probability, statistics, and other developments. Finally, it is hoped that the more general reader will find this book useful in exploring the endlessly fascinating and entertaining subject of probability.

On this occasion, I must take notice to such of my readers as are well versed in Vulgar Arithmetic, that it would not be difficult for them to make themselves Masters, not only of all the practical Rules in this book, but also of more useful Discoveries, if they would take the small Pains of being acquainted with the bare Notation of Algebra, which might be done in the hundredth part of the Time that is spent in learning to write Short-hand.

A. de Moivre, *The Doctrine of Chances*, 1717

Preface

This book begins with an introduction, chapter 1, to the basic ideas and methods of probability that are usually covered in a first course of lectures. The first part of the main text, subtitled Probability, comprising chapters 2–4, introduces the important ideas of probability in a reasonably informal and non-technical way. In particular, calculus is not a prerequisite.

The second part of the main text, subtitled Random Variables, comprising the final three chapters, extends these ideas to a wider range of important and practical applications. In these chapters it is assumed that the student has had some exposure to the small portfolio of ideas introduced in courses labelled 'calculus'. In any case, to be on the safe side and make the book as self-contained as possible, brief expositions of the necessary results are included at the ends of appropriate chapters.

The material is arranged as follows.

Chapter 1 contains an elementary discussion of what we mean by probability, and how our intuitive knowledge of chance will shape a mathematical theory.

Chapter 2 introduces some notation, and sets out the central and crucial rules and ideas of probability. These include independence and conditioning.

Chapter 3 begins with a brief primer on counting and combinatorics, including binomial coefficients. This is illustrated with examples from the origins of probability, including famous classics such as the gambler's ruin problem, and others.

Chapter 4 introduces the idea of a probability distribution. At this elementary level the idea of a probability density, and ways of using it, are most easily grasped by analogy with the discrete case. The chapter therefore includes the uniform, normal, and exponential densities, as well as the binomial, geometric, and Poisson distributions. We also discuss the idea of mean and variance.

Chapter 5 introduces the idea of a random variable; we discuss discrete random variables and those with a density. We look at functions of random variables, and at conditional distributions, together with their expected values.

Chapter 6 extends these ideas to several random variables, and explores all the above concepts in this setting. In particular, we look at independence, conditioning, covariance, and functions of several random variables (including sums). As in chapter 5 we treat continuous and discrete random variables together, so that students can learn by the use of analogy (a very powerful learning aid).

Chapter 7 introduces the ideas and techniques of generating functions, in particular probability generating functions and moment generating functions. This ingenious and elegant concept is applied to a variety of practical problems, including branching processes, random walks, and the central limit theorem.

In general the development of the subject is guided and illustrated by as many examples as could be packed into the text. Nevertheless, I have not shrunk from including proofs wherever they are important, or informative, or entertaining.

Naturally, some parts of the book are easier than others, and I would offer readers the advice, which is very far from original, that if they come to a passage that seems too difficult, then they should skip it, and return to it later. In many cases the difficulty will be found to have evaporated.

In general it is much easier and more pleasant to get to grips with a subject if you believe it to be of interest in its own right, rather than just a handy tool. I have therefore included a good deal of background material and illustrative examples to convince the reader that probability is one of the most entertaining and endlessly fascinating branches of mathematics. Furthermore, even in a long lecture course the time that can be devoted to examples and detailed explanations is necessarily limited. I have therefore endeavoured to ensure that the book can be read with a minimum of additional guidance.

Moreover, prerequisites have been kept to a minimum, and mathematical complexities have been rigorously excluded. You do need common sense, practical arithmetic, and some bits of elementary algebra. These are included in the core syllabus of all school mathematics courses.

Readers are strongly encouraged to attempt a respectable fraction of the exercises and problems. Tackling relevant problems (even when the attempt is not completely successful) always helps you to understand the concepts. In general, the exercises provide routine and transparent applications of ideas in the nearby text. Problems are often less routine; they may use ideas from further afield, and may put them in a new setting. Solutions and hints for most of the exercises and problems appear before the Index.

While all the exercises and problems have been kept as simple and straightforward as possible, it is inescapable that some may seem harder than others. I have resisted the temptation to magnify any slight difficulty by advertising it with an asterisk or equivalent decoration. You are at liberty to find any exercise easy, irrespective of any difficulties I may have anticipated.

It is certainly difficult to exclude every error from the text. I entreat readers to inform me of all those they discover.

Finally, you should note that the ends of examples, definitions, and proofs are denoted by the symbols $\bigcirc$, $\triangle$, and $\square$ respectively.

Oxford
January 1999

1
Introduction

I shot an arrow into the air
It fell to earth, I knew not where.

H.W. Longfellow

O! many a shaft at random sent
Finds mark the archer little meant.

W. Scott

1.1 PREVIEW

This chapter introduces probability as a measure of likelihood, which can be placed on a numerical scale running from 0 to 1. Examples are given to show the range and scope of problems that need probability to describe them. We examine some simple interpretations of probability that are important in its development, and we briefly show how the well-known principles of mathematical modelling enable us to progress. Note that in this chapter exercises and problems are chosen to motivate interest and discussion; they are therefore non-technical, and mathematical answers are not expected.

Prerequisites.   This chapter contains next to no mathematics, so there are no prerequisites. Impatient readers keen to get to an equation could proceed directly to chapter 2.

1.2 PROBABILITY

We all know what light is, but it is not easy to tell what it is.

Samuel Johnson

From the moment we first roll a die in a children's board game, or pick a card (*any* card), we start to learn what probability is. But even as adults, it is not easy to *tell* what it is, in the general way.

1

For mathematicians things are simpler, at least to begin with. We have the following:

Probability is a number between zero and one, inclusive.

This may seem a trifle arbitrary and abrupt, but there are many excellent and plausible reasons for this convention, as we shall show. Consider the following eventualities.

(i)   You run a mile in less than 10 seconds.
(ii) You roll two ordinary dice and they show a double six.
(iii) You flip an ordinary coin and it shows heads.
(iv) Your weight is less than 10 tons.

If you think about the relative likelihood (or chance or probability) of these eventualities, you will surely agree that we can compare them as follows.

The chance of running a mile in 10 seconds is *less* than the chance of a double six, which in turn is *less* than the chance of a head, which in turn is *less* than the chance of your weighing under 10 tons. We may write

$$\text{chance of 10 second mile} < \text{chance of a double six}$$
$$< \text{chance of a head}$$
$$< \text{chance of weighing under 10 tons.}$$

(Obviously it is assumed that you are reading this on the planet Earth, not on some asteroid, or Jupiter, that you are human, and that the dice are not crooked.)

It is easy to see that we can very often compare probabilities in this way, and so it is natural to represent them on a numerical scale, just as we do with weights, temperatures, earthquakes, and many other natural phenomena. Essentially, this is what numbers are *for*.

Of course, the two extreme eventualities are special cases. It is quite certain that you weigh less than 10 tons; nothing could be more certain. If we represent certainty by unity, then no probabilities exceed this. Likewise it is quite impossible for you to run a mile in 10 seconds or less; nothing could be less likely. If we represent impossibility by zero, then no probability can be less than this. Thus we can, if we wish, present this on a scale, as shown in figure 1.1.

The idea is that any chance eventuality can be represented by a point somewhere on this scale. Everything that is impossible is placed at zero – that the moon is made of

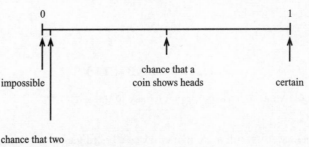

Figure 1.1. A probability scale.

cheese, formation flying by pigs, and so on. Everything that is certain is placed at unity – the moon is not made of cheese, Socrates is mortal, and so forth. Everything else is somewhere in [0, 1], i.e. in the interval between 0 and 1, the more likely things being closer to 1 and the more unlikely things being closer to 0.

Of course, if two things have the same chance of happening, then they are at the same point on the scale. That is what we mean by 'equally likely'. And in everyday discourse everyone, including mathematicians, has used and will use words such as very likely, likely, improbable, and so on. However, any detailed or precise look at probability requires the use of the numerical scale. To see this, you should ponder on just how you would describe a chance that is more than very likely, but less than very very likely.

This still leaves some questions to be answered. For example, the choice of 0 and 1 as the ends of the scale may appear arbitrary, and, in particular, we have not said exactly which numbers represent the chance of a double six, or the chance of a head. We have not even justified the claim that a head is more likely than double six. We discuss all this later in the chapter; it will turn out that if we regard probability as an extension of the idea of proportion, then we can indeed place many probabilities accurately and confidently on this scale.

We conclude with an important point, namely that the chance of a head (or a double six) is just a *chance*. The whole point of probability is to discuss uncertain eventualities *before* they occur. *After* this event, things are completely different. As the simplest illustration of this, note that even though we agree that if we flip a coin and roll two dice then the chance of a head is greater than the chance of a double six, nevertheless it may turn out that the coin shows a tail when the dice show a double six. Likewise, when the weather forecast gives a 90% chance of rain, or even a 99% chance, it may in fact not rain. The chance of a slip on the San Andreas fault this week is very small indeed, nevertheless it may occur today. The antibiotic is overwhelmingly likely to cure your illness, but it may not; and so on.

Exercises for section 1.2

1. Formulate your own definition of probability. Having done so, compare and contrast it with those in appendix I of this chapter.

2. (a) Suppose you flip a coin; there are two possible outcomes, head or tail. Do you agree that the probability of a head is $\frac{1}{2}$? If so, explain why.
   (b) Suppose you take a test; there are two possible outcomes, pass or fail. Do you agree that the probability of a pass is $\frac{1}{2}$? If not, explain why not.

3. In the above discussion we claimed that it was intuitively reasonable to say that you are more likely to get a head when flipping a coin than a double six when rolling two dice. Do you agree? If so, explain why.

1.3 THE SCOPE OF PROBABILITY

... nothing between humans is 1 to 3. In fact, I long ago come to the conclusion that all life is 6 to 5 against.

Damon Runyon, *A Nice Price*

> Life is a gamble at terrible odds; if it was a bet you wouldn't take it.
> Tom Stoppard, *Rosencrantz and Guildenstern are Dead,* Faber and Faber

In the next few sections we are going to spend a lot of time flipping coins, rolling dice, and buying lottery tickets. There are very good reasons for this narrow focus (to begin with), as we shall see, but it is important to stress that probability is of great use and importance in many other circumstances. For example, today seems to be a fairly typical day, and the newspapers contain articles on the following topics (in random order).

1. How are the chances of a child's suffering a genetic disorder affected by a grand-parent's having this disorder? And what difference does the sex of child or ancestor make?
2. Does the latest opinion poll reveal the true state of affairs?
3. The lottery result.
4. DNA profiling evidence in a trial.
5. Increased annuity payments possible for heavy smokers.
6. An extremely valuable picture (a Van Gogh) might be a fake.
7. There was a photograph taken using a scanning tunnelling electron microscope.
8. Should risky surgical procedures be permitted?
9. Malaria has a significant chance of causing death; prophylaxis against it carries a risk of dizziness and panic attacks. What do you do?
10. A commodities futures trader lost a huge sum of money.
11. An earthquake occurred, which had not been predicted.
12. Some analysts expected inflation to fall; some expected it to rise.
13. Football pools.
14. Racing results, and tips for the day's races.
15. There is a 10% chance of snow tomorrow.
16. Profits from gambling in the USA are growing faster than any other sector of the economy. (In connection with this item, it should be carefully noted that profits are made by the casino, not the customers.)
17. In the preceding year, British postmen had sustained 5975 dogbites, which was around 16 per day on average, or roughly one every 20 minutes during the time when mail is actually delivered. One postman had sustained 200 bites in 39 years of service.

Now, this list is by no means exhaustive; I could have made it longer. And such a list could be compiled every day (see the exercise at the end of this section). The subjects reported touch on an astonishingly wide range of aspects of life, society, and the natural world. And they all have the common property that chance, uncertainty, likelihood, randomness – call it what you will – is an inescapable component of the story. Conversely, there are few features of life, the universe, or anything, in which chance is not in some way crucial.

Nor is this merely some abstruse academic point; assessing risks and taking chances are inescapable facets of everyday existence. It is a trite maxim to say that life is a lottery; it would be more true to say that life offers a collection of lotteries that we can all, to some extent, choose to enter or avoid. And as the information at our disposal increases, it does not reduce the range of choices but in fact increases them. It is, for example, increasingly difficult successfully to run a business, practise medicine, deal in finance, or engineer things without having a keen appreciation of chance and probability. Of course you can make the attempt, by relying entirely on luck and uninformed guesswork, but in the long run you will probably do worse than someone who plays the odds in an informed way. This is amply confirmed by observation and experience, as well as by mathematics.

Thus, probability is important for all these severely practical reasons. And we have the bonus that it is also entertaining and amusing, as the existence of all those lotteries, casinos, and racecourses more than sufficiently testifies.

Finally, a glance at this and other section headings shows that chance is so powerful and emotive a concept that it is employed by poets, playwrights, and novelists. They clearly expect their readers to grasp jokes, metaphors, and allusions that entail a shared understanding of probability. (This feat has not been accomplished by algebraic struc-tures, or calculus, and is all the more remarkable when one recalls that the *literati* are not otherwise celebrated for their keen numeracy.) Furthermore, such allusions are of very long standing; we may note the comment attributed by Plutarch to Julius Caesar on crossing the Rubicon: 'Iacta alea est' (commonly rendered as 'The die is cast'). And the passage from Ecclesiastes: 'The race is not always to the swift, or the battle to the strong, but time and chance happen to them all'. The Romans even had deities dedicated to chance, Fors and Fortuna, echoed in Shakespeare's *Hamlet*: '... the slings and arrows of outrageous fortune ...'.

Many other cultures have had such deities, but it is notable that deification has not occurred for any other branch of mathematics. There is no god of algebra.

One recent stanza (by W.H. Henley) is of particular relevance to students of probability, who are often soothed and helped by murmuring it during difficult moments in lectures and textbooks:

> In the fell clutch of circumstance
> I have not winced or cried aloud:
> Under the bludgeonings of chance
> My head is bloody, but unbowed.

Exercise for section 1.3

1. Look at today's newspapers and mark the articles in which chance is explicitly or implicitly an important feature of the report.

1.4 BASIC IDEAS: THE CLASSICAL CASE

> The perfect die does not lose its usefulness or justification by the fact that real dice fail to live up to it.
>
> W. Feller

Our first task was mentioned above; we need to supply reasons for the use of the standard probability scale, and methods for deciding where various chances should lie on this scale. It is natural that in doing this, and in seeking to understand the concept of probability, we will pay particular attention to the experience and intuition yielded by flipping coins and rolling dice. Of course this is not a very bold or controversial decision;

any theory of probability that failed to describe the behaviour of coins and dice would be widely regarded as useless. And so it would be. For several centuries that we know of, and probably for many centuries before that, flipping a coin (or rolling a die) has been the epitome of probability, the paradigm of randomness. You flip the coin (or roll the die), and nobody can accurately predict how it will fall. Nor can the most powerful computer predict correctly how it will fall, if it is flipped energetically enough.

This is why cards, dice, and other gambling aids crop up so often in literature both directly and as metaphors. No doubt it is also the reason for the (perhaps excessive) popularity of gambling as entertainment. If anyone had any idea what numbers the lottery would show, or where the roulette ball will land, the whole industry would be a dead duck.

At any rate, these long-standing and simple gaming aids do supply intuitively convincing ways of characterizing probability. We discuss several ideas in detail.

I Probability as proportion

Figure 1.2 gives the layout of an American roulette wheel. Suppose such a wheel is spun once; what is the probability that the resulting number has a 7 in it? That is to say, what is the probability that the ball hits 7, 17, or 27? These three numbers comprise a proportion $\frac{3}{38}$ of the available compartments, and so the essential symmetry of the wheel (assuming it is well made) suggests that the required probability ought to be $\frac{3}{38}$. Likewise the

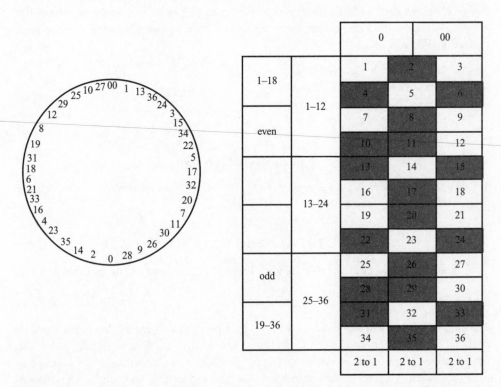

Figure 1.2. American roulette. Shaded numbers are black; the others are red except for the zeros.

probability of an odd compartment is suggested to be $\frac{18}{38} = \frac{9}{19}$, because the proportion of odd numbers on the wheel is $\frac{18}{38}$.

Most people find this proposition intuitively acceptable; it clearly relies on the fundamental symmetry of the wheel, that is, that all numbers are regarded equally by the ball. But this property of symmetry is shared by a great many simple chance activities; it is the same as saying that all possible outcomes of a game or activity are equally likely. For example:

- The ball is equally likely to land in any compartment.
- You are equally likely to select either of two cards.
- The six faces of a die are equally likely to be face up.

With these examples in mind it seems reasonable to adopt the following convention or rule. Suppose some game has n equally likely outcomes, and r of these outcomes correspond to your winning. Then the probability p that you win is r/n. We write

$$(1) \qquad p = \frac{r}{n} = \frac{\text{number of ways of winning the game}}{\text{number of possible outcomes of the game}}.$$

This formula looks very simple. Of course, it *is* very simple but it has many useful and important consequences. First note that we always have $0 \leqslant r \leqslant n$, and so it follows that

$$(2) \qquad\qquad 0 \leqslant p \leqslant 1.$$

If $r = 0$, so that it is impossible for you to win, then $p = 0$. Likewise if $r = n$, so that you are certain to win, then $p = 1$. This is all consistent with the probability scale introduced in section 1.2, and supplies some motivation for using it. Furthermore, this interpretation of probability as defined by proportion enables us to place many simple but important chances on the scale.

Example 1.4.1. Flip a coin and choose 'heads'. Then $r = 1$, because you win on the outcome 'heads', and $n = 2$, because the coin shows a head or a tail. Hence the probability that you win, which is also the probability of a head, is $p = \frac{1}{2}$. ◯

Example 1.4.2. Roll a die. There are six outcomes, which is to say that $n = 6$. If you win on an even number then $r = 3$, so the probability that an even number is shown is

$$p = \frac{3}{6} = \frac{1}{2}.$$

Likewise the chance that the die shows a 6 is $\frac{1}{6}$, and so on. ◯

Example 1.4.3. Pick a card at random from a pack of 52 cards. What is the probability of an ace? Clearly $n = 52$ and $r = 4$, so that

$$p = \frac{4}{52} = \frac{1}{13}.$$ ◯

Example 1.4.4. A town contains x women and y men; an opinion pollster chooses an adult at random for questioning about toothpaste. What is the chance that the adult is male? Here

$$n = x + y \quad \text{and} \quad r = y.$$

Hence the probability is

$$p = y/(x + y). \qquad \bigcirc$$

It may be objected that these results depend on an arbitrary imposition of the ideas of symmetry and proportion, which are clearly not always relevant. Nevertheless, such results and ideas are immensely appealing to our intuition; in fact the first probability calculations in Renaissance Italy take this framework more or less for granted. Thus Cardano (writing around 1520), says of a well-made die: 'One half of the total number of faces always represents equality ... I can as easily throw 1, 3, or 5 as 2, 4, or 6'.

Here we can clearly see the beginnings of the idea of probability as an expression of proportion, an idea so powerful that it held sway for centuries. However, there is at least one unsatisfactory aspect to this interpretation: it seems that we do not need ever to roll a die to say that the chance of a 6 is $\frac{1}{6}$. Surely actual experiments should have a role in our definitions? This leads to another idea.

II Probability as relative frequency

Figure 1.3 shows the proportion of sixes that appeared in a sequence of rolls of a die. The number of rolls is n, for $n = 0, 1, 2, \ldots$; the number of sixes is $r(n)$, for each n, and the proportion of sixes is

$$(3) \qquad\qquad p(n) = \frac{r(n)}{n}.$$

What has this to do with the probability that the die shows a six? Our idea of probability as a proportion suggests that the proportion of sixes in n rolls should not be too far from the theoretical chance of a six, and figure 1.3 shows that this seems to be true for large values of n. This is intuitively appealing, and the same effect is observed if you record such proportions in a large number of other repeated chance activities.

We therefore make the following general assertion. Suppose some game is repeated a large number n of times, and in $r(n)$ of these games you win. Then the probability p that

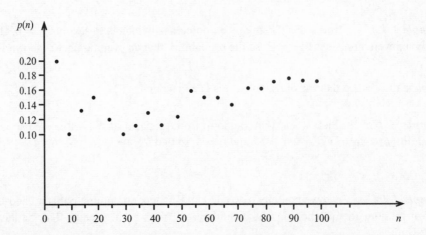

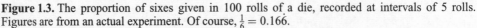

Figure 1.3. The proportion of sixes given in 100 rolls of a die, recorded at intervals of 5 rolls. Figures are from an actual experiment. Of course, $\frac{1}{6} = 0.16\dot{6}$.

you win some future similar repetition of this game is close to $r(n)/n$. We write

(4)
$$p \simeq \frac{r(n)}{n} = \frac{\text{number of wins in } n \text{ games}}{\text{number } n \text{ of games}}.$$

The symbol $\simeq$ is read as 'is approximately equal to'. Once again we note that $0 \leqslant r(n) \leqslant n$ and so we may take it that $0 \leqslant p \leqslant 1$.

Furthermore, if a win is impossible then $r(n) = 0$, and $r(n)/n = 0$. Also, if a win is certain then $r(n) = n$, and $r(n)/n = 1$. This is again consistent with the scale introduced in figure 1.1, which is very pleasant. Notice the important point that this interpretation supplies a way of approximately *measuring* probabilities rather than calculating them merely by an appeal to symmetry.

Since we can now calculate simple probabilities, and measure them approximately, it is tempting to stop there and get straight on with formulating some rules. That would be a mistake, for the idea of proportion gives another useful insight into probability that will turn out to be just as important as the other two, in later work.

III Probability and expected value

Many problems in chance are inextricably linked with numerical outcomes, especially in gambling and finance (where 'numerical outcome' is a euphemism for money). In these cases probability is inextricably linked to 'value', as we now show.

To aid our thinking let us consider an everyday concrete and practical problem. A plutocrat makes the following offer. She will flip a fair coin; if it shows heads she will give you \$1, if it shows tails she will give Jack \$1. What is this offer worth to you? That is to say, for what fair price \$$p$, should you sell it?

Clearly, whatever price \$$p$ this is worth to you, it is worth the same price \$$p$ to Jack, because the coin is fair, i.e. symmetrical (assuming he needs and values money just as much as you do). So, to the pair of you, this offer is altogether worth \$$2p$. But whatever the outcome, the plutocrat has given away \$1. Hence \$$2p = $ \$1, so that $p = \frac{1}{2}$ and the offer is worth \$$\frac{1}{2}$ to you.

It seems natural to regard this value $p = \frac{1}{2}$ as a measure of your chance of winning the money. It is thus intuitively reasonable to make the following general rule.

Suppose you receive \$1 with probability p (and otherwise you receive nothing). Then the *value* or fair price of this offer is \$$p$. More generally, if you receive \$$d$ with probability p (and nothing otherwise) then the fair price or expected value of this offer is given by

(5)
$$\text{expected value} = pd.$$

This simple idea turns out to be enormously important later on; for the moment we note only that it is certainly consistent with our probability scale introduced in figure 1.1. For example, if the plutocrat definitely gives you \$1 then this is worth exactly \$1 to you, and $p = 1$. Likewise if you are definitely given nothing, then $p = 0$. And it is easy to see that $0 \leqslant p \leqslant 1$, for any such offers.

In particular, for the specific example above we find that the probability of a head when a fair coin is flipped is $\frac{1}{2}$. Likewise a similar argument shows that the probability of a six when a fair die is rolled is $\frac{1}{6}$. (Simply imagine the plutocrat giving \$1 to one of six people selected by the roll of the die.)

The 'fair price' of such offers is often called the expected value, or *expectation*, to emphasize its chance nature. We meet this concept again, later on.

We conclude this section with another classical and famous manifestation of probability. It is essentially the same as the first we looked at, but is superficially different.

IV Probability as proportion again

Suppose a small meteorite hits the town football pitch. What is the probability that it lands in the central circle?

Obviously meteorites have no special propensity to hit any particular part of a football pitch; they are equally likely to strike any part. It is therefore intuitively clear that the chance of striking the central circle is given by the proportion of the pitch that it occupies. In general, if $|A|$ is the area of the pitch in which the meteorite lands, and $|C|$ is the area of some part of the pitch, then the probability p that C is struck is given by $p = |C|/|A|$.

Once again we formulate a general version of this as follows. Suppose a region A of the plane has area $|A|$, and C is some part of A with area $|C|$. If a point is picked at random in A, then the probability p that it lies in C is given by

(6) $$p = \frac{|C|}{|A|}.$$

As before we can easily see that $0 \leqslant p \leqslant 1$, where $p = 0$ if C is empty and $p = 1$ if $C = A$.

Example 1.4.5.   An archery target is a circle of radius 2. The bullseye is a circle of radius 1. A naive archer is equally likely to hit any part of the target (if she hits it at all) and so the probability of a bullseye for an arrow that hits the target is

$$p = \frac{\text{area of bullseye}}{\text{area of target}} = \frac{\pi \times 1^2}{\pi \times 2^2} = \frac{1}{4}.$$ ○

Exercises for section 1.4

1.   Suppose you read in a newspaper that the proportion of \$20 bills that are forgeries is 5%. If you possess what appears to be a \$20 bill, what is its expected value? Could it be more than \$19? Or could it be less? Explain! (Does it make any difference how you acquired the bill?)

2.   A point P is picked at random in the square $ABCD$, with sides of length 1. What is the probability that the distance from P to the diagonal AC is less than $\frac{1}{6}$?

1.5 BASIC IDEAS; THE GENERAL CASE

We must believe in chance, for how else can we account for the successes of those we detest?

Anon.

We noted that a theory of probability would be hailed as useless if it failed to describe the behaviour of coins and dice. But of course it would be equally useless if it failed to describe anything else, and moreover many real dice and coins (especially dice) have been known to be biased and asymmetrical. We therefore turn to the question of assigning probabilities in activities that do not necessarily have equally likely outcomes.

It is interesting to note that the desirability of doing this was implicitly recognized by Cardano (mentioned in the previous section) around 1520. In his *Book on Games of Chance*, which deals with supposedly fair dice, he notes that

'Every die, even if it is acceptable, has its favoured side.'

However, the ideas necessary to describe the behaviour of such biased dice had to wait for Pascal in 1654, and later workers. We examine the basic notions in turn; as in the previous section, these notions rely on our concept of probability as an extension of proportion.

I Probability as relative frequency

Once again we choose a simple example to illustrate the ideas, and a popular choice is the pin, or tack. Figure 1.4 shows a pin, called a Bernoulli pin. If such a pin is dropped onto a table the result is a success, *S*, if the point is not upwards; otherwise it is a failure, *F*.

What is the probability *p* of success? Obviously symmetry can play no part in fixing *p*, and Figure 1.5, which shows more Bernoulli pins, indicates that mechanical arguments will not provide the answer.

The only course of action is to drop many similar pins (or the same pin many times), and record the proportion that are successes (point down). Then if *n* are dropped, and *r(n)* are successes, we anticipate that the long-run proportion of successes is near to *p*, that is:

(1)
$$p \simeq \frac{r(n)}{n}, \quad \text{for large } n.$$

failure ≡ *F*   success ≡ *S*

Figure 1.4. A Bernoulli pin.

Figure 1.5. More Bernoulli pins.

If you actually obtain a pin and perform this experiment, you will get a graph like that of figure 1.6. It does seem from the figure that $r(n)/n$ is settling down around some number p, which we naturally interpret as the probability of success. It may be objected that the ratio changes every time we drop another pin, and so we will never obtain an exact value for p. But this gap between the real world and our descriptions of it is observed in all subjects at all levels. For example, geometry tells us that the diagonal of a unit square has length $\sqrt{2}$. But, as A. A. Markov has observed,

> If we wished to verify this fact by measurements, we should find that the ratio of diagonal to side is different for different squares, and is never $\sqrt{2}$.

It may be regretted that we have only this somewhat hit-or-miss method of measuring probability, but we do not really have any choice in the matter. Can you think of any other way of estimating the chance that the pin will fall point down? And even if you did think of such a method of estimation, how would you decide whether it gave the right answer, except by flipping the pin often enough to see? We can illustrate this point by considering a basic and famous example.

Example 1.5.1: sex ratio.   What is the probability that the next infant to be born in your local hospital will be male? Throughout most of the history of the human race it was taken for granted that essentially equal numbers of boys and girls are born (with some fluctuations, naturally). This question would therefore have drawn the answer $\frac{1}{2}$, until recently.

However, in the middle of the 16th century, English parish churches began to keep fairly detailed records of births, marriages, and deaths. Then, in the middle of the 17th century, one John Graunt (a draper) took the trouble to read, collate, and tabulate the numbers in various categories. In particular he tabulated the number of boys and girls whose births were recorded in London in each of 30 separate years.

To his, and everyone else's, surprise, he found that in every single year more boys were born than girls. And, even more remarkably, the ratio of boys to girls varied very little between these years. In every year the ratio of boys to girls was close to 14:13. The meaning and significance of this unarguable truth inspired a heated debate at the time. For us, it shows that the probability that the next infant born will be male, is approximately $\frac{14}{27}$. A few moments thought will show that there is no other way of answering the general question, other than by finding this relative frequency.

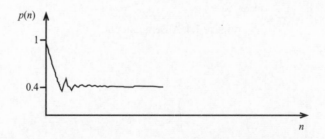

Figure 1.6. Sketch of the proportion $p(n)$ of successes when a Bernoulli pin is dropped n times. For this particular pin, p seems to be settling down at approximately 0.4.

It is important to note that the empirical frequency differs from place to place and from time to time. Graunt also looked at the births in Romsey over 90 years and found the empirical frequency to be 16:15. It is currently just under 0.513 in the USA, slightly less than $\frac{14}{27}$ ($\simeq 0.519$) and $\frac{16}{31}$ ($\simeq 0.516$).

Clearly the idea of probability as a relative frequency is very attractive and useful. Indeed it is generally the only interpretation offered in textbooks. Nevertheless, it is not always enough, as we now discuss.

II Probability as expected value

The problem is that to interpret probability as a relative frequency requires that we can repeat some game or activity as many times as we wish. Often this is clearly not the case. For example, suppose you have a Russian Imperial Bond, or a share in a company that is bankrupt and is being liquidated, or an option on the future of the price of gold. What is the probability that the bond will be redeemed, the share will be repaid, or the option will yield a profit? In these cases the idea of expected value supplies the answer. (For simplicity, we assume constant money values and no interest.)

The ideas and argument are essentially the same as those that we used in considering the benevolent plutocrat in section 1.4, leading to equation (5) in that section. For variety, we rephrase those notions in terms of simple markets. However, a word of warning is appropriate at this point. Real markets are much more complicated than this, and what we call the fair price or expected value will not usually be the actual or agreed market price in any case, or even very close to it. This is especially marked in the case of deals which run into the future, such as call options, put options, and other complicated financial derivatives. If you were to offer prices based on fairness or expected value as discussed here and above, you would be courting total disaster, or worse. See the discussion of bookmakers' odds in section 2.12 for further illustration and words of caution.

Suppose you have a bond with face value $1, and the probability of its being redeemed at par (that is, for $1) is p. Then, by the argument we used in section 1.4, the expected value μ, or fair price, of this bond is given by $\mu = p$. More generally, if the bond has face value d then the fair price is dp.

Now, as it happens, there are markets in all these things: you can buy Imperial Chinese bonds, South American Railway shares, pork belly futures, and so on. It follows that if the market gives a price μ for a bond with face value d, then it gives the probability of redemption as roughly

(2)
$$p = \frac{\mu}{d}.$$

Example 1.5.2. If a bond for a million roubles is offered to you for one rouble, and the sellers are assumed to be rational, then they clearly think the chance of the bond's being bought back at par is less than one in a million. If you buy it, then presumably you believe the chances are more than one in a million. If you thought the chances were less, you would reduce your offer. If you both agree that one rouble is a fair price for the bond, then you have assigned the value $p = 10^{-6}$ for the probability of its redemption. Of course this may vary according to various rumours and signals from the relevant banks and government (and note that the more ornate and attractive bonds now have some intrinsic value, independent of their chance of redemption).   ○

This example leads naturally to our final candidate for an interpretation of probability.

III Probability as an opinion or judgement

In the previous example we were able to assign a probability because the bond had an agreed fair price, *even though* this price was essentially a matter of opinion. What happens if we are dealing with probabilities that are purely personal opinions? For example, what is the probability that a given political party will win the next election? What is the probability that small green aliens regularly visit this planet? What is the probability that some accused person is guilty? What is the probability that a given, opaque, small, brick building contains a pig?

In each of these cases we could perhaps obtain an estimate of the probability by persuading a bookmaker to compile a number of wagers and so determine a fair price. But we would be at a loss if nobody were prepared to enter this game. And it would seem to be at best a very artificial procedure, and at worst extremely inappropriate, or even illegal. Furthermore, the last resort, betting with yourself, seems strangely unattractive.

Despite these problems, this idea of probability as a matter of opinion is often useful, though we shall not use it in this text.

Exercises for section 1.5

1. A picture would be worth $1000 000 if genuine, but nothing if a fake. Half the experts say it's a fake, half say it's genuine. What is it worth? Does it make any difference if one of the experts is a millionaire?

2. A machine accepts dollar bills and sells a drink for $1. The price is raised to 120c. Converting the machine to accept coins or give change is expensive, so it is suggested that a simple randomizer is added, so that each customer who inserts $1 gets nothing with probability $1/6$, or the can with probability $5/6$, and that this would be fair because the expected value of the output is $120 \times 5/6 = 100c = \$1$, which is exactly what the customer paid. Is it indeed fair?

   In the light of this, discuss how far our idea of a fair price depends on a surreptitious use of the concept of repeated experiments.

   Would you buy a drink from the modified machine?

1.6 MODELLING

If I wish to know the chances of getting a complete hand of 13 spades, I do not set about dealing hands. It would take the population of the world billions of years to obtain even a bad estimate of this.

John Venn

The point of the above quote is that we need a theory of probability to answer even the simplest of practical questions. Such theories are called *models*.

Example 1.6.1: cards. For the question above, the usual model is as follows. We assume that all possible hands of cards are equally likely, so that if the number of all possible hands is n, then the required probability is n^{-1}. $\bigcirc$

Experience seems to suggest that for a well-made, well-shuffled pack of cards, this answer is indeed a good guide to your chances of getting a hand of spades. (Though we must remember that such complete hands occur more often than this predicts, because humorists stack the pack, as a 'joke'.) Even this very simple example illustrates the following important points very clearly.

First, the model deals with abstract things. We cannot *really* have a perfectly shuffled pack of perfect cards; this 'collection of equally likely hands' is actually a fiction. We create the idea, and then use the rules of arithmetic to calculate the required chances. This is characteristic of all mathematics, which concerns itself only with rules defining the behaviour of entities which are themselves undefined (such as 'numbers' or 'points').

Second, the use of the model is determined by our interpretation of the rules and results. We do not need an interpretation of what chance is to calculate probabilities, but without such an interpretation it is rather pointless to do it.

Similarly, you do not need to have an interpretation of what lines and points are to do geometry and trigonometry, but it would all be rather pointless if you did not have one. Likewise chess is just a set of rules, but if checkmate were not interpreted as victory, not many people would play.

Use of the term 'model' makes it easier to keep in mind this distinction between theory and reality. By its very nature a model cannot include all the details of the reality it seeks to represent, for then it would be just as hard to comprehend and describe as the reality we want to model. At best, our model should give a reasonable picture of some small part of reality. It has to be a simple (even crude) description; and we must always be ready to scrap or improve a model if it fails in this task of accurate depiction. That having been said, old models are often still useful. The theory of relativity supersedes the Newtonian model, but all engineers use Newtonian mechanics when building bridges or motor cars, or probing the solar system.

This process of observation, model building, analysis, evaluation, and modification is called *modelling*, and it can be conveniently represented by a diagram; see figure 1.7. (This diagram is therefore in itself a model; it is a model for the modelling process.)

In figure 1.7, the top two boxes are embedded in the real world and the bottom two boxes are in the world of models. Box A represents our observations and experience of some phenomenon, together with relevant knowledge of related events and perhaps past experience of modelling. Using this we construct the rules of a model, represented by box B. We then use the techniques of logical reasoning, or mathematics, to deduce the way in which the model will behave. These properties of the model can be called theorems; this stage is represented by box C. Next, these characteristics of the model are interpreted in terms of predictions of the way the corresponding real system should work, denoted by box D. Finally, we perform appropriate experiments to discover whether these predictions agree with observation. If they do not, we change or scrap the model and go round the loop again. If they do, we hail the model as an engine of discovery, and keep using it to make predictions – until it wears out or breaks down. This last step is called using or checking the model or, more grandly, validation.

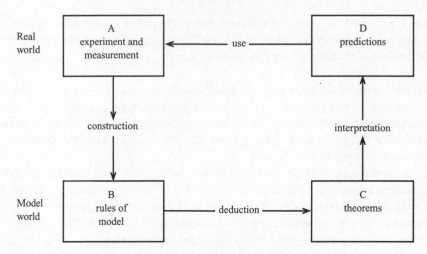

Figure 1.7. A model for modelling.

This procedure is so commonplace that we rather take it for granted. For example, it has been used every time you see a weather forecast. Meteorologists have observed the climate for many years. They have deduced certain simple rules for the behaviour of jet streams, anticyclones, occluded fronts, and so on. These rules form the model. Given any configuration of airflows, temperatures, and pressures, the rules are used to make a prediction; this is the weather forecast. Every forecast is checked against the actual outcome, and this experience is used to improve the model.

Models form extraordinarily powerful and economical ways of thinking about the world. In fact they are often so good that the model is confused with reality. If you ever think about atoms, you probably imagine little billiard balls; more sophisticated readers may imagine little orbital systems of elementary particles. Of course atoms are not 'really' like that; these visions are just convenient old models.

We illustrate the techniques of modelling with two simple examples from probability.

Example 1.6.2: setting up a lottery.   If you are organizing a lottery you have to decide how to allocate the prize money to the holders of winning tickets. It would help you to know the chances of any number winning and the likely number of winners. Is this possible? Let us consider a specific example.

Several national lotteries allow any entrant to select six numbers in advance from the integers 1 to 49 inclusive. A machine then selects six balls at random (without replacement) from an urn containing 49 balls bearing these numbers. The first prize is divided among entrants selecting these numbers.

Because of the nature of the apparatus, it seems natural to assume that any selection of six numbers is equally likely to be drawn. Of course this assumption is a mathematical model, not a physical law established by experiment. Since there are approximately 14 million different possible selections (we show this in chapter 3), the model predicts that your chance, with one entry, of sharing the first prize is about one in 14 million. Figure 1.8 shows the relative frequency of the numbers drawn in the first 1200 draws. It does not seem to discredit or invalidate the model so far as one can tell.

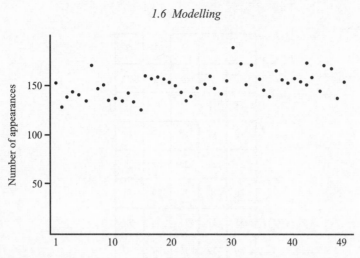

Figure 1.8. Frequency plot of an actual 6–49 lottery after 1200 drawings. The numbers do seem equally likely to be drawn.

The next question you need to answer is, how many of the entrants are likely to share the first prize? As we shall see, we need in turn to ask, how do lottery entrants choose their numbers?

This is clearly a rather different problem; unlike the apparatus for choosing numbers, gamblers choose numbers for various reasons. Very few choose at random; they use birthdays, ages, patterns, and so on. However, you might suppose that for any gambler chosen at random, that choice of numbers would be evenly distributed over the possibilities.

In fact this model would be wrong; when the actual choices of lottery numbers are examined, it is found that in the long run the chances that the various numbers will occur are very far from equal; see figure 1.9. This clustering of preferences arises because people choose numbers in lines and patterns which favour central squares, and they also favour the top of the card. Data like this would provide a model for the distribution of likely payouts to winners. ○

It is important to note that these remarks do not apply only to lotteries, cards, and dice. Venn's observation about card hands applies equally well to almost every other aspect of life. If you wished to design a telephone exchange (for example), you would first of all construct some mathematical models that could be tested (you would do this by making assumptions about how calls would arrive, and how they would be dealt with). You can construct and improve any number of mathematical models of an exchange very cheaply. Building a faulty real exchange is an extremely costly error.

Likewise, if you wished to test an aeroplane to the limits of its performance, you would be well advised to test mathematical models first. Testing a real aeroplane to destruction is somewhat risky.

So we see that, in particular, models and theories can save lives and money. Here is another practical example.

1	2	3	4	**5**
6	**7**	**8**	**9**	10
11	**12**	**13**	**14**	15
16	**17**	**18**	**19**	20
21	**22**	**23**	**24**	25
26	**27**	**28**	**29**	30
31	32	**33**	34	*35*
36	37	38	39	*40*
41	42	43	44	*45*
46	47	48	49	

Figure 1.9. Popular and unpopular lottery numbers: **bold**, most popular; roman, intermediate popularity; *italic*, least popular.

Example 1.6.3: first significant digit.   Suppose someone offered the following wager:

(i)   select any large book of numerical tables (such as a census, some company accounts, or an almanac);
(ii) pick a number from this book at random (by any means);
(iii) if the first significant digit of this number is one of $\{5, 6, 7, 8, 9\}$, then you win \$1; if it is one of $\{1, 2, 3, 4\}$, you lose \$1.

Would you accept this bet? You might be tempted to argue as follows: a reasonable intuitive model for the relative chances of each digit is that they are equally likely. On this model the probability p of winning is $\frac{5}{9}$, which is greater than $\frac{1}{2}$ (the odds on winning would be $5:4$), so it seems like a good bet. However, if you do some research and actually pick a large number of such numbers at random, you will find that the relative frequencies of each of the nine possible first significant digits are given approximately by

$$f_1 = 0.301, \quad f_2 = 0.176, \quad f_3 = 0.125,$$
$$f_4 = 0.097, \quad f_5 = 0.079, \quad f_6 = 0.067,$$
$$f_7 = 0.058, \quad f_8 = 0.051, \quad f_9 = 0.046.$$

Thus empirically the chance of your winning is

$$f_5 + f_6 + f_7 + f_8 + f_9 = 0.3$$

The wager offered is not so good for you! (Of course it would be quite improper for a mathematician to win money from the ignorant by this means.) This empirical distribution is known as Benford's law, though we should note that it was first recorded by S. Newcomb (a good example of Stigler's law of eponymy).   ○

We see that intuition is necessary and helpful in constructing models, but not sufficient; you also need experience and observations. A famous example of this arose in particle physics. At first it was assumed that photons and protons would satisfy the same statistical rules, and models were constructed accordingly. Experience and observations showed that in fact they behave differently, and the models were revised.

The theory of probability exhibits a very similar history and development, and we approach it in similar ways. That is to say, we shall construct a model that reflects our experience of, and intuitive feelings about, probability. We shall then deduce results and make predictions about things that have either not been explained or not been observed, or both. These are often surprising and even counter intuitive. However, when the predictions are tested against experiment they are almost always found to be good. Where they are not, new theories must be constructed.

It may perhaps seem paradoxical that we can explore reality most effectively by playing with models, but this fact is perfectly well known to all children.

Exercise for section 1.6

1. Discuss how the development of the high-speed computer is changing the force of Venn's observation, which introduced this section.

1.7 MATHEMATICAL MODELLING

> There are very few things which we know, which are not capable of being reduced to a mathematical reasoning; and when they cannot, it is a sign our knowledge of them is very small and confused; and where a mathematical reasoning can be had, it is as great a folly to make use of any other, as to grope for a thing in the dark, when you have a candle standing by you.
>
> John Arbuthnot, *Of the Laws of Chance*

The quotation above is from the preface to the first textbook on probability to appear in English. (It is in a large part a translation of a book by Huygens, which had previously appeared in Latin and Dutch.) Three centuries later, we find that mathematical reasoning is indeed widely used in all walks of life, but still perhaps not as much as it should be. A small survey of the reasons for using mathematical methods would not be out of place. The first question is, why be abstract at all? The blunt answer is that we have no choice, for many reasons.

In the first place, as several examples have made clear, practical probability is inescapably numerical. Betting odds can only be numerical, monetary payoffs are numerical, stock exchanges and insurance companies float on a sea of numbers. And even the simplest and most elementary problems in bridge and poker, or in lotteries, involve counting things. And this counting is often not a trivial task.

Second, the range of applications demands abstraction. For example, consider the following list of real activities:

- customers in line at a post office counter
- cars at a toll booth
- data in an active computer memory

- a pile of cans in a supermarket
- telephone calls arriving at an exchange
- patients arriving at a trauma clinic
- letters in a mail box

All these entail 'things' or 'entities' in one or another kind of 'waiting' state, before some 'action' is taken. Obviously this list could be extended indefinitely. It is desirable to abstract the essential structure of all these problems, so that the results can be interpreted in the context of whatever application happens to be of interest. For the examples above, this leads to a model called the theory of queues.

Third, we may wish to discuss the behaviour of the system without assigning specific numerical values to the rate of arrival of the objects (or customers), or to the rate at which they are processed (or serviced). We may not even know these values. We may wish to examine the way in which congestion depends generally on these rates. For all these reasons we are naturally forced to use all the mathematical apparatus of symbolism, logic, algebra, and functions. This is in fact very good news, and these methods have the simple practical and mechanical advantage of making our work very compact. This alone would be sufficient! We conclude this section with two quotations chosen to motivate the reader even more enthusiastically to the advantages of mathematical modelling. They illustrate the fact that there is also a considerable gain in understanding of complicated ideas if they are simply expressed in concise notation. Here is a definition of commerce.

> Commerce: a kind of transaction, in which A plunders from B the goods of C, and for compensation B picks the pocket of D of money belonging to E.
>
>   Ambrose Bierce, *The Devil's Dictionary*

The whole pith and point of the joke evaporates completely if you expand this from its symbolic form. And think of the expansion of effort required to write it. Using algebra is the reason – or at least one of the reasons – why mathematicians so rarely get writer's cramp or repetitive strain injury.

We leave the final words on this matter to Abraham de Moivre, who wrote the second textbook on probability to appear in English. It first appeared in 1717. (The second edition was published in 1738 and the third edition in 1756, posthumously, de Moivre having died on 27 November, 1754 at the age of 87.) He says in the preface:

> Another use to be made of this Doctrine of Chances is, that it may serve in conjunction with the other parts of mathematics as a fit introduction to the art of reasoning; it being known by experience that nothing can contribute more to the attaining of that art, than the consideration of a long train of consequences, rightly deduced from undoubted principles, of which this book affords many examples. To this may be added, that some of the problems about chance having a great appearance of simplicity, the mind is easily drawn into a belief, that their solution may be attained by the mere strength of natural good sense; which generally proving otherwise, and the mistakes occasioned thereby being not infrequent, it is presumed that a book of this kind, which teaches to distinguish truth from what seems so nearly to resemble it, will be looked on as a help to good reasoning.

These remarks remain as true today as when de Moivre wrote them around 1717.

1.8 MODELLING PROBABILITY

Rules and Models destroy genius and Art.

W. Hazlitt

First, we examine the real world and select the experiences and experiments that seem best to express the nature of probability, without too much irrelevant extra detail. You have already done this, if you have ever flipped a coin, or rolled a die, or wondered whether to take an umbrella.

Second, we formulate a set of rules that best describe these experiments and experiences. These rules will be mathematical in nature, for simplicity. (This is *not* paradoxical!) We do this in the next chapter.

Third, we use the structure of mathematics (thoughtfully constructed over the millenia for other purposes), to derive results of practical interest. We do this in the remainder of the book.

Finally, these results are compared with real data in a variety of circumstances: by scientists to measure constants, by insurance companies to avoid ruin, by actuaries to calculate your pension, by telephone engineers to design the network, and so on. This validates our model, and has been done by many people for hundreds of years. So we do not need to do it here.

This terse account of our program gives rise to a few questions of detail, which we address here, as follows. Do we in fact need to know what probability 'really' is? The answer here is, of course, no. We only need our model to describe what we observe. It is the same in physics; we do not need to know what mass really is to use Newton's or Einstein's theories. This is just as well, because we do *not* know what mass really is. We still do not know even what light 'really' is. Questions of reality are best left for philosophers to argue over, for ever.

Furthermore, in drawing up the rules, do we necessarily have to use the rather roundabout arguments employed in section 1.2? Is there not a more simple and straightforward way to say what probability does? After all, Newton only had to drop apples to see what gravity, force, and momentum did. Heat burns, electricity shocks, and light shines, to give some other trivial examples.

By contrast, probability is strangely intangible stuff; you cannot accumulate piles of it, or run your hands through it, or give it away. No meter will record its presence or absence, and it is not much used in the home. We cannot deny its existence, since we talk about it, but it exists in a curiously shadowy and ghost-like way. This difficulty was neatly pinpointed by John Venn in the 19th century:

> It is sometimes not easy to give a clear definition of a science at the outset, so as to set its scope and province before the reader in a few words. In the case of those sciences which are directly concerned with what are termed objects, this difficulty is not indeed so serious. If the reader is already familiar with the objects, a simple reference to them will give him a tolerably accurate idea of the direction and nature of his studies. Even if he is not familiar with them they will still be often to some extent connected and associated in his mind by a name, and the mere utterance of the name may thus convey a fair amount of preliminary information.

But when a science is concerned not so much with objects as with laws, the difficulty of giving preliminary information becomes greater.

The Logic of Chance

What Venn meant by this, is that books on subjects such as fluid mechanics need not ordinarily spend a great deal of time explaining the everyday concept of a fluid. The average reader will have seen waves on a lake, watched bathwater going down the plughole, observed trees bending in the wind, and been annoyed by the wake of passing boats. And anyone who has flown in an aeroplane has to believe that fluid mechanics demonstrably works. Furthermore, the language of the subject has entered into everyday discourse, so that when people use words like wave, or wing, or turbulence, or vortex, they think they know what they mean. Probability is harder to put your finger on.

1.9 REVIEW

In this chapter we have looked at chance and probability in a non-technical way. It seems obvious that we recognize the appearance of chance, but it is surprisingly difficult to give a comprehensive definition of probability. For this reason, and many others, we have begun to construct a theory of probability that will rely on mathematical models and methods.

Our first step on this path has been to agree that any probability is a number lying between zero and unity, inclusive. It can be interpreted as a simple proportion in situations with symmetry, or as a measure of long-run proportion, or as an estimate of expected value, depending on the context. The next task is to determine the rules obeyed by probabilities, and this is the content of the next chapter.

1.10 APPENDIX I. SOME RANDOMLY SELECTED DEFINITIONS OF PROBABILITY, IN RANDOM ORDER

One can hardly give a satisfactory definition of probability.

H. Poincaré

Probability is a degree of certainty, which is to certainty as a part is to the whole.

J. Bernoulli

Probability is the study of random experiments.

S. Lipschutz

Mathematical probability is a branch of mathematical analysis that has developed around the problem of assigning numerical measurement to the abstract concept of likelihood.

M. Munroe

Probability is a branch of logic which analyses nondemonstrative inferences, as opposed to demonstrative ones.

E. Nagel

I call that chance which is nothing but want of art.

J. Arbuthnot

The concept of probability is a generalization of the concepts of truth and falsehood.

J. Lucas

The probability of an event is the reason we have to believe that it has taken place or will take place.

S. Poisson

Probability is the science of uncertainty.

G. Grimmett

Probability is the reason that we have to think that an event will occur, or that a proposition is true.

G. Boole

Probability describes the various degrees of rational belief about a proposition given different amount of knowledge.

J. M. Keynes

Probability is likeness to be true.

J. Locke

An event will on a long run of trials tend to occur with a frequency proportional to its probability.

R. L. Ellis

One regards two events as equally probable when one can see no reason that would make one more probable than the other.

P. Laplace

The probability of an event is the ratio of the number of cases that are favourable to it, to the number of possible cases, when there is nothing to make us believe that one case should occur rather than any other.

P. Laplace

Probability is a feeling of the mind.

A. de Morgan

Probability is a function of two propositions.

H. Jeffreys

The probability of an event is the ratio between the value at which an expectation depending on the happening of the event ought to be computed, and the value of the thing expected upon its happening.

T. Bayes

To have p chances of a, and q chances of b, is worth $(ap + bq)/(p + q)$.

C. Huygens

Probability is a degree of possibility.

G. Leibniz

The limiting value of the relative frequency of an attribute will be called the probability of that attribute.

R. von Mises

The probability attributed by an individual to an event is revealed by the conditions under which he would be disposed to bet on that event.

B. de Finetti

Probability does not exist.

B. de Finetti

Personalist views hold that probability measures the confidence that a particular individual has in the truth of a particular proposition.

L. Savage

The probability of an outcome is our estimate for the most likely fraction of a number of repeated observations that will yield that outcome.

R. Feynman

It is likely that the word 'probability' is used by logicians in one sense and by statisticians in another.

F. P. Ramsey

1.11 APPENDIX II. REVIEW OF SETS AND FUNCTIONS

It is difficult to make progress in any branch of mathematics without using the ideas and notation of sets and functions. Indeed it would be perverse to try to do so, since these ideas and notation are very helpful in guiding our intuition and solving problems. (Conversely, almost the whole of mathematics can be constructed from these few simple concepts.) We therefore give a brief synopsis of what we need here, for completeness, although it is very likely that the reader will be familiar with all this already.

Sets

A *set* is a collection of things that are called the elements of the set. The elements can be any kind of entity: numbers, people, poems, blueberries, points, lines, and so on, endlessly.

For clarity, upper case letters are always used to denote sets. If the set S includes some element denoted by x, then we say x belongs to S, and write $x \in S$. If x does not belong to S, then we write $x \notin S$.

There are essentially two ways of defining a set, either by a *list* or by a *rule*.

Example 1.11.1.   If S is the set of numbers shown by a conventional die, then the *rule* is that S comprises the integers lying between 1 and 6 inclusive. This may be written formally as follows:
$$S = \{x: 1 \leqslant x \leqslant 6 \text{ and } x \text{ is an integer}\}.$$
Alternatively S may be given as a *list*:
$$S = \{1, 2, 3, 4, 5, 6\}. \qquad\qquad \bigcirc$$

One important special case arises when the rule is impossible; for example, consider the set of elephants playing football on Mars. This is impossible (there is no pitch on Mars) and the set therefore is empty; we denote the empty set by $\varnothing$. We may write $\varnothing$ as $\{\}$.

If S and T are two sets such that every element of S is also an element of T, then we say that T includes S, and write either $S \subseteq T$ or $S \subset T$. If $S \subset T$ and $T \subset S$ then S and T are said to be equal, and we write $S = T$.

Note that $\varnothing \subset S$ for every S. Note also that some books use the symbol '$\subseteq$' to denote inclusion and reserve '⊂' to denote strict inclusion, that is to say, $S \subset T$ if every element of S is in T, and some element of T is not in S. We do not make this distinction.

Combining sets

Given any non-empty set, we can divide it up, and given any two sets, we can join them together. These simple observations are important enough to warrant definitions and notation.

Definition.   Let A and B be sets. Their *union*, denoted by $A \cup B$, is the set of elements that are in A or B, or in both. Their *intersection*, denoted by $A \cap B$, is the set of elements in both A and B.   △

Note that in other books the union may be referred to as the *join* or *sum*; the intersection may be referred to as the *meet* or *product*. We do not use these terms. Note the following.

Definition.   If $A \cap B = \varnothing$, then A and B are said to be *disjoint*.   △

We can also remove bits of sets, giving rise to set differences, as follows.

Definition.   Let A and B be sets. That part of A which is not also in B is denoted by $A \setminus B$, called the *difference* of A from B. Elements which are in A or B but not both, comprise the *symmetric difference*, denoted by $A \vartriangle B$.   △

Finally we can combine sets in a more complicated way by taking elements in pairs, one from each set.

Definition.   Let A and B be sets, and let
$$C = \{(a, b): a \in A, \, b \in B\}$$
be the set of ordered pairs of elements from A and B. Then C is called the *product* of A and B and denoted by $A \times B$.   △

Example 1.11.2.   Let A be the interval $[0, a]$ of the x-axis, and B the interval $[0, b]$ of the y-axis. Then $C = A \times B$ is the rectangle of base a and height b with its lower left vertex at the origin, when $a, b > 0$.   ○

Venn diagrams

The above ideas are attractively and simply expressed in terms of *Venn diagrams*. These provide very expressive pictures, which are often so clear that they make algebra redundant. See figure 1.10.

In probability problems, all sets of interest A lie in a universal set Ω, so that $A \subset \Omega$ for all A. That part of Ω which is not in A is called the *complement* of A, denoted by A^c. Formally
$$A^c = \Omega \setminus A = \{x: x \in \Omega, x \notin A\}.$$
Obviously, from the diagram or by consideration of the elements
$$A \cup A^c = \Omega, \quad A \cap A^c = \varnothing, \quad (A^c)^c = A.$$
Clearly $A \cap B = B \cap A$ and $A \cup B = B \cup A$, but we must be careful when making more intricate combinations of larger numbers of sets. For example, we cannot write down simply $A \cup B \cap C$; this is not well defined because it is not always true that
$$(A \cup B) \cap C = A \cup (B \cap C).$$

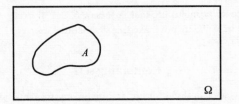

Figure 1.10. The set A is included in the universal set Ω.

We use the obvious notation

$$\bigcup_{r=1}^{n} A_r = A_1 \cup A_2 \cup \cdots \cup A_n,$$

$$\bigcap_{r=1}^{n} A_r = A_1 \cap A_2 \cap \cdots \cap A_n.$$

Definition.   If $A_j \cap A_k = \emptyset$, for $j \neq k$, and

$$\bigcup_{r=1}^{n} A_r = \Omega,$$

then the collection $(A_r; 1 \leq r \leq n)$ is said to form a *partition* of Ω. $\triangle$

Size

When sets are countable it is often useful to consider the number of elements they contain; this is called their *size* or *cardinality*. For any set A, we denote its size by $|A|$; when sets have a finite number of elements, it is easy to see that size has the following properties.

If sets A and B are disjoint then

$$|A \cup B| = |A| + |B|,$$

and more generally, when A and B are not necessarily disjoint,

$$|A \cup B| + |A \cap B| = |A| + |B|.$$

Naturally $|\emptyset| = 0$, and if $A \subseteq B$ then

$$|A| \leq |B|.$$

Finally, for the product of two such finite sets $A \times B$ we have

$$|A \times B| = |A| \times |B|.$$

When sets are infinite or uncountable, a great deal more care and subtlety is required in dealing with the idea of size. However, we can see intuitively that we can consider the length of subsets of a line, or areas of sets in a plane, or volumes in space, and so on. It is easy to see that if A and B are two subsets of a line, with lengths $|A|$ and $|B|$ respectively, then in general

$$|A \cup B| + |A \cap B| = |A| + |B|.$$

Therefore $|A \cup B| = |A| + |B|$ when $A \cap B = \emptyset$.

We can define the product of two such sets as a set in the plane with area $|A \times B|$, which satisfies the well-known elementary rule for areas and lengths

$$|A \times B| = |A| \times |B|$$

and is thus consistent with the finite case above. Volumes and sets in higher dimensions satisfy similar rules.

Functions

Suppose we have sets A and B, and a rule that assigns to each element a in A a unique element b in B. Then this rule is said to define a *function* from A to B; for the corresponding elements we write
$$b = f(a).$$
Here the symbol $f(\cdot)$ denotes the rule or function; often we just call it f. The set A is called the domain of f, and the set of elements in B that can be written as $f(a)$ for some a is called the range of f; we may denote the range by R.

Anyone who has a calculator is familiar with the idea of a function. For any function key, the calculator will supply $f(x)$, if x is in the domain of the function; otherwise it says 'error'.

Inverse function

If f is a function from A to B, we can look at any b in the range R of f and see how it arose from A. This defines a rule assigning elements of A to each element of R, so if the rule assigns a unique element a to each b this defines a function from R to A. It is called the *inverse function* and is denoted by $f^{-1}(\cdot)$:
$$a = f^{-1}(b).$$

Example 1.11.3: indicator function.   Let $A \subset \Omega$ and define the following function $I(\cdot)$ on Ω:
$$I(\omega) = 1 \quad \text{if } \omega \in A,$$
$$I(\omega) = 0 \quad \text{if } \omega \notin A.$$
Then I is a function from Ω to $\{0, 1\}$; it is called the *indicator* of A, because by taking the value 1 it indicates that $\omega \in A$. Otherwise it is zero. ○

This is about as simple a function as you can imagine, but it is surprisingly useful. For example, note that if A is finite you can find its size by summing $I(\omega)$ over all ω:
$$|A| = \sum_{\omega \in \Omega} I(\omega).$$

1.12 PROBLEMS

Note well: these are not necessarily mathematical problems; an essay may be a sufficient answer. They are intended to provoke thought about your own ideas of probability, which you may well have without realizing the fact.

1. Which of the definitions of probability in Appendix I do you prefer? Why? Can you produce a better one?

2. Is there any fundamental difference between a casino and an insurance company? If so, what is it? (Do not address moral issues.)

3. You may recall the classic paradox of Buridan's mule. Placed midway between two equally enticing bales of hay, it starved to death because it had no reason to choose one rather than the other. Would a knowledge of probability have saved it? (The paradox is first recorded by Aristotle.)

4. Suppose a coin showed heads 10 times consecutively. If it looked normal, would you nevertheless begin to doubt its fairness?

5.   Suppose Alf says his dice are fair, but Bill says they are crooked. They look OK. What would you do to decide the issue?

6.   What do you mean by risk? Many public and personal decisions seem to be based on the premise that the risks presented by food additives, aircraft disasters, and prescribed drugs are comparable with the risks presented by smoking, road accidents, and heart disease. In fact the former group present negligible risks compared with the latter. Is this rational? Is it comprehensible? Formulate your own view *accurately*.

7.   What kind of existence does chance have? (Hint: What kind of existence do numbers have?)

8.   It has been argued that seemingly chance events are not really random; the uncertainty about the outcome of the roll of a die is just an expression of our inability to do the mechanical calculations. This is the *deterministic* theory. Samuel Johnson remarked that determinism erodes free will. Do you think you have free will? Does it depend on the existence of chance?

9.   'Probability serves to determine our hopes and fears' – Laplace. Discuss what Laplace meant by this.

10. 'Probability has nothing to do with an isolated case' – A. Markov. What did Markov mean by saying this? Do you agree?

11. 'That the chance of gain is naturally overvalued, we may learn from the universal success of lotteries' – Adam Smith (1776). 'If there were no difference between objective and subjective probabilities, no rational person would play games of chance for money' – J. M. Keynes (1921).
     Discuss.

12. A proportion f of \$100 bills are forgeries. What is the value to you of a proffered \$100 bill?

13. Flip a coin 100 times and record the relative frequency of heads over five-flip intervals as a graph.

14. Flip a broad-headed pin 100 times and record the relative frequency of 'point up' over five-flip intervals.

15. Pick a page of the local residential telephone directory at random. Pick 100 telephone numbers at random (a column or so). Find the proportion p_2 of numbers whose last digit is odd, and also the proportion p_1 of numbers whose first digit is odd. (Ignore the area code.) Is there much difference?

16. Open a book at random and find the proportion of words in the first 10 lines that begin with a vowel. What does this suggest?

17. Show that $A = \emptyset$ if and only if $B = A \triangle B$.

18. Show that if $A \subseteq B$ and $B \subseteq A$ then $A = B$.

Part A

Probability

2

The rules of probability

Probability serves to define our hopes and fears.

P. Laplace

2.1 PREVIEW

In the preceding chapter we suggested that a model is needed for probability, and that this model would take the form of a set of rules. In this chapter we formulate these rules. When doing this, we shall be guided by the various intuitive ideas of probability as a relative of proportion that we discussed in Chapter 1. We begin by introducing the essential vocabulary and notation, including the idea of an event. After some elementary calculations, we introduce the addition rule, which is fundamental to the whole theory of probability, and explore some of its consequences.

Most importantly we also introduce and discuss the key concepts of conditional probability and independence. These are exceptionally useful and powerful ideas and work together to unlock many of the routes to solving problems in probability. By the end of this chapter you will be able to tackle a remarkably large proportion of the better-known problems of chance.

Prerequisites. We shall use the routine methods of elementary algebra, together with the basic concepts of sets and functions. If you have any doubts about these, refresh your memory by a glance at appendix II of chapter 1.

2.2 NOTATION AND EXPERIMENTS

From everyday experience, you are familiar with many ideas and concepts of probability; this knowledge is gained by observation of lotteries, board games, sport, the weather, futures markets, stock exchanges, and so on. You have various ways of discussing these random phenomena, depending on your personal experience. However, everyday discourse is too diffuse and vague for our purposes. We need to become routinely much more precise. For example, we have been happy to use words such as chance, likelihood, probability, and so on, more or less interchangeably. In future we shall confine ourselves

Table 2.1.

Procedure	Outcomes
Roll a die	One of 1, 2, 3, 4, 5, 6
Run a horse race	Some horse wins it, or there is a dead heat (tie)
Buy a lottery ticket	Your number either is or is not drawn

to using the word *probability*. The following are typical statements in this context.

The probability of a head is $\frac{1}{2}$.
The probability of rain is 90%.
The probability of a six is $\frac{1}{6}$.
The probability of a crash is 10^{-9}.

Obviously we could write down an endless list of probability statements of this kind; you should write down a few yourself (*exercise*). However, we have surely seen enough such assertions to realize that useful statements about probability can generally be cast into the following general form:

(1)   The probability of A is p.

In the above examples, A was 'a head', 'rain', 'a six', and 'a crash'; and p was '$\frac{1}{2}$', '90%', '$\frac{1}{6}$', and '10^{-9}' respectively. We use this format so often that, to save ink, wrists, trees, and time, it is customary to write (1) in the even briefer form

(2) $P(A) = p$.

This is obviously an extremely efficient and compact written representation; it is still pronounced as 'the probability of A is p'. A huge part of probability depends on equations similar to (2).

Here, the number p denotes the position of this probability on the probability scale discussed in chapter 1. It is most important to remember that on this scale

(3) $0 \leqslant p \leqslant 1$

If you ever calculate a probability outside this interval then it must be wrong!

We shall look at any event A, the probability function $P(\cdot)$, and probability $P(A)$ in detail in the next sections. For the moment we continue this section by noting that underlying every such probability statement is some procedure or activity with a random outcome; see table 2.1.

Useful probability statements refer to these outcomes. In everyday parlance this procedure and the possible outcomes are often implicit. In our new rigorous model this will not do. Every procedure and its possible outcomes must be completely explicit; we stress that if you do not follow this rule you will be very likely to make mistakes. (There are plenty of examples to show this.) To help in this task, we introduce some very convenient notation and jargon to characterize all such trials, procedures, and actions.

Definition.   Any activity or procedure that may give rise to a well-defined set of outcomes is called an *experiment*.   △

Definition. The set of all possible outcomes is denoted by Ω, and called the *sample space*. $\triangle$

The adjective 'well-defined' in the first definition just means that you know what all the possibilities of the experiment are, and could write them down if challenged to do so. Prior to the experiment you do not know for sure what the outcome will be; when you carry out the experiment it yields an outcome called the result. Often this result will have a specific label such as 'heads' or 'it rains'. In general, when we are not being specific, we denote the result of an experiment by ω. Obviously $\omega \in \Omega$; that is to say, the result always lies in the set of possible outcomes. For example, if Ω is the set of possible outcomes of a horse race in which Dobbin is a runner, then

$$\{\text{Dobbin wins}\} \subseteq \Omega$$

In this expression the curly brackets are used to alert you to the fact that what lies inside them is one (or more) of the possible outcomes.

We conclude with two small but important points. First, any experiment can have many different sample spaces attached to it.

Example 2.2.1. If you flip a coin twice, you may define

$$\Omega = \{HH,\ HT,\ TH,\ TT\}.$$

This lists everything that can arise from the experiment, so no larger sample space can be more informative. However, you may only be interested in the number of heads shown. In this case you may define

$$\Omega = \{0,\ 1,\ 2\}$$

and be quite satisfied. $\bigcirc$

The second point is in a sense complementary to the first. It is that you have little to lose by choosing a large enough sample space to be sure of including every possible outcome, even where some are implausible.

Example. 2.2.2 Suppose you are counting the number of pollen grains captured by a filter. A suitable sample space is the set of all non-negative integers

$$\Omega = \{0,\ 1,\ 2,\ 3,\ \ldots\}.$$

Obviously only a finite number of these are possible (since there is only a finite amount of pollen in existence), but any cut-off point would be unpleasantly arbitrary, and might be too small. $\bigcirc$

Furthermore it makes a calculation much easier if any number is allowed as a possibility, however unlikely. Otherwise it might be necessary to keep track of this bound throughout the calculation, a prospect which is both formidable and boring.

Exercise for section 2.2

1. Describe a suitable sample space Ω for the following experiments.
   (a) *m* balls are removed from an urn containing *j* jet (black) balls and *k* khaki balls.

(b)   The number of cars passing over a bridge in one week is counted.

(c)   Two players play the best of three sets at tennis.

(d)   You deal a poker hand to each of four players.

2.3 EVENTS

Suppose we have some experiment whose outcomes ω comprise Ω, the sample space. As we have noted above, the whole point of probability is to say how likely the outcomes are, either individually or collectively. We therefore make the following definition.

Definition.   An *event* is a subset of the sample space Ω.   △

Thus each event comprises one or more possible outcomes ω. By convention, events are always denoted by capital letters such as A, B, C, ..., with or without suffixes, super-fixes, or other adornments such as hats, bars, or stars. Here are a few simple but common examples.

Example 2.3.1: two dice are rolled

(i)   If we record their scores, then the sample space is

$$\Omega = \{(i, j): 1 \leqslant i, j \leqslant 6\}.$$

We may be interested in the event A that the first die is even and the second odd, which is

$$A = \{(i, j): i \in \{2, 4, 6\}, j \in \{1, 3, 5\}\}.$$

(ii) If we add their scores, then the sample space is

$$\Omega = \{k: 2 \leqslant k \leqslant 12\}.$$

The event B that the sum of the scores is 7 or 11 is

$$B = \{7, 11\} \qquad\qquad ○$$

Example 2.3.2.   Suppose you record the number of days this week on which it rains. The sample space is

$$\Omega = \{0, 1, 2, 3, 4, 5, 6, 7\}.$$

One outcome is that it rains on one day,

$$\omega_1 = 1.$$

The event that it rains more often than not is

$$A = \{4, 5, 6, 7\}, \qquad\qquad ○$$

comprising the outcomes 4, 5, 6, 7.

Example 2.3.3.   A doctor weighs a patient to the nearest pound. Then, to be on the safe side, we may agree that

$$\Omega = \{i: 0 \leqslant i \leqslant 20\,000\}.$$

Some outcomes here seem unlikely, or even impossible, but we lose little by including them. Then

$$C = \{i: 140 \leqslant i \leqslant 150\}$$

is the event that the patient weighed between 140 and 150 pounds. ○

Example 2.3.4. An urn contains a amber and b buff balls. Of these, c balls are removed without replacing any in the urn, where

$$c \leqslant \min\{a, b\} = a \wedge b.$$

Then Ω is the collection of all possible sequences of a's and b's of length c. We may define the event D that the number of a's and b's removed is the same. If c is odd, then this is the impossible event $\varnothing$. ○

Since events are sets, we can use all the standard ideas and notation of set theory as summarized in appendix II of Chapter 1. If the outcome ω of an experiment lies in the event A, then A is said to *occur*, or happen. In this case we have $\omega \in A$. We always have $A \subseteq \Omega$. If A does not occur, then obviously the complementary event A^c must occur, since ω lies in one of A or A^c.

The notation and ideas of set theory are particularly useful in considering combinations of events.

Example 2.3.5. Suppose you take one card from a conventional pack. Simple events include

$$A \equiv \text{the card is an ace,}$$

$$B \equiv \text{the card is red,}$$

$$C \equiv \text{the card is a club.}$$

More interesting events are denoted using the operations of union and intersection. For example

$$A \cap C \equiv \text{the card is the ace of clubs,}$$

$$A \cup B \equiv \text{the card is either red or an ace or both.}$$

Of course the card cannot be red and a club, so we have $B \cap C = \varnothing$, where $\varnothing$ denotes the impossible event, otherwise known as the empty set. ○

This leads to a useful definition.

Definition. Two events A and B are said to be *disjoint*, incompatible, or mutually exclusive, if they have no outcome in common. That is to say

$$A \cap B = \varnothing, \qquad\qquad \triangle$$

Obviously for any event A it is true that

$$A \cap A^c = \varnothing.$$

Notice that we have tacitly assumed that if we form any combination of events A and B, such as $C = A \cup B$ or $D = A \cap B$, then C and D are themselves events. This is almost completely obvious, but the point needs to be stressed: *unions, intersections, and complements of events are themselves events.*

Table 2.2. *Events and notation*

Certain event or sample space	Ω
Impossible event	$\varnothing$
The event A occurs	A
A does not occur	A^c
Both A and B occur	$A \cap B$
Either or both of A and B occur	$A \cup B$
If A occurs then B occurs	$A \subseteq B$
A occurs, but not B	$A \setminus B$
A and B are disjoint	$A \cap B = \varnothing$

Table 2.2 gives a brief summary of how set notation represents events and their relationships.

Definition. If $A_1, A_2, A_3, \ldots, A_n, \ldots$ is a collection of events such that $A_i \cap A_j = \varnothing$ for all $i \neq j$, and $\bigcup_i A_i = \Omega$, then the A's are said to form a *partition* of Ω. $\triangle$

As we have remarked above, many important relationships between events are very simply and attractively demonstrated by means of Venn diagrams. For example, figure 2.1 demonstrates very neatly that

$$(A \cup B) \cap C = (A \cap C) \cup (B \cap C) \quad \text{and} \quad A \cup (B \cap C) = (A \cup B) \cap (A \cup C).$$

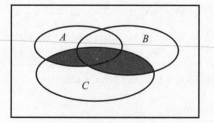

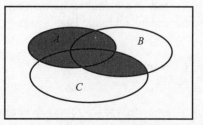

Figure 2.1. Venn diagrams. In the upper figure the shaded area is equal to $(A \cup B) \cap C$ and $(A \cap C) \cup (B \cap C)$. In the lower figure the shaded area is equal to $A \cup (B \cap C)$ and also to $(A \cup B) \cap (A \cup C)$.

Exercises for section 2.3

1. Here is a list of experiments, each with an associated event of interest. In each case write down a suitable sample space Ω, and define the event of interest in terms of the elements of Ω.
   (a) Two dice are rolled; their sum is 3.
   (b) 100 light bulbs are tested; at most 4 are defective.
   (c) A family comprises 3 children; they are all of the same sex.
   (d) The height of a tree is recorded to the nearest metre; it is between 10 and 15 metres high.
   (e) Rod and Fred play the best of three sets at tennis; Rod wins.
   (f) Tom and Jack are weighed to the nearest pound; Jack is heavier.

2. Show that for any events A and B
   (a) $(A \cup B)^c = A^c \cap B^c$
   (b) $A \cap (B \cup C) = (A \cap B) \cup (A \cap C)$.

2.4 PROBABILITY; ELEMENTARY CALCULATIONS

Now that we have defined events, we can discuss their probabilities. Suppose some experiment has outcomes that comprise Ω, and A is an event in Ω. Then the probability that A occurs is denoted by $P(A)$, where of course

$$(1) \qquad\qquad 0 \leqslant P(A) \leqslant 1.$$

Thus we can think of P as a rule that assigns a number $P(A) \in [0, 1]$ to each event A in Ω. The mathematical term for a rule like this is a function, as we discussed in appendix II of chapter 1. Thus $P(\cdot)$ is a function of the events in Ω, which takes values in the interval $[0, 1]$. Before looking at the general properties of P, it seems sensible to gain experience by looking at some simple special cases that are either familiar, or obvious, or both.

Example 2.4.1: Bernoulli trial. Many experiments have just two outcomes, for example: head or tail; even or odd; win or lose; fly or crash; and so on. To simplify matters these are often thought of as examples of an experiment with the two outcomes success or failure, denoted by S and F respectively. Then for the events S and F we write

$$P(S) = p \quad \text{and} \quad P(F) = q = 1 - p. \qquad\qquad\qquad \bigcirc$$

Example 2.4.2: equally likely outcomes. Suppose that an experiment has sample space Ω, such that each of the $|\Omega|$ outcomes in Ω is equally likely. This may be due to some physical symmetry or the conditions of the experiment. Now let A be any event; the number of outcomes in A is $|A|$. The equidistribution of probability among the outcomes implies that the probability that A occurs should be proportional to $|A|$ (we discussed this at length in section 1.4). In cases of this type we therefore write

$$(2) \qquad\qquad P(A) = \frac{|A|}{|\Omega|} = \frac{\text{number of outcomes in } A}{\text{number of outcomes in } \Omega}. \qquad\qquad \bigcirc$$

This is a large assumption, but it is very natural and compelling. It is so intuitively attractive that it was being used explicitly in the 16th century, and it is clearly implicit in the ideas and writings of several earlier mathematicians. Let us consider some further examples of this idea in action.

Example 2.4.3.   Two dice are rolled, one after the other. Let A be the event that the second number is greater than the first. Here

$$|\Omega| = 36$$

as there are 6×6 pairs of the form (i, j), with $1 \leqslant i, j \leqslant 6$. The pairs with $i < j$ are given by

$$A = \{(5, 6), (4, 5), (4, 6), (3, 4), \ldots, (1, 5), (1, 6)\};$$

it is easy to see that $|A| = 1 + 2 + \cdots + 5 = 15$. Hence

$$P(A) = \frac{15}{36} = \frac{5}{12}.$$

○

Example 2.4.4.   If a coin is flipped n times, what is the probability that each time the same face shows? Here Ω is the set of all possible sequences of H and T of length n, which we regard as equally likely. Hence $|\Omega| = 2^n$. The event A comprises all heads or all tails. Hence $|A| = 2$ and the required probability is

$$\frac{2}{2^n} = 2^{-(n-1)}.$$

○

Example 2.4.5: chain.   Suppose you are testing a chain to destruction. It has n links and is stretched between two shackles attached to a ram. The ram places the chain under increasing tension until a link fails. Any link is equally likely to be the one that snaps, and so if A is any group of links the probability that the failed link is in A is the proportion of the total number of links in A. Now $|\Omega| = n$, so

$$P(A) = \frac{|A|}{n}.$$

○

Example 2.4.6: lottery.   Suppose you have an urn containing 20 tickets marked 1, 2, $\ldots$, 20. A ticket is drawn at random. Thus

$$\Omega = \{1, 2, \ldots, 20\} = \{n: 1 \leqslant n \leqslant 20\}.$$

Events in Ω may include:

$$A \equiv \text{the number drawn is even};$$
$$B \equiv \text{the number drawn is divisible by 5};$$
$$C \equiv \text{the number drawn is less than 8}.$$

The implication of the words 'at random' is that any number is equally likely to be chosen. In this case our discussions above yield

$$P(A) = \frac{|A|}{|\Omega|} = \frac{10}{20} = \frac{1}{2},$$

$$P(B) = \frac{|B|}{|\Omega|} = \frac{4}{20} = \frac{1}{5},$$

and

$$P(C) = \frac{|C|}{|\Omega|} = \frac{7}{20}.$$

○

Here is a famous example.

Example 2.4.7: dice. Three dice are rolled and their scores added. Are you more likely to get 9 than 10, or vice versa?

Solution. There are $6^3 = 216$ possible outcomes of this experiment, as each die has six possible faces. You get a sum of 9 with outcomes such as (1, 2, 6), (2, 1, 6), (3, 3, 3) and so on. Tedious enumeration reveals that there are 25 such triples, so

$$P(9) = \frac{25}{216}.$$

A similar tedious enumeration shows that there are 27 triples, such as (1, 3, 6), (2, 4, 4) and so on, that sum to 10. So

$$P(10) = \frac{27}{216} > P(9).$$

This problem was solved by Galileo before 1642. ○

An interesting point about this example is that if your sample space does not distinguish between outcomes such as (1, 2, 6) and (2, 1, 6), then the possible sums 9 and 10 can each be obtained in the same number of different ways, namely six. However, actual experiments with real dice demonstrate that this alternative model is wrong.

Notice that symmetry is quite a powerful concept, and implies more than at first appears.

Example 2.4.8. You deal a poker hand of five cards face down. Now pick up the fifth and last card dealt; what is the probability that it is an ace? The answer is

$$P(A) = \frac{1}{13}.$$

Sometimes it is objected that the answer should depend on the first four cards, but of course if these are still face down they cannot affect the probability we want. By symmetry *any* card has probability $\frac{1}{13}$ of being an ace; it makes no difference whether the pack is dealt out or kept together, as long as only one card is actually inspected. ○

Our intuitive notions of symmetry and fairness enable us to assign probabilities in some other natural and appealing situations.

Example 2.4.9: rope. Suppose you are testing a rope to destruction: a ram places it under increasing tension until it snaps at a point S, say. Of course we suppose that the rope appeared uniformly sound before the test, so the failure point S is equally likely to be any point of the rope. If the rope is of length 10 metres, say, then the probability that it fails within 1 metre of each end is naturally P(S lies in those 2 metres) $= \frac{2}{10} = \frac{1}{5}$. Likewise the probability that S lies in any 2 metre length of rope is $\frac{1}{5}$, as is the probability that S lies in any 2 metres of the rope, however this 2 metres is made up. ○

Example 2.4.10: meteorite. Leaving your house one morning at 8 a.m. you find that a meteorite has struck your car. When did it do so? Obviously meteorites take no account of our time, so the time T of impact is equally likely to be any time between your leaving the car and returning to it. If this interval was 10 hours, say, then the probability that the meteorite fell between 1 a.m. and 2 a.m. is $\frac{1}{10}$. $\qquad\qquad\bigcirc$

These are special cases of experiments in which the outcome is equally likely to be any point of some interval $[a, b]$, which may be in time or space.

We can define a probability function for this kind of experiment quite naturally as follows. The sample space Ω is the interval $[a, b]$ of length $b - a$. Now let A be any interval in Ω of length $l = |A|$. Then the equidistribution of probability among the outcomes in Ω implies that the probability of the outcome being in A is

$$\mathrm{P}(A) = \frac{|A|}{|\Omega|} = \frac{l}{b - a}.$$

Example 2.4.9. revisited: rope. What is the probability that the rope fails nearer to the fixed point of the ram than the moving point? There are 5 metres nearer the fixed point, so this probability is $\frac{5}{10} = \frac{1}{2}$. $\qquad\qquad\bigcirc$

This argument is equally natural and appealing for points picked at random in regions of the plane. For definiteness, let Ω be some region of the plane with area $|\Omega|$. Let A be some region in Ω of area $|A|$. Suppose a point P is picked at random in Ω, with any point equally likely to be chosen. Then the equidistribution of probability implies that the probability of P being in A is

$$\mathrm{P}(A) = \frac{|A|}{|\Omega|}.$$

Example 2.4.10. The garden of your house is the region Ω, with area 100 square metres; it contains a small circular pond A of radius 1 metre. You are telephoned by a neighbour who tells you your garden has been struck by a small meteorite. What is the probability that it hit the pond? Obviously, by everything we have said above

$$\mathrm{P}(\text{hits pond}) = \frac{|A|}{|\Omega|} = \frac{\pi}{100}. \qquad\qquad\bigcirc$$

We return to problems of this type later on.

Let us conclude this section with an example which demonstrates a simple but important point.

Example 2.4.11: probabilistic equivalence. Consider the following three experiments.

(i)   A fair die is rolled, and the number shown is noted.
(ii) An urn contains 5 alabaster balls, and 1 beryl ball. They are removed until the beryl ball appears, and the number removed is noted.
(iii) Six different numbers are drawn for a lottery and the position in which the smallest of them is drawn is noted.

In each of these the sample space can be written

$$\Omega = \{1, 2, 3, 4, 5, 6\},$$

and by construction, in all three cases,

$$P(\omega_i) = \frac{1}{6}, \quad 1 \le i \le 6.$$

These experiments are probabilistically equivalent, since Ω and $P(\cdot)$ are essentially the same for each. ○

Since it makes no difference which experiment we use to yield these probabilities, there is something to be said for having a standard format to present the great variety of probability problems. For reasons of tradition, urns are often used.

Thus example 2.4.11(ii) would be a standard presentation of the probabilities above. There are three main reasons for this. The first is that urns are often useful in situations too complicated to be readily modelled by coins and dice. The second reason is that using urns (instead of more realistic descriptions) enables the student to see the probabilistic problems without being confused by false intuition. The third reason is historical: urns were widely used in conducting lotteries and elections (in both cases because they are opaque, thus preventing cheating in the first place and allowing anonymous voting in the second). It was therefore natural for early probabilists to use urns as models of real random behaviour.

Exercises for section 2.4

1. You make one spin of a roulette wheel. (Assume your wheel has 37 pockets numbered from 0 to 36 inclusive.) What is the probability that the outcome is odd?

2. You take two cards at random from a pack of 52. What is the probability that both are aces?

3. You pick a point P at random in a triangle ABC. What is the probability that is in ABD, where D lies in the side BC?

4. You pick a point at random in a circle. What is the probability that it is nearer to the centre than to the perimeter?

5. An urn contains a amber balls and b beryl balls. You remove them one by one. What is the probability that the last but one ball is beryl? (Hint: No one said you looked at them as you removed them.)

2.5 THE ADDITION RULES

Of course not all experiments have equally likely outcomes, so we need to fix rules that tell us about the properties of the probability function P, in general. Naturally we continue to require that for any event A

(1) $$0 \le P(A) \le 1,$$

and in particular the certain event has probability 1, so

(2) $$P(\Omega) = 1.$$

The most important rule is the following.

Addition rule. If A and B are disjoint events, then

(3) $$P(A \cup B) = P(A) + P(B).$$

This rule lies at the heart of probability. First let us note that we need such a rule, because $A \cup B$ *is* an event when A and B are events, and we therefore need to know its probability. Second, note that it follows from (3) (by induction) that if $A_1, A_2, \ldots, A_n$ is any collection of disjoint events then

(4) $P\left(\bigcup_{i=1}^{n} A_i\right) = P(A_1) + \cdots + P(A_n).$

The proof forms exercise 3, at the end of this section.

   Third, note that it is sometimes too restrictive to confine ourselves to a finite collection of events (we have seen several sample spaces with infinitely many outcomes), and we therefore need an extended version of (4).

Extended addition rule.   If $A_1, A_2, \ldots$ is a collection of disjoint events then

(5) $P(A_1 \cup A_2 \cup \cdots) = P(A_1) + P(A_2) + \cdots.$

Equation (5) together with (1) and (2),

                       $0 \leqslant P(A) \leqslant 1 \quad \text{and} \quad P(\Omega) = 1,$

are sometimes said to be the *axioms* of probability. They describe the behaviour of the probability function P defined on subsets of Ω. In fact, in everyday usage P is not referred to as a probability function but as a probability distribution. Formally we state the following.

Definition.   Let Ω be a sample space and suppose that $P(\cdot)$ is a probability function on a family of subsets of Ω satisfying (1), (2), and (5). Then P is called a *probability distribution* on Ω. $\triangle$

The word distribution is used because it is natural to think of probability as something that is distributed over the outcomes in Ω. The function P tells you just how it is distributed. In this respect probability behaves like distributed mass, and indeed in many books authors do speak of a unit of probability mass being distributed over the sample space, and refer to P as a probability mass function. This metaphor can be a useful aid to intuition because, of course, mass obeys exactly the same addition rule. If two distinct objects A and B have respective masses $m(A)$ and $m(B)$, then the mass of their union $m(A \cup B)$ satisfies

                            $m(A \cup B) = m(A) + m(B).$

Of course mass is non-negative also, which reinforces the analogy.

   We conclude this section by showing how the addition rule is consistent with, and suggested by, all our interpretations of probability as a proportion.

   First, consider an experiment with equally likely outcomes, for which we defined probability as the proportion

                                $P(A) = \frac{|A|}{|\Omega|}.$

If A and B are disjoint then, trivially,

                             $|A \cup B| = |A| + |B|.$

Hence in this case

$$P(A \cup B) = \frac{|A \cup B|}{|\Omega|} = \frac{|A|}{|\Omega|} + \frac{|B|}{|\Omega|} = P(A) + P(B).$$

Second, consider the interpretation of probability as reflecting relative frequency in the long run. Suppose an experiment is repeated N times. At each repetition, events A and B may, or may not, occur. If they are disjoint, they cannot both occur at the same repetition. We argued in section 1.8 that the relative frequency of any event should be not too far from its probability. Indeed, it is often the case that the relative frequency $N(A)/N$ of an event A is the only available guide to its probability $P(A)$. Now, clearly

$$N(A \cup B) = N(A) + N(B).$$

Hence, dividing by N, there is a powerful suggestion that we should have

$$P(A \cup B) = P(A) + P(B).$$

Third, consider probability as a measure of expected value. For this case we resurrect the benevolent plutocrat who is determined to give away \$1 at random. The events A and B are disjoint; if A occurs you get \$1 in your left hand, if B occurs you get \$1 in your right hand. If $(A \cup B)^c$ occurs, then Bob gets \$1. The value of this offer to you is \$$P(A \cup B)$, the value to your left hand is \$$P(A)$, and the value to your right hand is \$$P(B)$. But obviously it does not matter in which hand you get the money, so

$$P(A \cup B) = P(A) + P(B).$$

Finally, consider the case where we imagine a point is picked at random anywhere in some plane region Ω of area $|\Omega|$. If $A \subseteq \Omega$, we defined

$$P(A) = \frac{|A|}{|\Omega|}.$$

Since area also satisfies the addition rule, we have immediately, when $A \cap B = \varnothing$, that

$$P(A \cup B) = P(A) + P(B).$$

It is interesting and important to note that in this case the analogy with mass requires the unit probability mass to be distributed uniformly over the region Ω. We can envisage this distribution as a lamina of uniform density $|\Omega|^{-1}$ having total mass unity. This may seem a bizarre thing to imagine, but it turns out to be useful later on.

In conclusion, it seems that the addition rule is natural and compelling in every case where we have any insight into the behaviour of probability. Of course it is a big step to say that it should apply to probability in every other case, but it seems inevitable. And doing so has led to remarkably elegant and accurate descriptions of the real world.

Exercises for section 2.5

1. **Bernoulli trial.** Show that for a Bernoulli trial $P(S) + P(F) = 1$.

2. Why is the extended addition rule not necessary when Ω is finite?

3. Show that (4) follows from (3).

4. Show that $P(A \cap B) \leqslant \min\{P(A), P(B)\}$.

5. Show that for any events A and B, $P(A \cup B) \leqslant P(A) + P(B)$.

2.6 SIMPLE CONSEQUENCES

We have agreed that the probability distribution P satisfies

(1) $P(A) \geqslant 0,$ for any A,

(2) $P(\Omega) = 1,$

(3) $P(A \cup B) = P(A) + P(B),$ when $A \cap B = \varnothing$.

These rules are simple enough, but they have remarkably extensive consequences. Let us look at some simple applications and deductions.

Complements.   We know that for any event A, $\Omega = A \cup A^c$ and $A \cap A^c = \varnothing$. Hence by (3) and (2), and as illustrated in figure 2.2,

$$1 = P(\Omega) = P(A \cup A^c) = P(A) + P(A^c).$$

Thus the probability that A does not occur is

(4) $P(A^c) = 1 - P(A).$

In particular, for the impossible event $\varnothing$,

$$P(\varnothing) = P(\Omega^c) = 1 - P(\Omega) = 0.$$

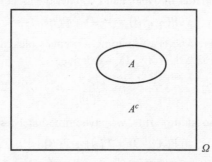

Figure 2.2. $P(A) + P(A^c) = P(\Omega) = 1.$

It is very pleasant to see this consistency with our intuitive probability scale. Note, however, that the converse is not true, that is, $P(A) = 0$ does not imply that $A = \varnothing$, as we now see in an example.

Example 2.6.1.   Pick a point at random in the unit square, say, and let A be the event that this point lies on a diagonal of the square. As we have seen above,

$$P(A) = \frac{|A|}{|\Omega|} = |A|,$$

where $|A|$ is the area of the diagonal. But lines have zero area; so $P(A) = 0$, even though the event A is clearly not impossible. $\bigcirc$

Let us consider some examples of the complement rule (4).

Example 2.6.2.   A die is rolled. How many rolls do you need, to have a better than evens chance of rolling at least one six?

Solution.   If you roll a die r times, there are 6^r possible outcomes. On each roll there are 5 faces that do not show a six, and there are therefore in all 5^r outcomes with no six. Hence P(no six in r rolls) $= 5^r/6^r$. Hence, by (4),

$$\text{P(at least one six)} = 1 - \text{P(no six)} = 1 - \left(\tfrac{5}{6}\right)^r.$$

For this to be better than evens, we need r large enough that $1 - \left(\tfrac{5}{6}\right)^r > \tfrac{1}{2}$. A short calculation shows that $r = 4$.   ○

Example 2.6.3: de Méré's problem.   Which of these two events is more likely?

(i)   Four rolls of a die yield at least one six.
(ii) Twenty-four rolls of two dice yield at least one (6, 6), i.e. double six.

Solution.   Let A denote the first event and B the second event. Then A^c is the event that no six is shown. There are 6^4 equally likely outcomes, and 5^4 of these show no six. Hence by (4)

$$P(A) = 1 - P(A^c) = 1 - \left(\tfrac{5}{6}\right)^4.$$

Likewise

$$P(B) = 1 - P(B^c) = 1 - \left(\tfrac{35}{36}\right)^{24}.$$

Now after a little calculation we find that

$$\tfrac{671}{1296} = P(A) > \tfrac{1}{2} > P(B) \simeq 0.491.$$

So the first event is more likely.   ○

Difference rule.   More generally we have the following rule for differences. Suppose that $B \subseteq A$. Then

$$A = B \cup (B^c \cap A) = B \cup (A \setminus B) \quad \text{and} \quad B \cap (B^c \cap A) = \emptyset.$$

Hence

$$P(A) = P(B) + P(B^c \cap A)$$

and so

(5) $$P(A \setminus B) = P(A) - P(B), \quad \text{if } B \subseteq A.$$

Figure 2.3 almost makes this argument unnecessary.

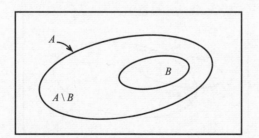

Figure 2.3. When $B \subseteq A$, then $P(A \setminus B) = P(A) - P(B)$.

Of course many events are not disjoint. What can we say of $P(A \cup B)$ when $A \cap B \neq \emptyset$? The answer is given by the following rule.

Inclusion–exclusion rule.   This says that, for any events A and B, the probability that either occurs is given by

(6) $$P(A \cup B) = P(A) + P(B) - P(A \cap B).$$

Proof.   Using the addition rule three times shows that
$$P(A) = P(A \cap B) + P(A \cap B^c),$$
$$P(B) = P(A \cap B) + P(B \cap A^c),$$
$$P(A \cup B) = P(A \cap B) + P(A \cap B^c) + P(B \cap A^c).$$

(Drawing the Venn diagram in Figure 2.4 makes these relations obvious.) Now, adding the first two equalities, and then subtracting that from the third, gives the result. $\square$

Let us consider some examples of this.

Example 2.6.4: wet and windy.   From meteorological records it is known that for a certain island at its winter solstice, it is wet with probability 30%, windy with probability 40%, and both wet and windy with probability 20%.

Using the above rules we can find the probability of other events of interest. For example:

(i) $P(\text{dry}) = P(\text{not wet}) = 1 - 0.3 = 0.7$, by (4);
(ii) $P(\text{dry and windy}) = P(\text{windy} \backslash \text{wet}) = P(\text{windy}) - P(\text{wet and windy}) = 0.2$, by (5);
(iii) $P(\text{wet or windy}) = 0.4 + 0.3 - 0.2 = 0.5$, by (6). $\bigcirc$

Example 2.6.5: alarms.   A kitchen contains two fire alarms; one is activated by smoke and the other by heat. Experience has shown that the probability of the smoke alarm sounding within one minute of a fire starting is 0.95, the probability of the heat alarm sounding within one minute of a fire starting is 0.91, and the probability of both sounding within a minute is 0.88. What is the probability of *at least one alarm* sounding within a minute?

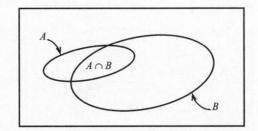

Figure 2.4. It can be seen that $P(A \cup B) = P(A) + P(B) - P(A \cap B)$.

Solution. We have, with an obvious notation,

$$P(H \cup S) = P(H) + P(S) - P(H \cap S)$$

$$= 0.91 + 0.95 - 0.88$$

$$= 0.98. \qquad \bigcirc$$

The inclusion–exclusion rule can be extended to cover more than two events (see the exercises and problems), but expressions become long and tedious. Sometimes it is enough to have bounds for probabilities; here is a famous example.

Boole's inequality. For any events $A_1, \ldots, A_n$, we have

(7) $$P(A_1 \cup A_2 \cup \cdots \cup A_n) \leqslant \sum_{r=1}^{n} P(A_r).$$

Proof. From (6) this is obviously true for $n = 2$. Suppose it is true for some $n \geqslant 2$; then

$$P(A_1 \cup A_2 \cup \cdots \cup A_{n+1}) \leqslant P(A_1 \cup A_2 \cup \cdots \cup A_n) + P(A_{n+1}) \leqslant \sum_{r=1}^{n+1} P(A_r).$$

The result follows by induction. $\qquad \square$

Exercises for section 2.6

1. Suppose that $A \subseteq B$. Show that $P(A) \leqslant P(B)$.

2. ***Wet, windy and warm.*** Show that for any events A (wet), B (windy), and C (warm),
   $$P(A \cup B \cup C) = P(A) + P(B) + P(C) - P(A \cap B) - P(B \cap C) - P(C \cap A) + P(A \cap B \cap C).$$

3. Two dice are rolled. How many rolls do you need for a better than evens chance of at least one double six?

4. ***Galileo's problem (example 2.4.7) revisited.*** Let S_k be the event that the sum of the three dice is k. Find $P(S_k)$ for all k.

5. ***Pepys' problem*** (1693)
   (a) Find the probability that at least 1 six is shown when 6 dice are rolled.
   (b) Find the probability that at least 2 sixes are shown when 12 dice are rolled, and compare with the answer to (a).

   Remark. Pepys put this problem to Isaac Newton, but was then very reluctant to accept Newton's (correct) answer.

6. Show that the probability that exactly one of the events A or B occurs is $P(A) + P(B) - 2P(A \cap B)$.

2.7 CONDITIONAL PROBABILITY; MULTIPLICATION RULE

In real life very few experiments amount to just one action with random outcomes; they usually have a more complicated structure in which some results may be known before the experiment is complete. Or conditions may change. We need a new rule to add to

Table 2.3. *Outcomes of 700 treatments grouped by size of stone*

	Outcomes		
Size	success (S)	failure	total
small	315	42	357
large (L)	247	96	343
total	562	138	700

those given in section 2.5; it is called the conditioning rule. Before we give the rule, here are some examples.

Example 2.7.1.   Suppose you are about to roll two dice, one from each hand. What is the probability that your right-hand die shows a larger number than the left-hand die?

There are 36 outcomes, and in 15 of these the right-hand score is larger. So

(1)   P(right-hand larger) $= \frac{15}{36}$

Now suppose you roll the left-hand die first, and it shows 5. What is the probability that the right-hand die shows more? It is clearly not $\frac{15}{36}$. In fact only one outcome will do: it must show 6. So the required probability is $\frac{1}{6}$.   ○

This is a special case of the general observation that if conditions change then results change. In particular, if the conditions under which an experiment takes place are altered, then the probabilities of the various outcomes may be altered. Here is another illustration.

Example 2.7.2: stones.   Kidney stones are either small, (i.e. <2 cm diameter) or large, (i.e. >2 cm diameter). Treatment can either succeed or fail. For a sequence of 700 patients the stone sizes and outcomes were as shown in table 2.3. Let L denote the event that the stone treated is large. Then, clearly, for a patient selected at random from the 700 patients,

(2)   P(L) $= \frac{343}{700}$.

Also, for a patient picked at random from the 700, the probability of success is

(3)   P(S) $= \frac{562}{700} \simeq 0.8$

However, suppose a patient is picked at random from those whose stone is large. The probability of success is *different* from that given in (3); it is obviously

$$\frac{247}{343} \simeq 0.72.$$

That is to say, the probability of success given that the stone is large is different from the probability of success given no knowledge of the stone.

This is natural and obvious, but it is most important and useful to have a distinct notation, in order to keep the conditions of an experiment clear and explicit in our minds and working. We therefore denote the probability of S given L by

(4) $$P(S|L) \simeq 0.72$$

and refer to $P(S|L)$ as the conditional probability of success, given L.

Likewise, we could imagine selecting a patient at random from those whose treatment was successful, and asking for the probability that such a patient's stone had been large. This is the conditional probability of L given S, and, from table 2.1, we easily see that

(5) $$P(L|S) = \tfrac{247}{562} \simeq 0.44 \qquad\qquad \bigcirc$$

After this preamble, we can now usefully state the rule governing conditional probability.

Conditioning rule. Let A and B be events. Then the *conditional probability* of A given that B occurs is defined as

(6) $$P(A|B) = \frac{P(A \cap B)}{P(B)}$$

whenever $P(B) > 0$.

This may seem a little arbitrary, but it is strongly motivated by our interpretation of probability as an extension of proportion. We may run through the usual examples in the usual way.

First, consider an experiment with equally likely outcomes, for which

$$P(A) = \frac{|A|}{|\Omega|} \quad \text{and} \quad P(B) = \frac{|B|}{|\Omega|}.$$

Given simply that B occurs, all the outcomes in B are still equally likely. Essentially, we now have an experiment with $|B|$ equally likely outcomes, in which A occurs if and only if $A \cap B$ occurs. Hence under these conditions

$$P(A|B) = \frac{|A \cap B|}{|B|}.$$

But

$$\frac{|A \cap B|}{|B|} = \frac{|A \cap B|}{|\Omega|} \bigg/ \frac{|B|}{|\Omega|} = \frac{P(A \cap B)}{P(B)},$$

which is what (6) says.

Second, we consider the argument from relative frequency. Suppose an experiment is repeated a large number n of times, yielding the event B on $N(B)$ occasions. Given that B occurs, we may confine our attention to these $N(B)$ repetitions. Now A occurs in just $N(A \cap B)$ of these, and so empirically

$$P(A|B) \simeq \frac{N(A \cap B)}{N(B)} = \frac{N(A \cap B)}{n} \frac{n}{N(B)}$$

$$\simeq \frac{P(A \cap B)}{P(B)}$$

which is consistent with (6).

Third, we return to the interpretation of probability as a fair price. Once again a plutocrat offers me a dollar. In this case I get the dollar only if both the events A and B

occur, so this offer is worth \$P($A \cap B$) to me. But we can also take it in stages: suppose that if B occurs, the dollar bill is placed on a table, and if A then occurs the bill is mine. Then

  (i)   the value of what will be on the table is \$P(B),
  (ii) the value of a dollar bill on the table is \$P($A|B$).

The value of the offer is the same, whether or not the dollar bill has rested on the table, so

$$P(A \cap B) = P(A|B)P(B),$$

which is (6) yet again.

Finally consider the experiment in which a point is picked at random in some plane region Ω. For any regions A and B, if we are given that B occurs then A can occur if and only if $A \cap B$ occurs. Naturally then it is reasonable to require that P($A|B$) is proportional to $|A \cap B|$, the area of $A \cap B$. That is, for some k,

$$P(A|B) = kP(A \cap B).$$

Now we observe that obviously P($\Omega|B$) = 1, so

$$1 = kP(\Omega \cap B) = kP(B),$$

as required. Figure 2.5 illustrates the rule (6).

Let us see how this rule applies to the simple examples at the beginning of this section.

Example 2.7.1 revisited.   According to the rule,

$$P(\text{right die larger}|\text{left shows 5}) = \frac{P(\text{right shows 6 and left shows 5})}{P(\text{left shows 5})}$$

$$= \frac{1}{36} \bigg/ \frac{1}{6} = \frac{1}{6}$$

as we saw in (1).   ○

Example 2.7.2 revisited.   According to the rule,

$$P(S|L) = \frac{P(S \cap L)}{P(L)} = \frac{247}{700} \bigg/ \frac{343}{700},$$

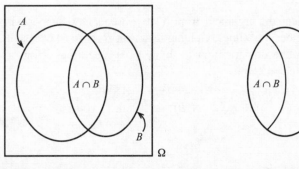

Figure 2.5. The left-hand diagram shows all possible outcomes Ω. The right-hand diagram corresponds to our knowing that the outcome must lie in B; P($A|B$) is thus the proportion P($A \cap B$)/P(B) of these possible outcomes.

which is (4), and

$$P(L|S) = \frac{P(L \cap S)}{P(S)} = \frac{247}{700} \Big/ \frac{562}{700},$$

which is (5). ○

Here are some fresh applications.

Example 2.7.3. A coin is flipped three times. Let A be the event that the first flip gives a head, and B the event that there are exactly two heads overall. We know that

$$\Omega = \{HHH, HHT, HTH, THH, TTH, THT, HTT, TTT\}$$

$$A = \{HTT, HHT, HHH, HTH\}$$

$$B = \{HHT, HTH, THH\}$$

$$A \cap B = \{HHT, HTH\}.$$

Hence by (6),

$$P(A|B) = \frac{P(A \cap B)}{P(B)} = \frac{|A \cap B|}{|\Omega|} \frac{|\Omega|}{|B|} = \tfrac{2}{3}$$

and

$$P(B|A) = \frac{P(A \cap B)}{P(A)} = \tfrac{1}{2}. \qquad ○$$

It is particularly important to be careful and painstaking in using conditional probability. This may seem a trite remark, but experience has shown that students are prone to attempt intuitive short cuts leading to wrong answers. Here is an example that demonstrates this.

Example 2.7.4. A box contains a double-headed coin, a double-tailed coin and a conventional coin. A coin is picked at random and flipped. It shows a head. What is the probability that it is the double-headed coin?

Solution. The three coins have 6 faces, so $|\Omega| = 6$.

Let D be the event that the coin is double-headed, and A the event that it shows a head. Then 3 faces yield A, so

$$P(A) = \tfrac{1}{2}$$

Two faces yield A and D (as it can be either way up) so

$$P(A \cap D) = \tfrac{1}{3}$$

Finally

$$P(D|A) = \frac{P(A \cap D)}{P(A)} = \tfrac{2}{3}. \qquad ○$$

The point of this example is that many people are prepared to argue as follows: 'If the coin shows a head, it is either double-headed or the conventional coin. Since the coin was picked at random, these are equally likely, so $P(D|A) = \tfrac{1}{2}$'.

This is superficially plausible but, as we have seen, it is totally wrong. Notice that this bogus argument avoids mentioning the sample space Ω; this is very typical of false reasoning in probability problems.

We conclude this section by looking at (6) again. It assumes that we know $P(A \cap B)$ and $P(B)$, and defines $P(A|B)$. However, in applications we quite often know $P(B)$ and $P(A|B)$, and wish to know $P(A \cap B)$. In cases like this, we use the following reformulation of (6).

Multiplication rule.   For any events A and B
(7) $P(A \cap B) = P(A|B)P(B)$.

Notice that we can avoid the tiresome reservation that $P(B) \neq 0$ in this version; when $P(B) = 0$ we just assign the value zero to both sides. This is not a very interesting case, of course.

Here is an example of (7) in use.

Example 2.7.5: socks.   A box contains 5 red socks and 3 blue socks. If you remove 2 socks at random, what is the probability that you are holding a blue pair?

Solution.   Let B be the event that the first sock is blue, and A the event that you have a pair of blue socks. If you have one blue sock, the probability that the second is blue is the chance of drawing one of the 2 remaining blues from the 7 remaining socks. That is to say

$$P(A|B) = \tfrac{2}{7}.$$

Here $A = A \cap B$ and so, by (7),

$$P(A) = P(A|B)P(B)$$

$$= \tfrac{2}{7} \times \tfrac{3}{8} = \tfrac{3}{28}.$$

Of course you could do this problem by enumerating the entire sample space for the experiment, but the above method is much easier.   ○

The multiplication rule can be usefully extended, thus.

Extended multiplication rule.   Let $A_1, A_2, \ldots, A_n$ be any events such that
(8) $P(A_1 \cap A_2 \cap \cdots \cap A_n) > 0$.

then we have the extended rule

$$P(A_1 \cap A_2 \cap \cdots \cap A_n) = P(A_n | A_{n-1} \cap \cdots \cap A_1)$$

$$\times P(A_{n-1} | A_{n-2} \cap \cdots \cap A_1)$$

$$\vdots$$

$$\times P(A_2 | A_1)P(A_1).$$

This follows immediately if we write, for all k,

$$P(A_k | A_{k-1} \cap \cdots \cap A_1) = \frac{P(A_k \cap \cdots \cap A_1)}{P(A_{k-1} \cap \cdots \cap A_1)}.$$

Then successive cancellation on the right-hand side gives the result required.   ○

Finally let us stress that conditional probability is not in any way an unnatural or purely theoretical concept. It is completely familiar and natural to you if you have ever bought insurance, played golf, or observed horse racing, to choose just three examples of the myriad available. Thus:

Insurance. Insurance companies require large premiums from young drivers to insure a car, because they know that P(claim|young driver) $\gg$ P(claim). For similarly obvious reasons, older customers must pay more for life insurance.

Golf. If you play against the Open Champion then P(you win) $\simeq 0$. However, given a sufficiently large number of strokes it can be arranged that P(you win|handicap) $\simeq \frac{1}{2}$. Thus any two players can have a roughly even contest.

Horse races. Similarly any horse race can be made into a much more open contest by requiring faster horses to carry additional weights. Much of the betting industry relies on the judgement of the handicappers in doing this.

The objective of the handicapper in choosing the weights is to equalize the chances to some extent and introduce more uncertainty into the result. The ante-post odds reflect the bookmakers' assessment of how far he has succeeded, and the starting prices reflect the gambler's assessment of the position. (See section 2.12 for an introduction to odds.)

Of course this is not the limit to possible conditions; if it rains heavily before a race then the odds will change to favour horses that run well in heavy conditions. And so on.

Clearly this idea of conditional probability is relevant in almost any experiment; you should think of some more examples (*exercise*).

Exercises for section 2.7

1. On any day the chance of rain is 25%. The chance of rain on two consecutive days is 10%. Given that it is raining today, what is the chance of rain tomorrow? Given that it will rain tomorrow, what is the chance of rain today?

2. Show that the conditional probability $P(A|B)$ satisfies the three axioms of probability:
   (a) $0 \leqslant P(A|B) \leqslant 1$,
   (b) $P(\Omega|B) = 1$,
   (c) $P(A_1 \cup A_2|B) = P(A_1|B) + P(A_2|B)$
   when A_1 and A_2 are disjoint. Show also that $P(A|B) = 1 - P(A^c|B)$.

3. ***Extended multiplication.*** Show that
$$P(A \cap B \cap C) = P(A|B \cap C)P(B|C)P(C).$$

4. Your kit bag contains 15 wrist bands, of which 6 are blue, 5 are red and 4 are green. You pick 3 at random. What is the probability that they are all (a) red? (b) the same colour?

5. Show that
$$P(A \cap B|A \cup B) \leqslant \min\{P(A \cap B|A), P(A \cap B|B)\}.$$

6. ***The prosecutor's fallacy.*** Let G be the event that some accused person is guilty, and T the event that some testimony or evidence presented is in fact true. It has been known for lawyers to argue on the assumption that $P(G|T) = P(T|G)$. Show that this holds if and only if $P(G) = P(T)$.

2.8 THE PARTITION RULE AND BAYES' RULE

In this section we look at some of the simple consequences of our definition of conditional probability. The first and most important thing to show is that conditional probability satisfies the rules for a probability function (otherwise its name would be very misleading).

Conditional probability is a probability

First, for any A and B,

$$0 \leqslant P(A \cap B) \leqslant P(B)$$

and so

(1) $$0 \leqslant P(A|B) \leqslant 1.$$

Second, for any B,

$$P(\Omega|B) = P(\Omega \cap B)/P(B) = 1.$$

Third, if A_1 and A_2 are disjoint then $A_1 \cap B$ and $A_2 \cap B$ are disjoint. Hence

(2) $$P(A_1 \cup A_2|B) = \{P(A_1 \cap B) + P(A_2 \cap B)\}/P(B)$$

$$= P(A_1|B) + P(A_2|B).$$

Next we consider some very important applications of conditional probability. Let us recall the multiplication rule for any events A and B,

(3) $$P(A \cap B) = P(A|B)P(B).$$

Since this holds true for any pair of events, it is also true for A and B^c that

(4) $$P(A \cap B^c) = P(A|B^c)P(B^c).$$

But B and B^c are disjoint events, so by the addition rule

(5) $$P(A \cap B) + P(A \cap B^c) = P((A \cap B) \cup (A \cap B^c))$$

$$= P(A).$$

Combining (3), (4), and (5) gives the extremely important

Partition rule

(6) $$P(A) = P(A|B)P(B) + P(A|B^c)P(B^c).$$

This has a conditional form as well: for any three events A, B, and C, we have the

Conditional partition rule

(7) $$P(A|C) = P(A|B \cap C)P(B|C) + P(A|B^c \cap C)P(B^c|C).$$

The proof of this is exercise 4 at the end of the section.

The partition rule may not seem very impressive at first sight, but it is in fact one of the most important and frequently used results in probability. Here are some examples of its use.

Example 2.8.1: pirates.   An expensive electronic toy made by Acme Gadgets Inc. is defective with probability 10^{-3}. These toys are so popular that they are copied and sold illegally but cheaply. Pirate versions capture 10% of the market, and any pirated copy is defective with probability $\frac{1}{2}$. If you buy a toy, what is the chance that it is defective?

Solution. Let A be the event that you buy a genuine article, and let D be the event that your purchase is defective. We know that

$$P(A) = \tfrac{9}{10}, \quad P(A^c) = \tfrac{1}{10}, \quad P(D|A) = \tfrac{1}{1000}, \quad P(D|A^c) = \tfrac{1}{2}.$$

Hence, by the partition rule,

$$P(D) = P(D|A)P(A) + P(D|A^c)P(A^c)$$

$$= \tfrac{9}{10\,000} + \tfrac{1}{20} \simeq 5\%. \qquad \bigcirc$$

Example 2.8.2: tests. In a population of individuals a proportion p are subject to a disease. A test is devised to indicate whether any given individual does have the disease; such an indication is called a positive test. No test is perfect, and in this case the probability that the test is positive for an individual with the disease is 95%, and the probability of a positive result for an individual who does not have the disease is 10%. If you test a randomly selected individual, what is the chance of a positive result?

Solution. Let R denote the event that the result is positive, and D the event that the individual has the disease. Then by (6)

$$P(R) = P(R|D)P(D) + P(R|D^c)P(D^c)$$

$$= 0.95p + 0.1(1 - p)$$

$$= 0.85p + 0.1.$$

For a test as bad as this you will get a lot of positive results even if the disease is rare; if it is rare, most of these will be false positives. $\bigcirc$

Example 2.8.3. Patients may be treated with any one of a number of drugs, each of which may give rise to side effects. A certain drug C has a 99% success rate in the absence of side effects, and side effects only arise in 5% of cases. However, if they do arise then C has only a 30% success rate. If C is used, what is the probability of the event A that a cure is effected?

Solution. Let B be the event that no side effects occur. We are given that

$$P(A|B \cap C) = \tfrac{99}{100},$$

$$P(B|C) = \tfrac{95}{100},$$

$$P(A^c|B^c \cap C) = \tfrac{30}{100},$$

$$P(B^c|C) = \tfrac{5}{100}.$$

Hence, by the conditional partition rule (7),

$$P(A|C) = \tfrac{99}{100} \times \tfrac{95}{100} + \tfrac{30}{100} \times \tfrac{5}{100} = \tfrac{9555}{10000} \simeq 95\%. \qquad \bigcirc$$

Of course many populations can be divided into more than two groups, and many experiments yield an arbitrary number of events. This requires a more general version of the partition rule.

Extended partition rule.   Let A be some event, and suppose that $(B_i; \; i \geqslant 1)$ is a collection of events such that.

$$A \subseteq B_1 \cup B_2 \cup \cdots = \bigcup_i B_i,$$

and, for $i \neq j$, $B_i \cap B_j = \emptyset$, that is to say, the B_i are disjoint. Then, by the extended addition rule (5) of section 2.5,

(8) $$P(A) = P(A \cap B_1) + P(A \cap B_2) + \cdots$$

$$= \sum_i P(A \cap B_i)$$

$$= \sum_i P(A|B_i)P(B_i).$$

This is the *extended partition rule*. Its conditional form is

(9) $$P(A|C) = \sum_i P(A|B_i \cap C)P(B_i|C).$$

Example 2.8.4: coins.   You have 3 double-headed coins, 1 double-tailed coin and 5 normal coins. You select one coin at random and flip it. What is the probability that it shows a head?

Solution.   Let D, T, and N denote the events that the coin you select is double-headed, double-tailed or normal, respectively. Then, if H is the event that the coin shows a head, by conditional probability we have

$$P(H) = P(H|D)P(D) + P(H|T)P(T) + P(H|N)P(N)$$

$$= 1 \times \tfrac{3}{9} + 0 \times \tfrac{1}{9} + \tfrac{1}{2} \times \tfrac{5}{9} = \tfrac{11}{18}. \qquad \bigcirc$$

Obviously the list of examples demonstrating the partition rule could be extended indefinitely; it is a crucial result. Now let us consider the examples given above from another point of view.

Example 2.8.1 revisited: pirates.   Typically, we are prompted to consider this problem when our toy proves to be defective. In this case we wish to know if it is an authentic product of Acme Gadgets Inc., in which case we will be able to get a replacement. Pirates, of course, are famous for not paying compensation. In fact we really want to know $P(A|D)$, which is an upper bound for the chance that you get a replacement. $\qquad \bigcirc$

Example 2.8.2 revisited: tests. Once again, for the individual the most important question is, given a positive result do you indeed suffer the disease? That is, what is $P(D|R)$? ○

Of course these questions are straightforward to answer by conditional probability, since

$$P(A|B) = P(A \cap B)/P(B).$$

The point is that in problems of this kind we are usually given $P(B|A)$ and $P(B|A^c)$. Expanding the denominator $P(B)$ by the partition rule gives an important result:

Bayes' rule

(10) $$P(A|B) = \frac{P(B|A)P(A)}{P(B|A)P(A) + P(B|A^c)P(A^c)}.$$

Here are some applications of this famous rule or theorem.

Example 2.8.2 continued: false positives. Now we can answer the question posed above: in the context of this test, what is $P(D|R)$?

Solution. By Bayes' rule,

$$P(D|R) = \frac{P(R|D)P(D)}{P(R)}$$

$$= \frac{0.95p}{0.85p + 0.1}.$$

On the one hand, if $p = \frac{1}{2}$ then we find

$$P(D|R) = \frac{19}{21}$$

and the test looks good. On the other hand, if $p = 10^{-6}$, so the disease is very rare, then

$$P(D|R) \simeq 10^{-5}$$

which is far from conclusive. Ordinarily one would hope to have further independent tests to use in this case. ○

Here is an example of Bayes' rule that has the merit of being very simple, albeit slightly frivolous.

Example 2.8.3: examinations. Suppose a multiple choice question has c available choices. A student either knows the answer with probability p, say, or guesses at random with probability $1 - p$. Given that the answer selected is correct, what is the probability that the student knew the answer?

Solution. Let A be the event that the question is answered correctly, and S the event that the student knew the answer. We require $P(S|A)$. To use Bayes' rule, we need to calculate $P(A)$, thus

$$P(A) = P(A|S)P(S) + P(A|S^c)P(S^c)$$

$$= p + c^{-1}(1 - p).$$

Now by conditional probability

$$P(S|A) = P(A|S)P(S)/P(A)$$

$$= \frac{p}{p + c^{-1}(1 - p)}$$

$$= \frac{cp}{1 + (c - 1)p}.$$

Notice that the larger c is, the more likely it is that the student knew the answer to the question, given that it is answered correctly. This is in accord with our intuition about such tests. Indeed if it were not true there would be little point in setting them.   ○

Exercises for section 2.8

1. An insurance company knows that the probability of a policy holder's having an accident in any given year is β if the insured is aged less than 25, and σ if the insured is 25 or over. A fraction ϕ of policy holders are less than 25. What is the probability that
   (a) a randomly selected policy holder has an accident?
   (b) a policy holder who has an accident is less than 25?

2. A factory makes tool bits; 5% are defective. A machine tests each bit. With probability 10^{-3} it incorrectly passes a defective bit; with probability 10^{-4} it incorrectly rejects a good bit. What is the probability that
   (a) a bit was good, given it was rejected?
   (b) the machine passes a randomly selected bit?

3. ***Red ace.***   A pack of four cards contains two clubs and two red aces.
   (a) Two cards are selected at random and a friend tells you that one is the ace of hearts. Can you say what the probability is that the other is the ace of diamonds?
   (b) Two cards are selected at random and inspected by a friend. You ask whether either of them is the ace of hearts and receive the answer 'Yes'. Can you say what the probability is that the other is the ace of diamonds? (Your friend always tells the truth.)

4. Prove the conditional partition rules (7) and (9).

2.9 INDEPENDENCE AND THE PRODUCT RULE

At the start of section 2.7 we noted that a change in the conditions of some experiment will often obviously change the probabilities of various outcomes. That led us to define conditional probability.

However, it is equally obvious that sometimes there are changes that make no difference whatever to the outcomes of the experiments, or to the probability of some event A of interest. For example, suppose you buy a lottery ticket each week; does the chance of your winning next week depend on whether you won last week? Of course not; the numbers chosen are independent of your previous history. What does this mean formally? Let A be the outcome of this week's lottery, and B the event that you won last week. Then we agree that obviously

(1) $$P(A|B) = P(A|B^c).$$

There are many events A and B for which, again intuitively, it seems natural that the chance that A occurs is not altered by any knowledge of whether B occurs or B^c occurs. For example, let A be the event that you roll a six and B the event that the dollar exchange rate fell. Clearly we must assume that (1) holds. You can see that this list of pairs A and B for which (1) is true could be prolonged indefinitely:

$$A \equiv \text{this coin shows a head}, \quad B \equiv \text{that coin shows a head};$$

$$A \equiv \text{you are dealt a flush}, \quad B \equiv \text{coffee futures fall}.$$

Think of some more examples yourself.

Now it immediately follows from (1) by the partition rule that

(2) $$P(A) = P(A|B)P(B) + P(A|B^c)P(B^c)$$

$$= P(A|B)\{P(B) + P(B^c)\}, \quad \text{by (1)}$$

$$= P(A|B).$$

That is, if (1) holds then

(3) $$P(A) = P(A|B) = P(A|B^c).$$

Furthermore, by the definition of conditional probability, we have in this case

(4) $$P(A \cap B) = P(A|B)P(B)$$

$$= P(A)P(B), \quad \text{by (2)}$$

This special property of events is called *independence* and, when (4) holds, A and B are said to be independent events. The final version (4) is usually taken to be definitive; thus we state the

Product rule. Events A and B are said to be *independent* if and only if

(5) $$P(A \cap B) = P(A)P(B).$$

Example 2.9.1. Suppose I roll a die and pick a card at random from a conventional pack. What is the chance of rolling a six and picking an ace?

Solution. We can look at this in two ways. The first way says that the events

$$A \equiv \text{roll a six}$$

and

$$B \equiv \text{pick an ace}$$

are obviously independent in the sense discussed above; that is, $P(A|B) = P(A)$ and of course $P(B|A) = P(B)$. Dice and cards cannot influence each other. Hence

$$P(A \cap B) = P(A)P(B)$$

$$= \tfrac{1}{6} \times \tfrac{1}{13} = \tfrac{1}{78}.$$

Alternatively, we could use the argument of chapter 1, and point out that by symmetry all $6 \times 52 = 312$ possible outcomes of die and card are equally likely. Four of them have an ace with a six, so

$$P(A \cap B) = \tfrac{4}{312} = \tfrac{1}{78}.$$

It is very gratifying that the two approaches yield the same answer, but not surprising. In fact, if you think about the argument from symmetry, you will appreciate that it tacitly assumes the independence of dice and cards. If there were any mutual influence it would break the symmetry.   ○

If A and B are not independent, then they are said to be *dependent*. Obviously dependence and independence are linked to our intuitive notions of cause and effect. There seems to be no way in which one coin can cause another to be more or less likely to show a head. However, you should beware of taking this too far. Independence is yet another assumption that we make in constructing our model of the real world. It is an extremely convenient assumption, but if it is inappropriate it will yield inaccurate and irrelevant results. Be careful.

The product rule (5) has an extended version, as follows:

Independence of n events.   The events $(A_r;\ r \geqslant 1)$ are independent if and only if

(6) $$P(A_{s_1} \cap A_{s_2} \cap \cdots \cap A_{s_n}) = P(A_{s_1}) \cdots P(A_{s_n})$$

for any selection $(s_1, \ldots, s_n)$ of the positive integers $\mathbb{Z}^+$.

We give various examples to demonstrate these ideas.

Example 2.9.2.   A sequence of fair coins is flipped. They each show a head or a tail independently, with probability $\tfrac{1}{2}$ in each case. Therefore the probability that any given set of n coins all show heads is 2^{-n}. Indeed, the probability that any given set of n coins shows a specified arrangement of heads and tails is 2^{-n}. Thus, for example, if you flip a fair coin 6 times,

$$P(HHHHHH) = P(HTTHTH) = 2^{-6}.$$

(The less experienced sometimes find this surprising.)   ○

Let us consider some everyday applications of the idea of independence.

Example 2.9.3: central heating.   Your heating system includes a pump and a boiler in a circuit of pipes. You might represent this as a diagram like figure 2.6.

Let F_p and F_b be the events that the pump or boiler fail, respectively. Then the event W that your system works is

$$W = F_p^c \cap F_b^c.$$

You might assume that pump and boiler break down independently, in which case, by (5),

(7) $$P(W) = P(F_p^c)P(F_b^c).$$

However, your plumber might object that if the power supply fails then both pump and boiler will fail, so the assumption of independence is invalid. To meet this objection we define the events

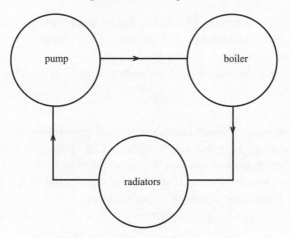

Figure 2.6. A central heating system.

$$F_e \equiv \text{power supply failure,}$$

$$M_p \equiv \text{mechanical failure of pump,}$$

$$M_b \equiv \text{mechanical failure of boiler.}$$

Then it is much more reasonable to suppose that F_e, M_p, and M_b are independent events. We represent this system in figure 2.7. Incidentally, this figure makes it clear that such diagrams are essentially formal in character; the water does not circulate through the power supply. We have now

$$W' = F_e^c \cap M_p^c \cap M_b^c,$$

$$F_b^c = M_b^c \cap F_e^c,$$

$$F_p^c = M_p^c \cap F_e^c.$$

Hence the probability that the system works is

(8) $$P(W') = P(F_e^c)P(M_p^c)P(M_b^c).$$

It is interesting to compare this with the answer obtained on the assumption that F_p and F_b are independent. From (7) this is

$$P(F_b^c)P(F_p^c) = P(M_b^c)P(M_p^c)[P(F_e^c)]^2 = P(W')P(F_e^c),$$

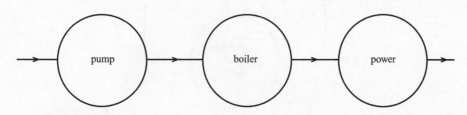

Figure 2.7. The system works only if all three elements in the sequence work.

on using independence, equation (6). This answer is smaller than that given in (8), showing how unjustified assumptions of independence can mislead.   ○

In the example above the elements of the system were in series. Sometimes elements are found in parallel.

Example 2.9.3 continued: central heating.   Your power supply is actually of vital importance (in a hospital, say) and you therefore fit an alternative generator for use in emergencies. The power system can now be represented as in figure 2.8. Let the event that the emergency power fails be E. If we assume that F_e and E are independent, then the probability that at least one source of power works is

$$P(F_e^c \cup E^c) = P((F_e \cap E)^c)$$

$$= 1 - P(F_e \cap E)$$

$$= 1 - P(F_e)P(E), \quad \text{by (5)}$$

$$\geqslant P(F_e^c).$$

Hence the probability that the system works is increased by including the reserve power unit, as you surely hoped.   ○

Many systems comprise blocks of independent elements in series or in parallel, and then $P(W)$ can be found by repeatedly combining blocks.

Example 2.9.4.   Suppose a system can be represented as in figure 2.9. Here each element works with probability p, independently of the others. Running through the blocks we can reduce this to figure 2.10, where the expression in each box is the probability of its working.   ○

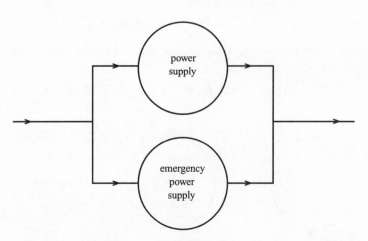

Figure 2.8. This system works if either of the two elements works.

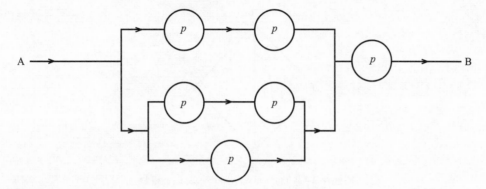

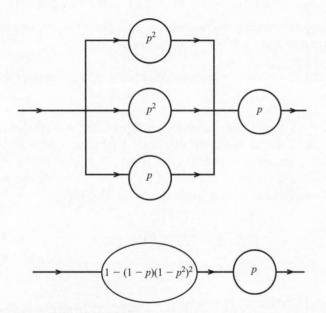

Figure 2.9. Each element works independently with probability p. The system works if a route exists from A to B that passes only through working elements.

Figure 2.10. Solution in stages, showing finally that $P(W) = p\{1 - (1 - p)(1 - p^2)^2\}$, where W is the event that the system works.

Sometimes elements are in more complicated configurations, in which case the use of conditional probability helps.

Example 2.9.5: snow. Four towns are connected by five roads, as shown in figure 2.11. Each road is blocked by snow independently with probability σ; what is the probability δ that you can drive from A to D?

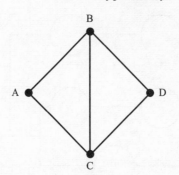

Figure 2.11. The towns lie at A, B, C, and D.

Solution.   Let R be the event that the road BC is open, and Q the event that you can drive from A to D. Then

$$P(Q) = P(Q|R)(1 - \sigma) + P(Q|R)\sigma = (1 - \sigma^2)^2(1 - \sigma) + [1 - \{1 - (1 - \sigma)^2\}^2]\sigma.$$

The last line follows on using the methods of example 2.9.4, because when R or R^c occurs the system is reduced to blocks in series and parallel.   ◯

Note that events can in fact be independent when you might reasonably expect them not to be.

Example 2.9.6.   Suppose three fair coins are flipped. Let A be the event that they all show the same face, and B the event that there is at most one head. Are A and B independent? Write 'yes' or 'no', and then read on.

Solution.   There are eight equally likely outcomes. We have

$$|A \cap B| = 1 = |\{TTT\}|,$$

$$|A| = 2 = |\{TTT, HHH\}|,$$

$$|B| = 4 = |\{TTT, HTT, THT, TTH\}|.$$

Hence

$$P(A \cap B) = \tfrac{1}{8} = P(TTT), \quad P(A) = \tfrac{1}{4} = P(TTT \cup HHH), \quad P(B) = \tfrac{1}{2},$$

and so $P(A \cap B) = P(A)P(B)$, and they *are* independent.   ◯

Very often indeed we need to use a slightly different statement of independence. Just as $P(A|C)$ is often different from $P(A)$, so also $P(A \cap B|C)$ may behave differently from $P(A \cap B)$. Specifically, A and B may be independent given C, even though they are not necessarily independent in general. This is called conditional independence; formally we state the following

Definition.   Events A and B are *conditionally independent* given C if

(9) $P(A \cap B|C) = P(A|C)P(B|C).$ △

Here is an example.

Example 2.9.7: high and low rolls. Suppose you roll a die twice. Let A_2 be the event that the first roll shows a 2, and B_5 the event that the second roll shows a 5. Also let L_2 be the event that the lower score is a 2, and H_5 the event that the higher score is a 5.

(i) Show that A_2 and B_5 are independent.
(ii) Show that L_2 and H_5 are *not* independent.
(iii) Let D be the event that one roll shows less than a 3 and one shows more than a 3. Show that L_2 and H_5 *are* conditionally independent given D.

Solution. For (i) we have easily

$$P(A_2 \cap B_5) = \tfrac{1}{36} = P(A_2)P(B_5)$$

For (ii), however,

$$P(L_2 \cap H_5) = \tfrac{2}{36}.$$

Also,

$$P(L_2) = \tfrac{9}{36} \text{ and } P(H_5) = \tfrac{9}{36}.$$

Therefore L_2 and H_5 are dependent, because

$$\tfrac{1}{18} \neq \tfrac{1}{16}.$$

For (iii) we now have $P(D) = \tfrac{12}{36} = \tfrac{1}{3}$. So, given D,

$$P(L_2 \cap H_5 | D) = P(L_2 \cap H_5 \cap D)/P(D) = \tfrac{1}{18}/\tfrac{1}{3} = \tfrac{1}{6}.$$

However, in this case

$$P(L_2 | D) = P(L_2 \cap D)/P(D) = \tfrac{6}{36}/\tfrac{12}{36} = \tfrac{1}{2}$$

and

$$P(H_5 | D) = P(H_5 \cap D)/P(D) = \tfrac{4}{36}/\tfrac{12}{36} = \tfrac{1}{3}$$

and of course $\tfrac{1}{6} = \tfrac{1}{2} \times \tfrac{1}{3}$, and so L_2 and H_5 *are* independent given D. ○

Conditional independence is an important idea that is frequently used surreptitiously, or taken for granted. It does not imply, nor is it implied by, independence; see the exercises below.

Exercises for section 2.9

1. Show that A and B are independent if and only if A^c and B^c are independent.

2. A and B are events such that $P(A) = 0.3$ and $P(A \cup B) = 0.5$. Find $P(B)$ when
   (a) A and B are independent,
   (b) A and B are disjoint,
   (c) $P(A|B) = 0.1$,
   (d) $P(B|A) = 0.4$.

3. A coin shows a head with probability p, or a tail with probability $1 - p = q$. It is flipped repeatedly until the first head occurs. Show that the probability that n flips are necessary, including the head, is $p_n = q^{n-1}p$.

4. Suppose that any child is equally likely to be male or female, and Anna has three children. Let
    A be the event that the family includes children of both sexes and B the event that the family
    includes at most one girl.
    (a)   Show that A and B are independent.
    (b)   Is this still true if boys and girls are not equally likely?
    (c)   What happens if Anna has four children?

5. Find events A, B, and C such that A and B are independent, but A and B are not conditionally
    independent given C.

6. Find events A, B, and C such that A and B are not independent, but A and B are conditionally
    independent given C.

7. Two conventional fair dice are rolled. Show that the event that their sum is 7 is independent of
    the score on the first die.

8. Some form of prophylaxis is said to be 90% effective at prevention during one year's treatment.
    If years are independent, show that the treatment is more likely than not to fail within seven
    years.

2.10 TREES AND GRAPHS

In real life you may be faced with quite a long sequence of uncertain contingent events.
For example, your computer may develop any of a number of faults, you may choose any
of a number of service agents, they may or may not correct it properly, the consequences
of an error are uncertain, and so on *ad nauseam*. (The same is true of bugs in software.)
  In such cases we need and use the extended form of the multiplication rule,

(1) $P(A_1 \cap A_2 \cap \cdots \cap A_n) = P(A_n|A_1 \cap A_2 \cap \cdots \cap A_n) \cdots P(A_2|A_1)P(A_1).$

The proof of (1) is trivial and is outlined after equation (8) in section 2.7.
  Now, if each of these events represents a stage in some system of multiple random
choices, it is not impossible that the student will become a little confused. In such cases it
is often helpful to use *tree diagrams* to illustrate what is going on. (This is such a natural
idea that it was first used by C. Huygens in the 17th century, in looking at the earliest
problems in probability.) These diagrams will not enable you to avoid the arithmetic and
algebra, but they do help in keeping track of all the probabilities and possibilities. The
basic idea is best explained by an example.

Example 2.10.1: faults.   (i) A factory has two robots producing capeks. (A capek is
not unlike a widget or a gubbins, but it is more colourful.) One robot is old and one is
new; the newer one makes twice as many capeks as the old. If you pick a capek at
random, what is the probability that it was made by the new machine? The answer is
obviously $\frac{2}{3}$, and we can display all the possibilities in a natural and appealing way in
Figure 2.12. The arrows in a tree diagram point to possible events, in this example N
(new) or N^c (old). The probability of the event is marked beside the relevant arrow.
  (ii) Now we are told that 5% of the output of the old machine is defective (D), but 10%
of the output of the new machine is defective. What is the probability that a randomly
selected capek is defective? This time we draw a diagram first, figure 2.13. Now we begin
to see why this kind of picture is called a tree diagram. Again the arrows point to possible

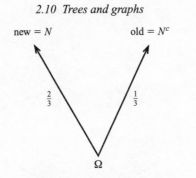

Figure 2.12. A small tree.

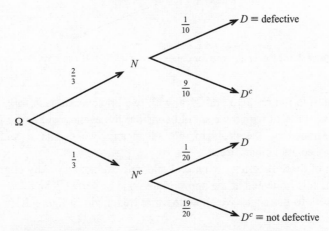

Figure 2.13. A tree drawn left to right: classifying capeks.

events. However, the four arrows on the right are marked with conditional probabilities, because they originate in given events. Thus

$$P(D|N) = \tfrac{1}{10}, \quad P(D^c|N^c) = \tfrac{19}{20},$$

and so on.

The probability of traversing any route in the tree is the product of the probabilities on the route, by (1). In this case two routes end at a defective capek, so the required probability is

$$P(D) = \tfrac{2}{3} \times \tfrac{1}{10} + \tfrac{1}{3} \times \tfrac{1}{20}$$

$$= P(D|N)P(N) + P(D|N^c)P(N^c),$$

$$= \tfrac{1}{12},$$

which is the partition rule, of course. You always have a choice of writing the answer down by routine algebra, but drawing a diagram often helps.

It is interesting to look at this same example from a slightly different viewpoint, as follows. By conditional probability we have easily that

$$P(N|D) = \tfrac{4}{5}, \, P(N^c|D) = \tfrac{1}{5},$$

$$P(N|D^c) = \tfrac{36}{55}, \, P(N^c|D^c) = \tfrac{19}{55}.$$

Then we can draw what is known as the *reversed tree*; see figure 2.14. ○

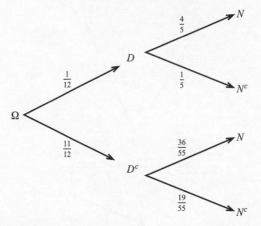

Figure 2.14. Reversed tree for capeks: D or D^c is followed by N or N^c.

Trees like those in figures 2.13 and 2.14, with two branches at each fork, are known as *binary trees*. The order in which we should consider the events, and hence draw the tree, is usually determined by the problem, but given any two events A and B there are obviously two associated binary trees.

The notation of these diagrams is natural and self-explanatory. Any edge corresponds to an event, which is indicated at the appropriate node or vertex. The relevant probability is written adjacent to the edge. We show the first tree again in figure 2.15, labelled with symbolic notation.

The edges may be referred to as branches, and the final node may be referred to as a leaf. The probability of the event at any node, or leaf, is obtained by multiplying the probabilities labelling the branches leading to it. For example,

$$(2) \qquad\qquad P(A^c \cap B) = P(B|A^c)P(A^c).$$

Furthermore, since event B occurs at the two leaves marked with an asterisk, the diagram

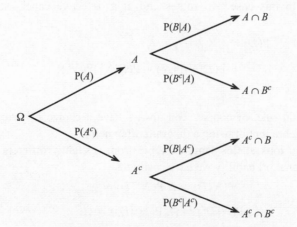

Figure 2.15. A or A^c is followed by B or B^c.

shows that

(3) $$P(B) = P(B|A)P(A) + P(B|A^c)P(A^c)$$

as we know.

Figure 2.16 is the reversed tree. If we have the entries on either tree we can find the entries on the other by using Bayes' rule.

Similar diagrams arise quite naturally in knock-out tournaments, such as Wimbledon. The diagram is usually displayed the other way round in this case, so that the root of such a tree is the winner in the final.

Example 2.8.2 revisited: tests. Recall that a proportion p of a population is subject to a disease. A test may help to determine whether any individual has the disease. Unfortunately unless the test is extremely accurate, this will result in false-positive results. One would like accurate tests, but reliable tests usually involve invasive biopsy. This itself can lead to undesirable results; if you have not got some disease you would regret a biopsy, to find out, that resulted in your death. It is therefore customary, where possible, to use a two-stage procedure; the population of interest is first given the non-invasive but less reliable test. Only those testing positive are subject to biopsy. The tree may take the form shown in figure 2.17; T denotes that the result of the biopsy is positive. ○

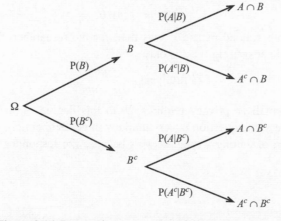

Figure 2.16. Reversed tree: B or B^c is followed by A or A^c.

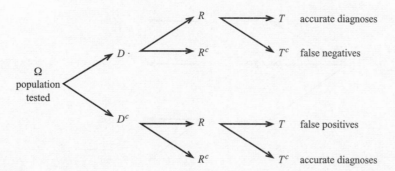

Figure 2.17. Sequential tests. D, disease present; R, first test positive; T, second test positive.

Example 2.10.2: embarrassment.   Privacy is important to many people. For example, suppose you were asked directly whether you dye your hair, 'borrow' your employer's stationery, drink to excess, or suffer from some stigmatized illness. You might very well depart from strict truth in your reply, or simply decline to respond.

Nevertheless, there is no shortage of researchers and other busybodies who would like to know the correct answer to these and other embarrassing questions. They have invented the following scheme.

A special pack of cards is shuffled; of these cards a proportion n say 'Answer "no"', a proportion p say 'Answer "yes"', and a proportion q say 'Answer truthfully'. You draw a card at random, secretly, the question is put, and you obey the instructions on the card.

The point is that you can afford to tell the truth when instructed to do so, because no one can say whether your answer actually is true, even when embarrassing. However, in this way, in the long run the researchers can estimate the true incidence of alcoholism, or pilfering in the population, using Bayes' rule. The tree is shown in figure 2.18. The researcher wishes to know m, the true proportion of the population that is embarrassed. Now, using the partition rule as usual, we can say that the probability of the answer being 'yes' is

$$(4) \qquad\qquad P(\text{yes}) = p + mq.$$

Then the busybody can estimate m by finding the proportion y who answer 'yes' and setting

$$(5) \qquad\qquad m = (y - p)/q.$$

Conversely, the embarrassed person knows that, for the researcher, the probability that they were truthful in answering 'yes' is only

$$(6) \qquad\qquad P(\text{truth}|\text{yes}) = \frac{qm}{p + qm},$$

which can be as small as privacy requires. With this degree of security conferred by randomizing answers, the question is very unlikely to get deliberately wrong answers. (In fact, the risk is run of getting wrong answers because the responder fails to understand the procedure.)   ○

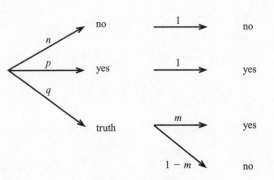

Figure 2.18. Evasive tree.

We note that trees can be infinite in many cases.

Example 2.10.3: craps, an infinite tree.   In this well-known game two dice are rolled and their scores added. If the sum is 2, 3, or 12 the roller loses, if it is 7 or 11 the roller wins, if it is any other number, say n, the dice are rolled again. On this next roll, if the sum is n then the roller wins, if it is 7 the roller loses, otherwise the dice are rolled again. On this and all succeeding rolls the roller loses with 7, wins with n, or rolls again otherwise. The corresponding tree is shown in figure 2.19.   ○

We conclude this section by remarking that sometimes diagrams other than trees are useful.

Example 2.10.4: tennis.   Rod and Fred are playing a game of tennis, and have reached deuce. Rod wins any point with probability r or loses it with probability $1 - r$. Let us denote the event that Rod wins a point by R. Then if they share the next two points the game is back to deuce; an appropriate diagram is shown in figure 2.20.   ○

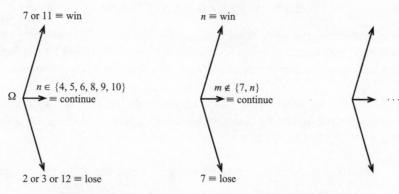

Figure 2.19. Tree for craps. The game continues indefinitely.

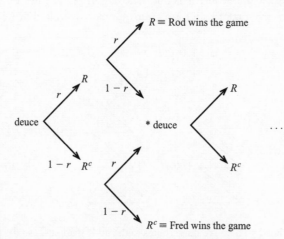

Figure 2.20. The diagram is not a tree because the edges rejoin at *.

Larger diagrams of this kind can be useful.

Example 2.10.5: coin tossing.   Suppose you have a biased coin that yields a head
with probability p and a tail with probability q. Then one is led to the diagram in figure
2.21 as the coin is flipped repeatedly; we truncate it at three flips.   ○

Exercises for section 2.10

1.   Prove the multiplication rule (1).

2.   In tennis a tie-break is played when games are 6–6 in a set. Draw the diagram for such a tie-
     break. What is the probability that Rod wins the tie-break 7–1, if he wins any point with
     probability p?

3.   In a card game you cut for the deal, with the convention that if the cards show the same value,
     you recut. Draw the graph for this experiment.

4.   You flip the coin of example 2.10.5 four times. What is the probability of exactly two tails?

2.11 WORKED EXAMPLES

The rules of probability (we have listed them in subsection 2.14.II), especially the ideas
of independence and conditioning, are remarkably effective at working together to
provide neat solutions to a wide range of problems. We consider a few examples.

Example 2.11.1.   A coin shows a head with probability p, or a tail with probability
$1 - p = q$. It is flipped repeatedly until the first head appears. Find $P(E)$, the probability
of the event E that the first head appears at an even number of flips.

Solution.   Let H and T denote the outcomes of the first flip. Then, by the partition
rule,

$$(1) \qquad P(E) = P(E|H)P(H) + P(E|T)P(T).$$

Now of course $P(E|H) = 0$, because 1 is odd. Turning to $P(E|T)$, we now require an odd
number of flips after the first to give an even number overall. Furthermore, flips are
independent and so

$$(2) \qquad P(E|T) = P(E^c) = 1 - P(E)$$

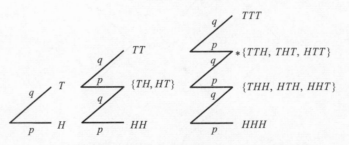

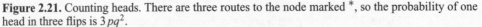

Figure 2.21. Counting heads. There are three routes to the node marked *, so the probability of one
head in three flips is $3pq^2$.

Hence, using (1) and (2),

$$P(E) = \{1 - P(E)\}q,$$

so $P(E) = q/p$.

You can check this, and appreciate the method, by writing down the probability of a head after $2r$ flips, that is, $q^{2r-1}p$, and then summing over r,

$$\sum_r q^{2r-1}p = \frac{pq}{1-q^2} = \frac{q}{p}. \qquad \bigcirc$$

Here is the same problem for a die.

Example 2.11.2. You roll a die repeatedly. What is the probability of rolling a six for the first time at an odd number of rolls?

Solution. Let A be the event that a six appears for the first time at an odd roll. Let S be the event that the first roll is a six. Then by the partition rule, with an obvious notation,

$$P(A) = P(A|S)\tfrac{1}{6} + P(A|S^c)\tfrac{5}{6}.$$

But obviously $P(A|S) = 1$. Furthermore, the rolls are all independent, and so

$$P(A|S^c) = 1 - P(A)$$

Therefore

$$P(A) = \tfrac{1}{6} + \tfrac{5}{6}\{1 - P(A)\}$$

which yields

$$P(A) = \tfrac{6}{11}. \qquad \bigcirc$$

Let us try something trickier.

Example 2.11.3: Huygens' problem. Two players take turns at rolling dice; they each need a different score to win. If they do not roll the required score, play continues. At each of their attempts A wins with probability α, whereas B wins with probability β. What is the probability that A wins if he rolls first? What is it if he rolls second?

Solution. Let p_1 be the probability that A wins when he has the first roll, and p_2 the probability that A wins when B has the first roll. By conditioning on the outcome of the first roll we see that, when A is first,

$$p_1 = \alpha + (1 - \alpha)p_2.$$

When B is first, conditioning on the first roll gives

$$p_2 = (1 - \beta)p_1.$$

Hence solving this pair gives

$$p_1 = \frac{\alpha}{1 - (1 - \alpha)(1 - \beta)}$$

and

$$p_2 = \frac{(1 - \beta)\alpha}{1 - (1 - \alpha)(1 - \beta)}. \qquad \bigcirc$$

Example 2.11.4: Huygen's problem again.   Two coins, A and B, show heads with respective probabilities α and β. They are flipped alternately, giving $ABABAB \ldots$. Find the probability of the event E that A is first to show a head.

Solution.   Consider the three events

$$\{H\} \equiv \text{1st flip heads,}$$

$$\{TH\} \equiv \text{1st flip tails, 2nd flip heads,}$$

$$\{TT\} \equiv \text{1st and 2nd flips tails.}$$

These form a partition of Ω, so, by the extended partition rule (8) of section 2.8,

$$P(E) = P(E|H)\alpha + P(E|TH)(1 - \alpha)\beta + P(E|TT)(1 - \alpha)(1 - \beta).$$

Now obviously

$$P(E|H) = 1 \quad \text{and} \quad P(E|TH) = 0.$$

Furthermore

$$P(E|TT) = P(E).$$

To see this, just remember that after two tails, everything is essentially back to the starting position, and all future flips are independent of those two. Hence

$$P(E) = \alpha + (1 - \alpha)(1 - \beta)P(E)$$

and so

$$P(E) = \frac{\alpha}{\alpha + \beta - \alpha\beta}. \qquad \qquad \bigcirc$$

Example 2.11.5: deuce.   Rod and Fred are playing a game of tennis, and the game stands at deuce. Rod wins any point with probability p, independently of any other point. What is the probability γ that he wins the game?

Solution.   We give two methods of solution.

Method I. Recall that Rod wins as soon as he has won two more points in total than Fred. Therefore he can win only when an even number $2n + 2$ of points have been played. Of these Rod has won $n + 2$ and Fred has won n. Let W_{2n+2} be the event that Rod wins at the $(2n + 2)$th point. Now at each of the first n deuces, there are two possibilities: either Rod gains the advantage and loses it, or Fred gains the advantage and loses it. At the last deuce Rod wins both points. Thus there are 2^n different outcomes in W_{2n+2}, and each has probability $p^{n+2}(1 - p)^n$. Hence, by the addition rule, $P(W_{2n+2}) = 2^n p^{n+2}(1 - p)^n$. If $P(W_{2n})$ is the probability that Rod wins the game at the $2n$th point then, by the extended addition rule (5) of section 2.5,

$$\gamma = P(W_2) + P(W_3) + \cdots$$

$$= p^2 + 2p^3(1 - p) + 2^2 p^4(1 - p)^2 + \cdots$$

$$= \frac{p^2}{1 - 2p(1 - p)}.$$

The possible progress of the game is made clearer by the tree diagram in Figure 2.22. Clearly after an odd number of points either the game is over, or some player has the advantage. After an even number, either the game is over or it is deuce.

Method II. The tree diagram suggests an alternative approach. Let α be the probability that Rod wins the game eventually given he has the advantage, and β the probability that Rod wins the game eventually given that Fred has the advantage.

Further, R be the event that Rod wins the game and W_i be the event that he wins the ith point. Then, by the partition rule,

$$\gamma = P(R)$$

$$= P(R|W_1 \cap W_2)P(W_1 \cap W_2)$$

$$\quad + P(R|W_1^c \cap W_2^c)P(W_1^c \cap W_2^c)$$

$$\quad + P\big(R|(W_1 \cap W_2^c) \cup (W_1^c \cap W_2)\big)P((W_1 \cap W_2^c) \cup (W_1^c \cap W_2))$$

$$= p^2 + 0 + \gamma 2p(1 - p).$$

This is the same as we obtained by the first method. ◯

Example 2.11.6.   Three players, known as A, B, and C, roll a die repeatedly in the order $ABCABCA$ .... The first to roll a six is the winner; find their respective probabilities of winning.

Solution.   Let the players' respective probabilities of winning be α, β, and $1 - \alpha - \beta$, and let the event that the first roll shows a six be S. Then by conditional probability

$$\alpha = P(A \text{ wins}) = P(A \text{ wins}|S)\tfrac{1}{6} + P(A \text{ wins}|S^c)\tfrac{5}{6}.$$

Now

$$P(A \text{ wins}|S) = 1.$$

If S^c occurs, then by independence the game takes the same form as before, except that

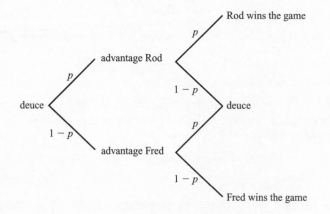

Figure 2.22. Deuce.

the rolls are in the order $BCABCA \dots$ and A is third to roll. Hence, *starting from this point*, the probability that A wins is now $1 - \alpha - \beta$, and we have that

$$\alpha = \tfrac{1}{6} + (1 - \alpha - \beta)\tfrac{5}{6}$$

Applying a similar argument to the sequence of rolls beginning with B, we find.

$$1 - \alpha - \beta = \tfrac{5}{6}\beta$$

because B must fail to roll a six for A to have a chance of winning, and then the sequence takes the form $CABCAB \dots$, in which A is second, with probability β of winning.

Applying the same argument to the sequence of rolls beginning with C yields

$$\beta = \tfrac{5}{6}\alpha$$

because C must fail to roll a six, and then A is back in first place. Solving these three equations gives

$$\alpha = \tfrac{36}{91}, \quad \beta = \tfrac{30}{91}, \quad 1 - \alpha - \beta = \tfrac{25}{91}. \qquad\qquad \bigcirc$$

Another popular and extremely useful approach to many problems in probability entails using conditioning and independence to yield difference equations. We shall see more of this in chapter 3 and later; for the moment here is a brief preview. We start with a trivial example.

Example 2.11.7. A biased coin is flipped repeatedly until the first head is shown. Find the probability $p_n = P(A_n)$ of the event A_n that n flips are required.

Solution. By the partition rule, and conditioning on the outcome of the first flip,

$$P(A_n) = P(A_n|H)p + P(A_n|T)q$$

$$= \begin{cases} p & \text{if } n = 1 \\ 0 + qP(A_{n-1}) & \text{otherwise,} \end{cases}$$

by independence. Hence

$$p_n = qp_{n-1} = q^{n-1}p_1 = q^{n-1}p, \quad n \geqslant 1. \qquad\qquad \bigcirc$$

Of course this result is trivially obvious anyway, but it illustrates the method. Here is a trickier problem.

Example 2.11.8. A biased coin is flipped up to and including the flip on which it has first shown two successive tails. Let A_n be the event that n flips are required. Show that, if $p_n = P(A_n)$, p_n satisfies

$$p_n = pp_{n-1} + pqp_{n-2}, \quad n > 2.$$

Solution. As usual we devise a partition; in this case H, TH, TT are three appropriate disjoint events. Then

$$p_n = P(A_n) = P(A_n|H)p + P(A_n|TH)pq + P(A_n|TT)q^2$$

$$= \begin{cases} q^2, & n = 2 \\ pp_{n-1} + pqp_{n-2} & \text{otherwise,} \end{cases}$$

by independence of flips. ○

Here is another way of using conditional probability.

Example 2.11.9: degraded signals. A digital communication channel transmits data using the two symbols 0 and 1. As a result of noise and other degrading influences, any symbol is incorrectly transmitted with probability q, independently of the rest of the symbols. Otherwise it is correctly transmitted with probability $p = 1 - q$.

On receiving a signal R comprising n symbols, you decode it by assuming that the sequence S which was sent is such that $P(R|S)$ is as large as possible.

For example, suppose you receive the signal 101 and the possible sequences sent are 111 and 000, then

$$P(101|S) = \begin{cases} p^2q & \text{if } S = 111 \\ q^2p & \text{if } S = 000. \end{cases}$$

Thus if $p > q$ the signal 101 is decoded as 111. ○

Next we turn to a problem that was considered (and solved) by many 18th century probabilists, and later generalized by Laplace and others. It arose in Paris with the rather shadowy figure of a Mr Waldegrave, a friend of Montmort. He is described as an English gentleman, and proposed the problem to Montmort sometime before 1711. de Moivre studied the same problem in 1711 in his first book on probability. It seems unlikely that these events were independent; there is no record of Waldegrave visiting the same coffee house as de Moivre, but this seems a very likely connection. (de Moivre particularly favoured Slaughter's coffee house, in St Martin's Lane). de Moivre also worked as a mathematics tutor to the sons of the wealthy, so an alternative hypothesis is that Waldegrave was a pupil or a parent.

The problem is as follows.

Example 2.11.10: Waldegrave's problem. There are $n + 1$ players of some game, $A_0, A_1, \ldots, A_n$, who may be visualized as sitting around a circular table. They play a sequence of rounds in pairs as follows. First A_0 plays against A_1; then the winner plays against A_2; after that the new winner plays against A_3, and so on. The first player to win n rounds consecutively (thus beating all other players) is the overall victor, and the game stops. One may ask several questions, but a natural one is to seek the probability that the game stops at the rth round. Each round is equally likely to be won by either player.

Solution. As so often in probability problems, it is helpful to restate the problem before solving it. Each round is played by a challenger and a fresh player, the challenged. Since each round is equally likely to be won by either player, we might just as well flip a coin or roll a die. The game is then rephrased as follows.

The first round is decided by rolling a die; if it is even A_0 wins, if it is odd A_1 wins. All following rounds are decided by flipping a coin. If it shows heads the challenger wins, if it shows tails the challenged wins. Now it is easy to see that if the coin shows $n-1$ consecutive heads then the game is over. Also, the game can only finish when this occurs. Hence the first round does not count towards this, and so the required result is given by the probability p_r that the coin first shows $n-1$ consecutive heads at the $(r-1)$th flip. But this is a problem we know how to solve; it is just an extension of example 2.11.8.

First we note that (using an obvious notation) the following is a partition of the sample space:

$$\{T, HT, H^2T, \ldots, H^{n-2}T, H^{n-1}\}.$$

Using conditional probability and independence of flips, this gives

(3) $$p_r = \tfrac{1}{2}p_{r-1} + \left(\tfrac{1}{2}\right)^2 p_{r-2} + \cdots + \left(\tfrac{1}{2}\right)^{n-1} p_{r-n+1}, \quad r > n$$

with

$$p_n = \left(\tfrac{1}{2}\right)^{n-1}$$

and

$$p_1 = p_2 = \cdots = p_{n-1} = 0$$

In particular, when $n = 3$, (3) becomes

(4) $$p_r = \tfrac{1}{2}p_{r-1} + \left(\tfrac{1}{2}\right)^2 p_{r-2}, \quad r \geqslant 4.$$

Solving this constitutes problem 26 in section 2.16.   ○

Exercises for section 2.11

1. A biased coin is flipped repeatedly. Let p_n be the probability that n flips have yielded an even number of heads, with $p_0 = 1$. As usual $P(H) = p = 1 - q$, on any flip. Show that
    $$p_n = p(1 - p_{n-1}) + qp_{n-1}, \quad n \geqslant 1$$
    and find p_n (for a list of the basic rules, see section 2.15).

2. A die is 'fixed' so that when rolled the score cannot be the same as the previous score, all other scores having equal probability $\tfrac{1}{5}$. If the first score is 6, what is the probability p_n that the nth score is 6? What is the probability q_n that the nth score is j for $j \neq 6$?

2.12 ODDS

> ... and this particular season the guys who play the horses are being murdered by the bookies all over the country, and are in terrible distress. ... But personally I consider all horse players more or less daffy anyway. In fact, the way I see it, if a guy is not daffy he will not be playing the horses.
>
> Damon Runyon, *Dream Street Rose*

Occasionally, statements about probability are made in terms of odds. This is universally true of bookmakers who talk of 'long odds', '100–1 odds', 'the 2–1 on favourite', and so on. Many of these phrases and customs are also used colloquially, so it is as well to make it clear what all this has to do with our theory of probability.

In dealing with these ideas we must distinguish very carefully between fair odds and bookmakers' payoff odds. *These are not the same.* First, we define fair odds.

Definition.   If an event A has probability $P(A)$, then the fair odds against A are

(1) $$\phi_a(A) = \frac{1 - P(A)}{P(A)} \equiv \{1 - P(A)\} : P(A)$$

and the fair odds on A are

(2) $$\phi_o(A) = \frac{P(A)}{1 - P(A)} \equiv P(A) : \{1 - P(A)\}$$

The ratio notation on the right is often used for odds.

For example, for a fair coin the odds on and against a head are

$$\phi_o(H) = \frac{1/2}{1/2} = \phi_a(H) \equiv 1 : 1$$

These are equal, so these odds are said to be *evens*. If a die is rolled, the odds on and against a six are

$$\phi_o(6) = \frac{1/6}{1 - 1/6} \equiv 1 : 5,$$

$$\phi_a(6) = \frac{1 - 1/6}{1/6} \equiv 5 : 1.$$

You should note that journalists and reporters (on the principle that ignorance is bliss) will often refer to 'the odds on A', when in fact they intend to state the odds *against A*. Be careful.

Now although the fair odds against a head when you flip a coin are 1:1, no bookmaker would pay out at evens for a bet on heads. The reason is that in the long run she would pay out just as much in winnings as she would take from losers. Nevertheless, book-makers and casinos offer odds; where do they come from? First let us consider casino odds.

When a casino offers odds of 35 to 1 against an event A, it means that if you stake \$1 and then A occurs, you will get your stake back plus \$35. If A^c occurs then you forfeit your stake. For this reason such odds are often called *payoff odds*. How are they fixed?

In fact, 35:1 is exactly the payoff odds for the event that a single number you select comes up at roulette. In the American roulette wheel there are 38 compartments. In a well-made wheel they should be equally likely, by symmetry, so the chance that your number comes up is $\frac{1}{38}$.

Now, as we have discussed above, if you get \$d with probability $P(A)$ and otherwise you get nothing, then \$P(A)d is the value of this offer to you.

We say that a bet is fair if the value of your return is equal to the value of your stake. To explain this terminology, suppose you bet \$1 at the fair odds given in (1) against A. You get

$$\$1 + \$\frac{1 - P(A)}{P(A)}$$

with probability $P(A)$, so the value of this to you is

$$\$P(A)\left\{1 + \frac{1 - P(A)}{P(A)}\right\} = \$1.$$

This is the same as your stake, which seems fair. Now consider the case of roulette. Here you get $\$(1 + 35)$ with probability $\frac{1}{38}$. The value of the return is $\$\frac{18}{19}$, which is less than your stake. Of course the difference, $\$\frac{1}{19}$, is what the casino charges you for the privilege of losing your money, so this (in effect) is the value you put on playing roulette. If you get more than $\$\frac{1}{19}$ worth of pleasure out of wagering $\$1$, then for you the game is worth playing.

It is now obvious that in a casino, if the fair odds against your winning are ϕ_a, then the payoff odds π_a will always satisfy

(3) $$\pi_a < \phi_a.$$

In this way the casino ensures that the expected value of the payoff is always less than the value of the stake. It is as well to stress that if the casino has arranged that (3) is true for every available bet, then *no* system of betting can be fair, or favourable to the gambler. Such systems can only change the rate at which you lose money.

Now let us consider bookmakers' odds; for definiteness let us consider a horse race. There are two main ways of betting on horse races. We will consider that known as the Tote system; this is also known as *pari-mutuel* betting. When you place your bet you do not know the payoff odds; they are not fixed until betting stops just before the race begins. For this reason they are known as starting prices, and are actually determined by the bets placed by the gamblers. Here is how it works. Suppose there are n horses entered for the race, and a total of $\$b_j$ is wagered on the jth horse, yielding a total of $\$b$ bet on the race, where

$$b = \sum_{j=1}^{n} b_j.$$

Then the Tote payoff odds for the jth horse are quoted as

(4) $$\pi_a(j) = \frac{1 - p_j}{p_j}$$

where

$$p_j = \frac{b_j}{(1 - t)b}$$

for some positive number t, less than 1.

What does all this mean? For those who together bet a total of $\$b_j$ on the jth horse the total payoff if it wins is

(5) $$b_j\left(1 + \frac{1 - p_j}{p_j}\right) = \frac{b_j}{p_j} = (1 - t)b = b - tb$$

which is $\$tb$ less than the total stake, and independent of j. That is to say, the bookmaker will enjoy a profit of $\$tb$, the 'take', no matter which horse wins. (Bets on places and other events are treated in a similar but slightly more complicated way.)

Now suppose that the *actual* probability that the jth horse will win the race is h_j. (Of course we can never *know* this probability.) Then the value to the gamblers of their bets on this horse is $h_j b(1 - t)$, and the main point of betting on horse races is that this *may* be greater than b_j. But usually it will not be.

It is clear that you should avoid using payoff odds (unless you are a bookmaker). You should also avoid using fair odds, as the following example illustrates.

Example 2.12.1. Find the odds on $A \cap B$ in terms of the odds on A and the odds on B, when A and B are independent.

Solution. From the definition (2) of odds we have

$$\phi_o(A \cap B) = \frac{P(A \cap B)}{1 - P(A \cap B)}$$

$$= \frac{P(A)P(B)}{1 - P(A)P(B)}, \quad \text{by independence}$$

$$= \frac{\phi_o(A)\phi_o(B)}{1 + \phi_o(A) + \phi_o(B)}.$$

Compare this horrible expression with $P(A \cap B) = P(A)P(B)$, to see why the use of odds is best avoided in algebraic work. (Of course it suits bookmakers to obfuscate matters.) $\bigcirc$

Finally we note that when statisticians refer to an 'odds ratio', they mean a quantity such as

$$R(A{:}B) = \frac{P(A)}{P(A^c)} \bigg/ \frac{P(B)}{P(B^c)}.$$

More loosely, people occasionally call any quotient of the form $P(A)/P(B)$ an odds ratio. Be careful.

Exercises for section 2.12

1. Suppose the fair odds against an event are ϕ_a and the casino payoff odds are π_a. Show that the casino's percentage take is

$$100\left(\frac{\phi_a - \pi_a}{\phi_a + 1}\right)\%.$$

2. Suppose you find a careless bookmaker offering payoff odds of $\pi(j)$ against the jth horse in an n-horse race, $1 \leqslant j \leqslant n$, and

$$\sum_{j=1}^{n} \frac{1}{1 + \pi(j)} < 1.$$

Show that if you bet $\${1 + \pi(j)}^{-1}$ on the jth horse, for all n horses, then you surely win.

3. Headlines recently trumpeted that the Earth had a one in a thousand chance of being destroyed by an asteroid shortly. The story then revealed that these were bookmakers' payoff odds. Criticize the reporters. (Hint: do you think a 'chance' is given by ϕ_a or π_a?)

2.13 POPULAR PARADOXES

Probability is the only branch of mathematics in which good mathematicians frequently get results which are entirely wrong.

C. S. Pierce

This section contains a variety of material that, for one reason or another, seems best placed at the end of the chapter. It comprises a collection of 'paradoxes', which probability supplies in seemingly inexhaustible numbers. These could have been included earlier, but the subject is sufficiently challenging even when not paradoxical; it seems unreasonable for the beginner to be asked to deal with gratuitously tricky ideas as well. They are not really paradoxical, merely examples of confused thinking, but, as a by-now experienced probabilist, you may find them entertaining. Many of them arise from false applications of Bayes' rule and conditioning. You can now use these routinely and appropriately, of course, but in the hands of amateurs, Bayes' rule is deadly.

Probability has always attracted more than its fair share of disputes in the popular press; and several of the hardier perennials continue to enjoy a zombie-like existence on the internet (or web). One may speculate about the reasons for this; it may be no more than the fact that anyone can roll dice, or pick numbers, but rather fewer take the trouble to get the algebra right. At any rate we can see that, from the very beginning of the subject, amateurs were very reluctant to believe what the mathematicians told them. We observe Pepys badgering Newton, de Méré pestering Pascal, and so on. Recall the words of de Moivre: 'Some of the problems about chance having a great appearance of simplicity, the mind is easily drawn into a belief that their solution may be attained by the mere strength of natural good sense; which generally proves otherwise ...'; so still today.

In the following examples '*Solution*' denotes a false argument, and *Resolution* or *Solution* denotes a true argument.

Most of the early paradoxes arose through confusion and ignorance on the part of non-mathematicians. One of the first mathematicians who chose to construct paradoxes was Lewis Carroll. When unable to sleep, he was in the habit of solving mathematical problems in his head (that is to say, without writing anything); he did this, as he put it, 'as a remedy for the harassing thoughts that are apt to invade a wholly unoccupied mind'. The following was resolved on the night of 8 September 1887.

Carroll's paradox.   A bag contains two counters, as to which nothing is known except that each is either black or white. Show that one is black and the other white.

'Solution'.   With an obvious notation, since colours are equally likely, the possibilities have the following distribution:

$$P(BB) = P(WW) = \tfrac{1}{4}, \quad P(BW) = \tfrac{1}{2}.$$

Now add a black counter to the bag, then shake the bag, and pick a counter at random. What is the probability that it is black? By conditioning on the three possibilities we have

$$P(B) = 1 \times P(BBB) + \tfrac{2}{3} \times P(BWB) + \tfrac{1}{3} \times P(WWB)$$

$$= 1 \times \tfrac{1}{4} + \tfrac{2}{3} \times \tfrac{1}{2} + \tfrac{1}{3} \times \tfrac{1}{4} = \tfrac{2}{3}.$$

But if a bag contains three counters, and the chance of drawing a black counter is $\frac{2}{3}$, then there must be two black counters and one white counter, by symmetry. Therefore, before we added the black counter, the bag contained BW, *viz.*, one black and one white.

Resolution.   The two experiments, and hence the two sample spaces, are different. The fact that an event has the same probability in two experiments cannot be used to deduce that the sample spaces are the same. And in any case, if the argument were valid, and you applied it to a bag with one counter in it, you would find that the counter had to be half white and half black, that is to say, random, which is what we knew already.   ○

Galton's paradox (1894).   Suppose you flip three fair coins. At least two are alike, and it is an evens chance whether the third is a head or a tail, so the chance that all three are the same is $\frac{1}{2}$.

Solution.   In fact
$$P(\text{all same}) = P(TTT) + P(HHH) = \tfrac{1}{8} + \tfrac{1}{8} = \tfrac{1}{4}.$$
What is wrong?

Resolution.   Again this paradox arises from fudging the sample space. This 'third' coin is not identified initially in Ω, it is determined by the others. The chance whether the 'third' is a head or a tail is a conditional probability, not an unconditional probability. Easy calculations show that

$\left. \begin{array}{l} P(\text{3rd is } H \,|\, HH) = \tfrac{1}{4} \\ P(\text{3rd is } T \,|\, HH) = \tfrac{3}{4} \end{array} \right\}$ *HH* denotes the event that there are at least two heads.

$\left. \begin{array}{l} P(\text{3rd is } T \,|\, TT) = \tfrac{1}{4} \\ P(\text{3rd is } H \,|\, TT) = \tfrac{3}{4} \end{array} \right\}$ *TT* denotes the event that there are at least two tails.

In *no circumstances* therefore is it true that it is an evens chance whether the 'third' is a head or a tail; the argument collapses.   ○

Bertrand's other paradox.   There are three boxes. One contains two black counters, one contains two white counters and one contains a black and a white counter. Pick a box at random and remove a counter without looking at it; it is equally likely to be black or white. The other counter is equally likely to be black or white. Therefore the chance that your box contains identical counters is $\frac{1}{2}$. But this is clearly false: the correct answer is $\frac{2}{3}$.

Resolution.   This is very similar to Galton's paradox. Having picked a box and counter, the probability that the other counter is the same is a conditional probability, not an unconditional probability. Thus easy calculations give (with an obvious notation)

(1) $$P(\text{both black} \,|\, B) = \tfrac{2}{3} = P(\text{both white} \,|\, W);$$

in neither case is it true that the *other* counter is equally likely to be black or white.   ○

Simpson's paradox.   A famous clinical trial compared two methods of treating kidney stones, either by surgery or nephrolithotomy; we denote these by S and N respectively. In all, 700 patients were treated, 350 by S and 350 by N. Then it was found that for cure rates

$$P(\text{cure}|S) = \frac{273}{350} \simeq 0.78,$$

$$P(\text{cure}|N) = \frac{289}{35} \simeq 0.83.$$

Surgery seems to have an inferior rate of success at cures. However, the size of the stones removed was also recorded in two categories:

$$L = \text{diameter more than 2 cm,}$$

$$T = \text{diameter less than 2 cm.}$$

When patients were grouped by stone size as well as treatment, the following results emerged:

$$P(\text{cure}|S \cap T) \simeq 0.93,$$
$$P(\text{cure}|N \cap T) \simeq 0.87,$$

and

$$P(\text{cure}|S \cap L) \simeq 0.73,$$
$$P(\text{cure}|N \cap L) \simeq 0.69.$$

In both these cases surgery has the *better* success rate; but when the data are pooled to ignore stone size, surgery has an *inferior* success rate. This seems paradoxical, which is why it is known as Simpson's paradox. However, it is a perfectly reasonable property of a probability distribution, and occurs regularly. Thus it is not a paradox.

Another famous example arose in connection with the admission of graduates to the University of California at Berkeley. Women in fact had a better chance than men of being admitted to individual faculties, but when the figures were pooled they seemed to have a smaller chance. This situation arose because women applied in much greater numbers to faculties where *everyone* had a slim chance of admission. Men tended to apply to faculties where everyone had a good chance of admission.   ○

The switching paradox: goats and cars, the Monty Hall problem.   Television has dramatically expanded the frontiers of inanity, so you are not too surprised to be faced with the following decision. There are three doors; behind one there is a costly car, behind two there are cheap (non-pedigree) goats. You will win whatever is behind the door you finally choose. You make a first choice, but the presenter does not open this door, but a different one (revealing a goat), and asks you if you would like to change your choice to the final unopened door that you did not choose at first. Should you accept this offer to switch? Or to put it another way: what is the probability that the car is behind your first choice compared to the probability that it lies behind this possible fresh choice?

Answer.   The blunt answer is that you cannot calculate this probability as the question stands. You can only produce an answer if you assume that you know how the presenter is running the show. Many people find this unsatisfactory, but it is important to realize why it is the unpalatable truth. We discuss this later; first we show why there is no one answer.

I The 'usual' solution. The usual approach *assumes* that the presenter is attempting to make the 'game' longer and less dull. He is therefore *assumed* to behaving as follows.

Rules. Whatever your first choice, he will show you a goat behind a different door; with a choice of two goats he picks either at random.

Let the event that the car is behind the door you chose first be C_f, let the event that the car is behind your alternative choice be C_a, and let the event that the host shows you a goat be G. We require $P(C_a|G)$, and of course we *assume* that initially the car is equally likely to be anywhere. Call your first choice D_1, the presenter's open door D_2, and the alternative door D_3. Then

$$(2) \qquad P(C_a|G) = \frac{P(C_a \cap G)}{P(G)}$$

$$= \frac{P(G|C_a)P(C_a)}{P(G|C_f)P(C_f) + P(G|C_a)P(C_a)}$$

$$= \frac{P(G|C_a)}{P(G|C_f) + P(G|C_a)},$$

because $P(C_a) = P(C_f)$, by assumption.

Now by the presenter's rules

$$P(G|C_a) = 1$$

because he must show you the goat behind D_3. However,

$$P(G|C_f) = \tfrac{1}{2}$$

because there are two goats to choose from, behind D_2 and D_3, and he picks the one behind D_3 with probability $\tfrac{1}{2}$. Hence

$$P(C_a|G) = \frac{1}{1 + \tfrac{1}{2}} = \tfrac{2}{3}.$$

II The 'cheapskate' solution. Suppose we make a different set of assumptions. *Assume* the presenter is trying to save some money (the show has given away too many cars lately). He thus behaves as follows.

Rules.

(i)   If there is a goat behind the first door you choose, then he will open that door with no further ado.

(ii) If there is a car behind the first door, then he will open another door (D_3), and hope you switch to D_2.

In this case obviously $P(C_a|G) = 0$, because you only get the opportunity to switch when the first door conceals the car.

III The 'mafia' solution. There are other possible assumptions; here is a very realistic set-up. Unknown to the producer, you and the presenter are members of the same family. If the car is behind D_1, he opens the door for you; if the car is behind D_2 or D_3, he opens the other door concealing a goat. You then choose the alternative because obviously

$$P(C_a|G) = 1. \qquad \qquad \bigcirc$$

Remark. This problem is also sometimes known as the Monty Hall problem, after the presenter of a programme that required this type of decision from participants. It appeared in this form in *Parade* magazine, and generated a great deal of publicity and follow-up articles. It had, however, been around in many other forms for many years before that.

Of course this is a trivial problem, albeit entertaining, but it is important. This importance lies in the lesson that, in any experiment, the procedures and rules that define the sample space and all the probabilities must be explicit and fixed before you begin. This predetermined structure is called a *protocol*. Embarking on experiments without a complete protocol has proved to be an extremely convenient method of faking results over the years. And will no doubt continue to be so.

There are many more 'paradoxes' in probability. As we have seen, few of them are genuinely paradoxical. For the most part such results attract fame simply because someone once made a conspicuous error, or because the answer to some problem is contrary to uninformed intuition. It is notable that many such errors arise from an incorrect use of Bayes' rule, despite the fact that as long ago as 1957, W. Feller wrote this warning:

> Unfortunately Bayes' rule has been somewhat discredited by metaphysical applications of the type described by Laplace. In routine practice this kind of argument can be dangerous .... Plato used this type of argument to prove the existence of Atlantis, and philosophers used it to prove the absurdity of Newtonian mechanics.

Of course Atlantis never existed, and Newtonian mechanics are not absurd. But despite all this experience, the popular press and even, sometimes, learned journals continue to print a variety of these bogus arguments in one form or another.

Exercises for section 2.13

1. ***Prisoners paradox.*** Three prisoners, A, B, and C, are held in solitary confinement. The warder W tells each of them that two are to be freed, the third is to be flogged. Prisoner A, say, then knows his chance of being released is $\frac{2}{3}$. At this point the warder reveals to A that one of those to be released is B; this warder is known to be truthful. Does this alter A's chance of release? After all, he already knew that one of B or C was to be released. Can it be that knowing the name changes the probability?

2. ***Goats and cars revisited.*** The 'incompetent' solution. Due to a combination of indolence and incompetence the presenter has failed to find out which door the car is actually behind. So when you choose the first door, he picks another at random and opens it (hoping it does not conceal the car). Show that in this case $P(C_a|G) = \frac{1}{2}$.

2.14 REVIEW: NOTATION AND RULES

In this chapter we have used our intuitive ideas about probability to formulate rules that probability must satisfy in general. We have introduced some simple standard notation to help us in these tasks; we summarize the notation and rules here.

I Notation

Ω: sample space of outcomes

$A, B, C, \ldots$: possible events included in Ω

$\varnothing$: impossible event

$P(\cdot)$: the probability function

$P(A)$: the probability that A occurs

$A \cup B$: union; either A or B occurs or both occur

$A \cap B$: intersection; both A and B occur

A^c: complementary event

$A \subseteq B$: inclusion; B occurs if A occurs

$A \backslash B$: difference; A occurs and B does not

II Rules

Range: $0 \leqslant P(A) \leqslant 1$

Impossible event: $P(\varnothing) = 0$

Certain event: $P(\Omega) = 1$

Addition: $P(A \cup B) = P(A) + P(B)$ when $A \cap B = \varnothing$

Countable addition: $P(\cup_i A_i) = \sum_i P(A_i)$ when $(A_i; \ i \geqslant 1)$ are disjoint events

Inclusion–exclusion: $P(A \cup B) = P(A) + P(B) - P(A \cap B)$

Complement: $P(A^c) = 1 - P(A)$

Difference: when $B \subseteq A$, $P(A \backslash B) = P(A) - P(B)$

Conditioning: $P(A|B) = P(A \cap B)/P(B)$

Addition: $P(A \cup B|C) = P(A|C) + P(B|C)$ when $A \cap C$ and $B \cap C$ are disjoint

Multiplication: $P(A \cap B \cap C) = P(A|B \cap C)P(B|C)P(C)$

The partition rule: $P(A) = \sum_i P(A|B_i)P(B_i)$ when $(B_i; \ i \geqslant 1)$ are disjoint events, and
$A \subseteq \cup_i B_i$

Bayes' rule: $P(B_i|A) = P(A|B_i)P(B_i)/P(A)$

Independence: A and B are independent if and only if $P(A \cap B) = P(A)P(B)$

   This is equivalent to $P(A|B) = P(A)$ and to $P(B|A) = P(B)$

Conditional independence: A and B are conditionally independent given C when
$P(A \cap B|C) = P(A|C)P(B|C)$

Value and expected value: If an experiment yields the numerical outcome a with
probability p, or zero otherwise, then its value (or expected value) is ap

2.15 APPENDIX. DIFFERENCE EQUATIONS

On a number of occasions above, we have used conditional probability and independence to show that the answer to some problem of interest is the solution of a difference equation. For example, in example 2.11.7 we considered

$$(1) \qquad\qquad p_n = qp_{n-1},$$

in example 2.11.8 we derived

$$(2) \qquad\qquad p_n = pp_{n-1} + pqp_{n-2}, \quad pq \neq 0,$$

and in exercise 1 at the end of section 2.11 you derived

$$(3) \qquad\qquad p_n = (q - p)p_{n-1} + p.$$

We need to solve such equations systematically. Note that any sequence $(x_r; r \geq 0)$ in which each term is a function of its predecessors, so that

$$(4) \qquad\qquad x_{r+k} = f(x_r, x_{r+1}, \ldots, x_{r+k-1}), \quad r \geq 0,$$

is said to satisfy the *recurrence relation* (4). When f is linear this is called a difference equation of order k:

$$(5) \qquad\qquad x_{r+k} = a_0 x_r + a_1 x_{r+1} + \cdots + a_{k-1} x_{r+k-1} + g(r), \quad a_0 \neq 0.$$

When $g(r) = 0$, the equation is homogeneous:

$$(6) \qquad\qquad x_{r+k} = a_0 x_r + a_1 x_{r+1} + \cdots + a_{k-1} x_{r+k-1}, \quad a_0 \neq 0.$$

Solving (1) is easy because $p_{n-1} = qp_{n-2}$, $p_{n-2} = qp_{n-3}$ and so on. By successive substitution we obtain

$$p_n = q^n p_0.$$

Solving (3) is nearly as easy when we notice that

$$p_n = \tfrac{1}{2}$$

is a particular solution. Now writing $p_n = \tfrac{1}{2} + x_n$ gives

$$x_n = (q - p)x_{n-1} = (q - p)^n x_0.$$

Hence

$$p_n = \tfrac{1}{2} + (q - p)^n x_0.$$

Equation (2) is not so easy but, after some work which we omit, it turns out that (2) has solution

$$(7) \qquad\qquad p_n = c_1 \lambda_1^n + c_2 \lambda_2^n$$

where λ_1 and λ_2 are the roots of

$$x^2 - px - pq = 0$$

and c_1 and c_2 are arbitrary constants. You can verify this by substituting (7) into (2).

Having seen these preliminary results, you will not now be surprised to see the general solution to the second-order difference equation: let

$$(8) \qquad\qquad x_{r+2} = a_0 x_r + a_1 x_{r+1} + g(r), \quad r \geq 0.$$

Suppose that $\pi(r)$ is any function such that

$$\pi(r + 2) = a_0 \pi(r) + a_1 \pi(r + 1) + g(r)$$

and suppose that λ_1 and λ_2 are the roots of

$$x^2 = a_0 + a_1 x.$$

Then the solution of (8) is given by

$$x_r = \begin{cases} c_1 \lambda_1^r + c_2 \lambda_2^r + \pi(r), & \lambda_1 \neq \lambda_2 \\ (c_1 + c_2 r)\lambda_1^r + \pi(r), & \lambda_1 = \lambda_2, \end{cases}$$

where c_1 and c_2 are arbitrary constants. Here $\pi(r)$ is called a *particular solution*, and you should note that λ_1 and λ_2 may be complex, as then may c_1 and c_2.

The solution of higher-order difference equations proceeds along similar lines; there are more λ's and more c's.

2.16 PROBLEMS

1. The classic slot machine has three wheels each marked with 20 symbols. You rotate the wheels by means of a lever, and win if each wheel shows a bell when it stops. Assume that the outside wheels each have one bell symbol, the central wheel carries 10 bells, and that wheels are independently equally likely to show any of the symbols (academic licence). Find:
    (a) the probability of getting exactly two bells;
    (b) the probability of getting three bells.

2. You deal two cards from a conventional pack. What is the probability that their sum is 21? (Court cards count 10, and aces 11.)

3. You deal yourself two cards, and your opponent two cards. Your opponent reveals that the sum of those two cards is 21; what is the probability that the sum of your two cards is 21? What is the probability that you both have 21?

4. A weather forecaster says that the probability of rain on Saturday is 25%, and the probability of rain on Sunday is 25%. Can you say the chance of rain at the weekend is 50%? What *can* you say?

5. My lucky number is 3, and your lucky number is 7. Your PIN is equally likely to be any number between 1001 and 9998. What is the probability that it is divisible by at least one of our two lucky numbers?

6. You keep rolling a die until you first roll a number that you have rolled before. Let A_k be the event that this happens on the kth roll.
    (a) What is $P(A_{12})$?   (b) Find $P(A_3)$ and $P(A_6)$.

7. Ann aims three darts at the bullseye and Bob aims one. What is the probability that Bob's dart is nearest the ball? Given that one of Ann's darts is nearest, what is the probability that Bob's dart is next nearest? (They are equally skilful.)

8. In the lottery of 1710, one in every 40 tickets yielded a prize. It was widely believed at the time that you needed to buy 40 tickets at least, to have a better than evens chance of a prize. Was this belief correct?

9. (a) You have two red cards and two black cards. Two cards are picked at random; show that the probability that they are the same colour is $\frac{1}{3}$.
    (b) You have one red card and two black cards; show that the probability that two cards picked at random are the same colour is $\frac{1}{3}$. Are you surprised?
    (c) Calculate this probability when you have
        (i) three red cards and three black cards, (ii) two red cards and three black cards.

10. A box contains three red socks and two blue socks. You remove socks at random one by one until you have a pair. Let T be the event that you need only two removals, R the event that the first sock is red and B the event that the first sock is blue. Find
    $$(a)\ P(B|T),\quad (b)\ P(R|T),\quad (C)\ P(T).$$

11. Let A, B and C be events. Show that
    $$A \cap B = (A^c \cup B^c)^c,$$
    and
    $$A \cup B \cup C = (A^c \cap B^c \cap C^c)^c.$$

12.   Let $(A_r; r \geq 1)$ be events. Show that for all $n \geq 1$,

$$\left(\bigcup_{r=1}^{n} A_r \right)^c = \bigcap_{r=1}^{n} A_r^c \quad \text{and} \quad \left(\bigcap_{r=1}^{n} A_r \right)^c = \bigcup_{r=1}^{n} A_r^c.$$

Let A and B be events with $P(A) = \frac{3}{5}$ and $P(B) = \frac{1}{2}$. Show that

$$\tfrac{1}{10} \leq P(A \cap B) \leq \tfrac{1}{2}$$

and give examples to show that both extremes are possible. Can you find bounds for $P(A \cup B)$?

14.   Show that if $P(A|B) > P(A)$, then

$$P(B|A) > P(B) \quad \text{and} \quad P(A^c|B) < P(A^c).$$

15.   Show that if A is independent of itself, then either $P(A) = 0$ or $P(A) = 1$.

16.   A pack contains n cards labelled $1, 2, 3, \ldots, n$ (one number on each card). The cards are dealt out in random order. What is the probability that
      (a)   the kth card shows a larger number than its $k - 1$ predecessors?
      (b)   each of the first k cards shows a larger number than its predecessors?
      (c)   the kth card shows n, given that the kth card shows a larger number than its $k - 1$ predecessors?

17.   Show that $P(A \backslash B) \leq P(A)$.

18.   Show that

$$P\left(\bigcup_{r=1}^{n} A_r \right) = \sum_{r} P(A_r) - \sum_{r<s} P(A_r \cap A_s) + \cdots + (-)^{n+1} P\left(\bigcap_{r=1}^{n} A_r \right).$$

Is there a similar formula for $P(\bigcap_{r=1}^{n} A_r)$?

19.   Show that

$$P(A \cap B) - P(A)P(B) = P((A \cup B)^c) - P(A^c)P(B^c).$$

20.   An urn contains a amber balls and b buff balls. A ball is removed at random.
      (a)   What is the probability α that it is amber?
      (b)   Whatever colour it is, it is returned to the urn with a further c balls of the same colour as the first. Then a second ball is drawn at random from the urn. Show that the probability that it is amber is α.

21.   In the game of *antidarts* a player shoots an arrow into a rectangular board measuring six metres by eight metres. If the arrow is within one metre of the centre it scores 1 point, between one and two metres away it scores 2, between two and three metres it scores 3, between three and four metres and yet still on the board it scores 4, and further than four metres but still on the board it scores 5. William Tell always lands his arrows on the board but otherwise they are purely random.
      (a)   Show that the probability that his first arrow scores more than 3 points is $1 - \frac{3}{16}\pi$.

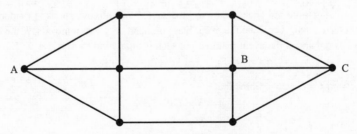

Figure 2.23. The mole's burrows.

(b) Find the probability that he scores a total of exactly 4 points in his first two arrows.

(c) Show that the probability that he scores exactly 15 points in three arrows is given by

$$\left(1 - \frac{2}{3}\sin^{-1}\left(\frac{3}{4}\right) - \frac{1}{8}\sqrt{7}\right)^3.$$

22. A mole has a network of burrows as shown in figure 2.23. Each night he sleeps at one of the junctions. Each day he moves to a neighbouring junction but he chooses a passage randomly, all choices being equally likely from those available at each move.

(a) He starts at A. Find the probability that two nights later he is at B.

(b) Having arrived at B, find the probability that two nights later he is again at B.

(c) A second mole is at C at the same time as the first mole is at A. What is the probability that two nights later the two moles share the same junction?

23. Three cards in an urn bear pictures of ants and bees; one card has ants on both sides, and one card has bees on both sides, and one has an ant on one side and a bee on the other.

A card is removed at random and placed flat. If the upper face shows a bee, what is the probability that the other side shows an ant?

24. You pick a card at random from a conventional pack and note its suit. With an obvious notation define the events

$$A_1 = S \cup H, \quad A_2 = S \cup D, \quad A_3 = S \cup C.$$

Show that A_j and A_k are independent when $j \neq k$, $1 \leqslant j$, $k \leqslant 3$.

25. A fair die is rolled repeatedly. Find

(a) the probability that the number of sixes in k rolls is even,

(b) the probability that in k rolls the number of sixes is divisible by 3.

26. ***Waldegrave's problem, example 2.11.10.***   Show that, with four players, equation (4) in this example has the solution

$$p_r = \frac{1}{2\sqrt{5}}\left(\frac{1+\sqrt{5}}{4}\right)^{r-2} - \frac{1}{2\sqrt{5}}\left(\frac{1-\sqrt{5}}{4}\right)^{r-2}.$$

27. Karel flips $n + 1$ fair coins and Newt flips n fair coins. Karel wins if he has more heads than Newt, otherwise he loses Show that P(Karel wins) $= \frac{1}{2}$.

28. Arkle (A) and Dearg (D) are connected by roads as in figure 2.24. Each road is independently blocked by snow with probability p. Find the probability that it is possible to travel by road from A to D.

Funds are available to snow-proof just one road. Would it be better to snow-proof AB or BC?

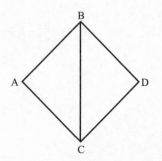

Figure 2.24. Roads.

29. You are lost on Mythy Island in the summer, when tourists are two-thirds of the population. If you ask a tourist for directions the answer is correct with probability $\frac{3}{4}$; answers to repeated questions are independent even if the question is the same. If you ask a local for directions, the answer is always false.

    (a) You ask a passer-by whether Mythy City is East or West. The answer is East. What is the probability that it is correct?

    (b) You ask her again, and get the same reply. Show that the probability that it is correct is $\frac{1}{2}$.

    (c) You ask her one more time, and the answer is East again. What is the probability that it is correct?

    (d) You ask her for the fourth and last time and get the answer West. What is the probability that East is correct?

    (e) What if the fourth answer were also East?

30. A bull is equally likely to be anywhere in the square field ABCD, of side 1. Show that the probability that it is within a distance x from A is

$$p_x = \begin{cases} \dfrac{\pi x^2}{4}, & 0 \leqslant x \leqslant 1 \\[2mm] (x^2 - 1)^{1/2} + \dfrac{\pi x^2}{4} - x^2 \cos^{-1}\left(\dfrac{1}{x}\right), & 1 \leqslant x \leqslant \sqrt{2}. \end{cases}$$

The bull is now tethered to the corner A by a chain of length 1. Find the probability that it is nearer to the fence AB than the fence CD.

31. A theatre ticket is in one of three rooms. The event that it is in the ith room is B_i, and the event that a cursory search of the ith room fails to find the ticket is F_i, where

$$0 \leqslant P(F_i|B_i) < 1.$$

Show that $P(B_i|F_i) < P(B_i)$, that is to say, if you fail to find it in the ith room on one search, then it is less likely to be there. Show also that $P(B_i|F_j) > P(B_i)$ for $i \neq j$, and interpret this.

32. 10% of the surface of a sphere S is coloured blue, the rest is coloured red. Show that, however the colours are distributed, it is possible to inscribe a cube in S with 8 red vertices. (Hint: Pick a cube at random from the set of all possible inscribed cubes, let $B(r)$ be the event that the rth vertex is blue, and consider the probability that any vertex is blue.)

3

Counting and gambling

It is clear that the enormous variety which can be seen both in nature and in the actions of mankind, and which makes up the greater part of the beauty of the universe, arises from the many different ways in which objects are arranged or chosen. But it often happens that even the cleverest and best-informed men are guilty of that error of reasoning which logicians call the insufficient, or incomplete, enumeration of cases.

J. Bernoulli (*ca.* 1700)

3.1 PREVIEW

We have seen in the previous chapter that many chance experiments have equally likely outcomes. In these problems many questions can be answered by merely counting the outcomes in events of interest. Moreover, quite often simple counting turns out to be useful and effective in more general circumstances.

In the following sections, therefore, we review the basic ideas about how to count things. We illustrate the theory with several famous examples, including birthday problems and lottery problems. In particular we solve the celebrated problem of the points. This problem has the honour of being the first to be solved using modem methods (by Blaise Pascal in 1654), and therefore marks the official birth of probability. A natural partner to it is the even more famous gambler's ruin problem. We conclude with a brief sketch of the history of chance, and some other famous problems.

Prerequisites. You need only the usual basic knowledge of elementary algebra. We shall often use the standard factorial notation

$$r! = r(r-1) \times \cdots \times 3 \times 2 \times 1.$$

Remember that $0! = 1$, by convention.

3.2 FIRST PRINCIPLES

Recall that many chance experiments have equally likely outcomes. In these cases the probability of any event A is just

$$P(A) = \frac{|A|}{|\Omega|}$$

and we 'only' have to count the elements of A and Ω. For example, suppose you are dealt five cards at poker; what is the probability of a full house? You first need the number of ways of being dealt five cards, assumed equally likely. Next you need the number of such hands that comprise a full house (three cards of one kind and two of another kind, e.g. QQQ33). We shall give the answer to this problem shortly; first we remind ourselves of the basic rules of counting. No doubt you know them informally already, but it can do no harm to collect them together explicitly here.

The first is obvious but fundamental.

Correspondence rule.   Suppose we have two finite sets A and B. Let the numbers of objects in A and B be $|A|$ and $|B|$ respectively. Then if we can show that each element of A corresponds to one and only one element of B, and vice versa, then $|A| = |B|$.

Example 3.2.1.   Let $A = \{11, 12, 13\}$ and $B = \{\heartsuit, \diamondsuit, \clubsuit\}$. Then $|A| = |B| = 3$.   ○

Example 3.2.2: reflection.   Let A be a set of distinct real numbers. Define the set B such that

$$B = \{b: -b \in A\}.$$

Then $|A| = |B|$.   ○

Example 3.2.3: choosing.   Let A be a set of size n. Let $c(n, k)$ be the number of ways of choosing k of the n elements in A. Then

$$c(n, k) = c(n, n - k),$$

because to each choice of k elements there corresponds one and only one choice of the remaining $n - k$ elements.   ○

Our next rule is equally obvious.

Addition rule.   Suppose that A and B are disjoint finite sets, so that $A \cap B = \emptyset$. Then

$$|A \cup B| = |A| + |B|.$$

Example 3.2.4: choosing.   Let A be a set containing n elements, and recall that $c(n, k)$ is the number of ways of choosing k of these elements. Show that

(1) $c(n, k) = c(n - 1, k) + c(n - 1, k - 1).$

Solution.   We can label the elements of A as we please; let us label one of them the *first* element. Let B be the collection of all subsets of A that contain k elements. This can be divided into two sets: $B(f)$, in which the first element always appears, and $B(\sim f)$, in which the first element does not appear. Now on the one hand

$$|B(f)| = c(n - 1, k - 1)$$

because the first element is guaranteed to be in all these. On the other hand

$$|B(\sim f)| = c(n - 1, k)$$

because the first element is not in these, and we still have to choose k from the $n - 1$ remaining. Obviously $|B| = c(n, k)$, by definition. Hence, by the addition rule, (1) follows. ○

The addition rule has an obvious extension to the union of several disjoint sets; write this down yourself (*exercise*).

The third counting rule will come as no surprise. As we have seen several times in chapter 2, we often combine simple experiments to obtain more complicated sample spaces. For example, we may roll several dice, or flip a sequence of coins. In such cases the following rule is often useful.

Multiplication rule. Let A and B be finite sets, and let C be the set obtained by choosing any element of A and any element of B. Thus C is the collection of ordered pairs

$$C = \{(a, b): a \in A, b \in B\}.$$

Then

(2)
$$|C| = |A\|B|.$$

This rule is often expressed in other words; one may speak of *decisions*, or *operations*, or *selections*. The idea is obvious in any case. To establish (2) it is sufficient to display all the elements of C in an array:

$$C \equiv \begin{vmatrix} (a_1, b_1) & \dots & (a_1, b_n) \\ \vdots & & \vdots \\ (a_m, b_1) & \dots & (a_m, b_n) \end{vmatrix}$$

Here $m = |A|$ and $n = |B|$. The rule (2) is now obvious by the addition rule. Again, this rule has an obvious extension to the product of several sets.

Example 3.2.5: sequences. Let A be a finite set. A *sequence* of length r from A is an ordered set of elements of A (which may be repeated as often as required). We denote such a sequence by $(a_1, a_2, \dots, a_r)$; suppose that $|A| = n$. By the multiplication rule we find that there are n^r such sequences of length r. ○

Example 3.2.6: crossing a cube. Let A and B be diametrically opposite vertices of a cube. How many ways are there of traversing edges from A to B using exactly three edges?

Solution. There are three choices for the first step, then two for the second, then one for the last. The required number is 3! ○

For our final rule we consider the problem of counting the elements of $A \cup B$, when A and B are not disjoint. This is given by the inclusion–exclusion rule, as follows.

Inclusion–exclusion rule.   For any finite sets A and B,

(3) $$|A \cup B| = |A| + |B| - |A \cap B|.$$

To see this, note that any element in $A \cup B$ appears just once on each side unless it is in $A \cap B$. In that case it appears in all three terms on the right, and so contributes $1 + 1 - 1 = 1$ to the total, as required. The three-set version is of course

(4) $$|A \cup B \cup C| = |A| + |B| + |C| - |A \cap B| - |A \cap C|$$
$$- |B \cap C| + |A \cap B \cap C|.$$

A straightforward induction yields the general form of this rule for n sets $A_1, \ldots, A_n$ in the form

(5) $$|A_1 \cup \cdots \cup A_n| = \sum_i |A_i| - \sum_{i<j} |A_i \cap A_j| + \cdots + (-1)^{n+1} |A_1 \cap \cdots \cap A_n|$$

Example 3.2.7: derangements.   An urn contains three balls numbered 1, 2, 3. They are removed at random, without replacement. What is the probability p that none of the balls is drawn in the same position as the number it bears?

Solution.   By the multiplication rule, $|\Omega| = 6$. Let A_i be the set of outcomes in which the ball numbered i is drawn in the ith place. Then we have

$$|A_i| = 2,$$

$$|A_i \cap A_j| = 1, \quad i < j,$$

$$|A_1 \cap A_2 \cap A_3| = 1.$$

Hence, by (4),

$$|A_1 \cup A_2 \cup A_3| = 2 + 2 + 2 - 1 - 1 - 1 + 1 = 4.$$

Therefore, by the addition rule, the number of ways of getting no ball in the same position as its number is two. Therefore, since the six outcomes in Ω are assumed to be equally likely, $p = \frac{2}{6} = \frac{1}{3}$.   ○

Exercises for section 3.2

1. Let A and B be diametrically opposite vertices of a cube. How many ways are there of traversing edges from A to B, without visiting any vertex twice, using exactly (a) five edges? (b) six edges? (c) seven edges?

2. Show that equation (1), $c(n, k) = c(n - 1, k) + c(n - 1, k - 1)$, is satisfied by
$$c(n, k) = \frac{n!}{k!(n-k)!}.$$

3. A die is rolled six times. Show that the probability that all six faces are shown is 0.015, approximately.

4. A die is rolled 1000 times. Show that the probability that the sum of the numbers shown is 1100 is the same as the probability that the sum of the numbers shown is 5900.

3.3 ARRANGING AND CHOOSING

The sequences considered in section 3.2 sometimes allowed the possibility of repetition; for example, in rolling a die several times you might get two or more sixes. However, a great many experiments supply outcomes with no repetition; for example, you would be very startled (to say the least) to find more than one ace of spades in your poker hand. In this section we consider problems involving selections and arrangements without repetition.

We begin with arrangements or orderings.

Example 3.3.1.   You have five books on probability. In how many ways can you arrange them on your bookshelf?

Solution.   Any of the five can go on the left. This leaves four possibilities for the second book, and so by the multiplication rule there are $5 \times 4 = 20$ ways to put the first two on your shelf. That leaves three choices for the third book, yielding $5 \times 4 \times 3 = 60$ ways of shelving the first three. Then there are two possibilities for the penultimate book, and only one choice for the last book, so there are altogether

$$5 \times 4 \times 3 \times 2 \times 1 = 5! = 120$$

ways of arranging them.

Incidentally, in the course of showing this we have shown that the number of ways of arranging a selection of r books, $0 \le r \le 5$, is

$$5 \times \cdots \times (5 - r + 1) = \frac{5!}{(5-r)!}.$$ ○

It is quite obvious that the same argument works if we seek to arrange a selection of r things from n things. We can choose the first in n ways, the second in $n - 1$ ways, and so on, with the last chosen in $n - r + 1$ ways. By the product rule, this gives $n(n - 1) \cdots (n - r + 1)$ ways in total. We display this result, and note that the conventional term for such an ordering or arrangement is a *permutation*. (Note also that algebraists use it differently.)

Permutations.   The number of permutations of r things from n things is

(1) $$n(n - 1) \cdots (n - r + 1) = \frac{n!}{(n - r)!}, \quad 0 \le r \le n.$$

This is a convenient moment to turn aside for a word on notation and conventions. We are familiar with the factorial notation

$$n! = n(n - 1) \times \cdots \times 2 \times 1,$$

defined for positive integers n, with the convention that $0! = 1$. The number of permutations of r things from n things, given in (1), crops up so frequently that it is often given a special symbol. We write

(2) $$x(x - 1) \cdots (x - r + 1) = x^{\underline{r}},$$

which is spoken as 'the rth falling factorial power of x', which is valid for any real number x. By convention $x^{\underline{0}} = 1$. When x is a positive integer,

$$x^{\underline{r}} = \frac{x!}{(x-r)!}.$$

In particular, $r^{\underline{r}} = r!$

Note that various other notations are used for this, most commonly $(x)_r$ in the general case, and $^x P_r$ when x is an integer.

Next we turn to the problem of counting arrangements when the objects in question are not all distinguishable.

In the above example involving books, we naturally assumed that the books were all distinct. But suppose that, for whatever strange reason, you happen to have two new copies of some book. They are unmarked, and therefore indistinguishable. How many different permutations of all five are possible now? There are in fact 60 different arrangements. To see this we note that in the 120 arrangements in example 3.3.1 there are 60 pairs in which each member of the pair is obtained by exchanging the positions of the two identical books. But these pairs are indistinguishable, and therefore the same. So there are just 60 different permutations.

We can generalize this result as follows. If there are n objects of which n_1 form one indistinguishable group, n_2 another, and so on up to n_r, where

(3) $$n_1 + n_2 + \cdots + n_r = n,$$

then there are

(4) $$M(n_1, \ldots, n_r) = \frac{n!}{n_1! n_2! \cdots n_r!}$$

distinct permutations of these n objects. It is easy to prove this, as follows. For each of the M such arrangements suppose that the objects in each group are then numbered, and hence distinguished. Then the objects in the first group can now be arranged in $n_1!$ ways, and so on for all r groups. By the multiplication rule there are hence $n_1! n_2! \cdots n_r! M$ permutations. But we already know that this number is $n!$. Equating these two gives (4).

This argument is simpler than it may appear at first sight; the following example makes it obvious.

Example 3.3.2. Consider the word 'dada'. In this case $n = 4$, $n_1 = n_2 = 2$, and (4) gives

$$M(2, 2) = \frac{4!}{2! 2!} = 6,$$

as we may verify by exhaustion:

aadd, adad, daad, dada, adda, ddaa.

Now, as described above we can number the a's and d's, and permute these now distinguishable objects for each of the six cases. Thus

$$aadd \text{ yields } \begin{cases} a_1 a_2 d_1 d_2 \\ a_2 a_1 d_1 d_2 \\ a_1 a_2 d_2 d_1 \\ a_2 a_1 d_2 d_1 \end{cases}$$

and likewise for the other five cases. There are therefore $6 \times 4 = 24$ permutations of 4 objects, as we know already since $4! = 24$. ○

Once again we interject a brief note on names and notation. The numbers $M(n_1, \ldots, n_r)$ are called *multinomial coefficients*. An alternative notation is

$$M(n_1, \ldots, n_r) = \binom{n_1 + n_2 + \cdots + n_r}{n_1, n_2, \ldots, n_r}.$$

The most important case, and the one which we see most often, is the *binomial coefficient*

$$\binom{n}{r} = M(n - r, r) = \frac{n!}{(n - r)!r!}$$

This is also denoted by ^nC_r. We can also write it as

$$\binom{x}{r} = \frac{x^{\underline{r}}}{r!},$$

which makes sense when x is any real number. For example

$$\binom{-1}{r} = (-1)^r.$$

Binomial coefficients arise very naturally when we count things without regard to their order, as we shall soon see.

In counting permutations the idea of order is essential. However, it is often the case that we choose things and pay no particular regard to their order.

Example 3.3.3: quality control. You have a box of numbered components, and you have to select a fixed quantity (r, say) for testing. If there are n in the box, how many different selections are possible?

If $n = 4$ and $r = 2$, then you can see by exhaustion that from the set $\{a, b, c, d\}$ you can pick six pairs, namely

$$ab, \; ac, \; ad, \; bc, \; bd, \; cd. \qquad\qquad\qquad\qquad\qquad \bigcirc$$

There are many classical formulations of this basic problem; perhaps the most frequently met is the hand of cards, as follows.

You are dealt a hand of r cards from a pack of n. How many different possible hands are there? Generally $n = 52$; for poker $r = 5$, for bridge $r = 13$.

The answer is called the number of *combinations* of r objects from n objects. The key result is the following.

Combinations. The number of ways of choosing r things from n things (taking no account of order) is

$$(5) \qquad\qquad \binom{n}{r} = \frac{n!}{r!(n - r)!}, \qquad 0 \leqslant r \leqslant n.$$

Any such given selection of r things is called a *combination*, or an *unordered sample*. It is of course just a *subset* of size r, in more workaday terminology.

This result is so important and useful that we are going to establish it in several different ways. This provides insight into the significance of the binomial coefficients, and also illustrates important techniques and applications.

First derivation of (5).   We know from (1) that the number of permutations of r things from n things is $n^{\underline{r}}$. But any permutation can also be fixed by performing two operations:

(i)   choose a subset of size r;
(ii)   choose an order for the subset.

Suppose that step (i) can be made in $c(n, r)$ ways; this number is what we want to find. We know step (ii) can be made in $r!$ ways. By the multiplication rule (2) of section 3.2 the product of these two is $n^{\underline{r}}$, so

(6) $$c(n, r)r! = n^{\underline{r}} = \frac{n!}{(n - r)!}.$$

Hence

(7) $$c(n, r) = \binom{n}{r} = \frac{n!}{r!(n - r)!} \qquad \square$$

This argument is very similar to that used to establish (4); and this remark suggests an alternative proof.

Second derivation of (5).   Place the n objects in a row, and mark the r selected objects with the symbol S. Those not selected are marked F. Therefore, by construction, there is a one–one correspondence between the combinations of r objects from n and the permutations of r S-symbols and $n - r$ F-symbols. But, by (4), there are

(8) $$M(r, n - r) = \frac{n!}{r!(n - r)!}$$

permutations of these S- and F-symbols. Hence using the correspondence rule (see the start of section 3.2) proves (5). $\square$

Another useful method of counting a set is to split it up in some useful way. This supplies another derivation.

Third derivation of (5).   As above we denote the number of ways of choosing a subset of size r from a set of size n objects by $c(n, r)$. Now suppose one of the n objects is in some way distinctive; for definiteness we shall say it is pink. Now there are two distinct methods of choosing subsets of size r:

(i)   include the pink one and choose $r - 1$ more objects from the remaining $n - 1$;
(ii)   exclude the pink one and choose r of the $n - 1$ others.

There are $c(n - 1, r - 1)$ ways to choose using method (i), and $c(n - 1, r)$ ways to choose using method (ii). By the addition rule their sum is $c(n, r)$, which is to say

(9) $$c(n, r) = c(n - 1, r) + c(n - 1, r - 1).$$

Of course we always have $c(n, 0) = c(r, n) = 1$, and it is an easy matter to check that the solution of (9) is

$$c(n, r) = \binom{n}{r};$$

we just plod through a little algebra:

(10)
$$\binom{n}{r} = \frac{n!}{r!(n-r)!} = \frac{n(n-1)!}{r(r-1)!(n-r)(n-r-1)!}$$

$$= \frac{(n-1)!}{(r-1)!(n-r-1)!}\left\{\frac{1}{r} + \frac{1}{n-r}\right\}$$

$$= \binom{n-1}{r} + \binom{n-1}{r-1}. \qquad \Box$$

Exercises for section 3.3

1. Show in three different ways that
$$\binom{n}{r} = \binom{n}{n-r}.$$

2. Show that the multinomial coefficient can be written as a product of binomial coefficients:
$$M(n_1, \ldots, n_r) = \binom{s_r}{s_{r-1}}\binom{s_{r-1}}{s_{r-2}}\cdots\binom{s_2}{s_1}$$
    where $s_r = \sum_{i=1}^{r} n_i$.

3. Four children are picked at random (with no replacement) from a family which includes exactly two boys. The chance that neither boy is chosen is half the chance that both are chosen. How large is the family?

4. You flip a fair coin n times. What is the probability that
    (a) there have been exactly three heads?
    (b) there have been at least two heads?
    (c) there have been equal numbers of heads and tails?
    (d) there have been twice as many tails as heads?

3.4 BINOMIAL COEFFICIENTS AND PASCAL'S TRIANGLE

The binomial coefficients
$$c(n, r) = \binom{n}{r}$$

can be simply and memorably displayed as an array. There are of course many ways to organize such an array; let us place them in the nth row and rth column like this:

```
0th row →    1
             1    1
             1    2    1
             1    3    3    1
             1    4    6    4    1
             1    5   10   10    5    1
             1    6   15   20   15    6    1
             1    7   21   35   35   21    7    1
                   . . .
             ↑
        0th column
```

Thus for example, in the 5th column and 7th row we find

$$\binom{7}{5} = 21$$

This array is called Pascal's triangle in honour of Blaise Pascal, who wrote a famous book *Treatise on the Arithmetic Triangle* in 1654, published a decade later. In this he brought together most of what was known about this array of numbers at the time, together with many significant contributions of his own.

Any entry in Pascal's triangle can be calculated individually from the fact that

(1)
$$\binom{n}{r} = \frac{n!}{r!(n-r)!},$$

but it is also convenient to observe that rows can be calculated recursively from the identity we proved in (10) of section 3.3, namely

(2)
$$\binom{n}{r} = \binom{n-1}{r-1} + \binom{n-1}{r}.$$

This says that any entry in the triangle is the sum of the entry in the row above and its neighbour on the left.

It is easy to see that any entry is also related to its neighbour in the same row by the relation

(3)
$$\binom{n}{r} = \frac{n-r+1}{r} \binom{n}{r-1}.$$

This offers a very easy way of calculating

$$\binom{n}{k},$$

by starting with

$$\binom{n}{0} = 1$$

and then applying (3) recursively for $r = 1, 2, \ldots, k$. Equation (3) is most easily shown by direct substitution of (1), but as with many such identities there is an alternative combinatorial proof. (We have already seen an example of this in our several derivations of (5) in section 3.3.)

***Example 3.4.1: demonstration of* (3)**.   Suppose that you have n people and a van with $k \leq n$ seats. In how many ways w can you choose these k travellers, with one driver?

(i)   You can choose k to go in $\binom{n}{k}$ ways, and choose one of these k to drive in k ways. So

$$w = k\binom{n}{k}.$$

(ii)   You can choose $k-1$ passengers in

$$\binom{n}{k-1}$$

ways, and then pick the driver in $n - (k-1)$ ways. So

$$w = (n - k + 1)\binom{n}{k - 1}.$$

Now (3) follows. ○

We give one more example of this technique: its use to prove a famous formula.

Example 3.4.2: Van der Monde's formula. Remarkably, it is true that for integers m, n, and $r \leqslant m \wedge n$,

$$\sum_k \binom{m}{k}\binom{n}{r - k} = \binom{m + n}{r}.$$

Solution. Suppose there are m men and n women, and you wish to form a team with r members. In how many distinct ways can this be done? Obviously in

$$\binom{m + n}{r}$$

ways if you choose directly from the whole group. But now suppose you choose k men from those present and $r - k$ women from those present. This may be done in

$$\binom{m}{k}\binom{n}{r - k}$$

ways, by the multiplication rule. Now summing over all possible k gives the left-hand side, by the addition rule. ○

Exercises for section 3.4

1. Show that the number of subsets of a set of n elements is 2^n. (Do not forget to include the empty set $\varnothing$.)

2. Prove that

$$(x + y)^n = \sum_{k=0}^{n}\binom{n}{k}x^k y^{n-k}.$$

3. Show that

$$\sum_{k=0}^{n}\binom{n}{k} = 2^n.$$

4. Use the correspondence and addition rules (section 3.2) to show that

$$\binom{n}{k} = \sum_{r=1}^{n}\binom{r - 1}{k - 1}.$$

    (Hint: How many ways are there to choose k numbers in a lottery when the largest chosen is r?)

5. **Ant.** An ant walks on the non-negative plane integer lattice starting at $(0, 0)$. When at (j, k) it can step either to $(j + 1, k)$ or $(j, k + 1)$. In how many ways can it walk to the point (r, s)?

3.5 CHOICE AND CHANCE

We now consider some probability problems, both famous and commonplace, to illustrate the use of counting methods. Remember that the fundamental situation is a sample space Ω, all of whose outcomes are equally likely. Then for any event A

$$P(A) = \frac{|A|}{|\Omega|} = \frac{\text{number of outcomes in } A}{\text{number of outcomes in } \Omega}.$$

We begin with some simple examples. As will become apparent, the golden rule in tackling all problems of this kind is:

> Make sure you understand exactly what the sample space Ω and the event of interest A actually are.

Example 3.5.1: personal identifier numbers.   Commonly PINs have four digits. A computer assigns you a PIN at random. What is the probability that all four are different?

Solution.   Conventionally PINs do not begin with zero (though there is no technical reason why they should not). Therefore, using the multiplication rule,

$$|\Omega| = 9 \times 10 \times 10 \times 10.$$

Now A is the event that no digit is repeated, so

$$|A| = 9 \times 9 \times 8 \times 7.$$

Hence

$$P(A) = \frac{|A|}{|\Omega|} = \frac{9^{\underline{3}}}{10^3} = 0.504. \qquad \bigcirc$$

Example 3.5.2: poker dice.   A set of poker dice comprises five cubes each showing $\{9, 10, J, Q, K, A\}$, in an obvious notation. If you roll such a set of dice, what is the probability of getting a 'full house' (three of one kind and two of another)?

Solution.   Obviously $|\Omega| = 6^5$, because each die may show any one of the six faces. A particular full house is chosen as follows:

- choose a face to show three times;
- choose another face to show twice;
- choose three dice to show the first face.

By the multiplication rule, and (5) of section 3.3, we have therefore that

$$|A| = 6 \times 5 \times \binom{5}{3}.$$

Hence

$$P(\text{full house}) = 2\binom{6}{4}\binom{5}{3}6^{-5} \simeq 0.039 \qquad \bigcirc$$

Here is a classic example of this type of problem.

Example 3.5.3: birthdays. For reasons that are mysterious, some (rather vague) significance is sometimes attached to the discovery that two individuals share a birthday. Given a collection of people, for example a class or lecture group, it is natural to ask for the chance that at least two do share the same birthday.

We begin by making some assumptions that greatly simplify the arithmetic, without in any way sacrificing the essence of the question or the answer. Specifically, we assume that there are r individuals (none of whom was born on 29 February) who are all independently equally likely to have been born on any of the 365 days of a non-leap year.

Let s_r be the probability that at least two of the r share a birthday. Then we ask the following two questions:

(i) How big does r need to be to make $s_r > \frac{1}{2}$? That is, how many people do we need to make a shared birthday more likely than not?
(ii) In particular, what is s_{24}?

(In fact births are slightly more frequent in the late summer, multiple births do occur, and some births occur on 29 February. However, it is obvious, and it can be proved, that the effect of these facts on our answers is practically negligible.)

Before we tackle these two problems we can make some elementary observations. First, we can see easily that

$$s_2 = \frac{1}{365} \simeq 0.003$$

because there are $(365)^2$ ways for two people to have their birthdays, and in 365 cases they share it. With a little more effort we can see that

$$s_3 = \frac{1093}{133225} \simeq 0.008$$

because there are $(365)^3$ ways for three people to have their birthdays, there are $365 \times 364 \times 363$ ways for them to be different, and so there are $(365)^3 - 365 \times 364 \times 363$ ways for at least one shared day. Hence, as required,

(1) $$s_3 = \frac{(365)^3 - 365 \times 364 \times 363}{(365)^3}.$$

These are rather small probabilities but, at the other extreme, we have $s_{366} = 1$, which follows from the pigeonhole principle. That is, even if 365 people have different birthdays then the 366th person must share one. At this point, before we give the solution, you should write down your intuitive guesses (very roughly) at the answers to (i) and (ii).

Solution. The method for finding s_r has already been suggested by our derivation of s_3. We first find the number of ways in which all r people have different birthdays. There are 365 possibilities for the first, then 364 different possibilities for the second, then 363 possibilities different from the first two, and so on. Therefore, by the multiplication rule, there are

$$365 \times 364 \times \cdots \times (365 - r + 1)$$

ways for all r birthdays to be different.

Also, by the multiplication rule, there are $(365)^r$ ways for the birthdays to be distributed. Then by the addition rule there are

$$(365)^r - 365 \times \cdots \times (365 - r + 1)$$

ways for at least one shared day. Thus

(2)
$$s_r = \frac{(365)^r - 365 \times \cdots \times (365 - r + 1)}{(365)^r}$$

$$= 1 - \frac{364 \times \cdots \times (365 - r + 1)}{(365)^{r-1}}$$

Now after a little calculation (with a calculator) we find that approximately

$$s_{24} \simeq 0.54, \quad s_{23} \simeq 0.51, \quad s_{22} \simeq 0.48.$$

Thus a group of 23 randomly selected people is sufficiently large to ensure that a shared birthday is more likely than not.

This is generally held to be surprisingly low, and at variance with uninformed intuition. How did your guesses compare with the true answer?   ○

At this point we pause to make a general point. You must have noticed that in all the above examples the sample space Ω has the property that

$$|\Omega| = n^r, \quad r \geqslant 1$$

for some n and r. It is easy to see that this is so because the n objects could each supply any one of r outcomes independently. This is just the same as the sample space you get if from an urn containing n distinct balls you remove one, inspect it, and replace it, and do this r times altogether. This situation is therefore generally called *sampling with replacement*.

If you did not replace the balls at any time then

$$|\Omega| = n^{\underline{r}} = \frac{n!}{(n - r)!}, \quad 1 \leqslant r \leqslant n.$$

Naturally this is called *sampling without replacement*.

We now consider some classic problems of this latter kind.

Example 3.5.4: bridge hands.   You are dealt a hand at bridge. What is the probability that it contains s spades, h hearts, d diamonds, and c clubs?

Solution.   A hand is formed by choosing 13 of the 52 cards, and so

$$|\Omega| = \binom{52}{13}.$$

Then the s spades may be chosen in

$$\binom{13}{s}$$

ways, and so on for the other suits. Hence by the multiplication rule, using an obvious notation,

$$|A(s, h, d, c)| = \binom{13}{s}\binom{13}{h}\binom{13}{d}\binom{13}{c}.$$

With a calculator and some effort you can show that, for example, the probability of 4 spades and 3 of each of the other three suits is

$$P(A(4, 3, 3, 3)) = \frac{\binom{13}{4}\binom{13}{3}^3}{\binom{52}{13}} \simeq 0.026.$$

In fact, no other specified hand is more likely. ○

Example 3.5.5: bridge continued; shapes. You are dealt a hand at bridge. Find the probability of the event B that the hand contains s_1 of one suit, s_2 of another suit and so on, where $s_1 + s_2 + s_3 + s_4 = 13$ and

$$s_1 \geqslant s_2 \geqslant s_3 \geqslant s_4.$$

Solution. There are three cases.

(i) The s_i are all different. In this case the event B in question arises if $A(s_1, s_2, s_3, s_4)$ occurs, or if $A(s_2, s_1, s_3, s_4)$ occurs, or if $A(s_4, s_1, s_2, s_3)$ occurs, and so on. That is, any permutation of (s_1, s_2, s_3, s_4) will do. There are 4! such permutations, so

$$P(B) = 4!P(A(s_1, s_2, s_3, s_4)).$$

(ii) Exactly 2 of s_1, s_2, s_3, s_4 are the same. In this case there are $4!/2! = 12$ distinct permutations of (s_1, s_2, s_3, s_4), so by the same argument as in (i),

$$P(B) = 12P(A).$$

(iii) Exactly 3 of s_1, s_2, s_3, s_4 are the same. In this case there are $4!/3! = 4$ distinct permutations, so $P(B) = 4P(A)$.

With a calculator and some effort you can show that the probability that the shape of your hand is $(4, 4, 3, 2)$ is 0.22 approximately. And in fact no other shape is more likely. ○

Notice how this differs from the case when suits are specified. The shape $(4, 3, 3, 3)$ has probability 0.11, approximately, even though it was the most likely hand when suits were specified.

Example 3.5.6: poker. You are dealt a hand of 5 cards from a conventional pack. A *full house* comprises 3 cards of one value and 2 of another (e.g. 3 twos and 2 fours). If the hand has 4 cards of one value (e.g. 4 jacks), this is called *four of a kind*. Which is more likely?

Solution. (i) First we note that Ω comprises all possible choices of 5 cards from 52 cards. Hence

$$|\Omega| = \binom{52}{5}.$$

(ii) For a full house you can choose the value of the triple in 13 ways, and then you can choose their 3 suits in

$$\binom{4}{3}$$

ways. The value of the double can then be chosen in 12 ways, and their suits in

$$\binom{4}{2}$$

ways. Hence

$$P(\text{full house}) = 13\binom{4}{3}12\binom{4}{2}\bigg/\binom{52}{5}$$

$$\simeq 0.0014.$$

(iii) Four of a kind allows 13 choices for the quadruple and then 48 choices for the other card. Hence

$$P(\text{four of a kind}) = 13 \times 48 \bigg/ \binom{52}{5}$$

$$\simeq 0.00024. \qquad \qquad \bigcirc$$

Example 3.5.7: tennis.   Rod and Fred are playing a game of tennis. The scoring is conventional, which is to say that scores run through (0, 15, 30, 40, game), with the usual provisions for deuce at 40–40.

Rod wins any point with probability p. What is the probability g that he wins the game? We assume that all points are won or lost independently.

You can use the result of example 2.11.5.

Solution.   Let A_k be the event that Rod wins the game and Fred wins exactly k points during the game; let A_d be the event that Rod wins from deuce. Clearly

$$g = P(A_0) + P(A_1) + P(A_2) + P(A_d).$$

Let us consider these terms in order.

(i) For A_0 to occur, Rod wins 4 consecutive points; $P(A_0) = p^4$.

(ii) For A_1 to occur Fred wins a point at some time before Rod has won his 4 points. There are

$$\binom{4}{1} = 4$$

occasions for Fred to win his point, and in each case the probability that Rod wins 4 and Fred 1 is $p^4(1 - p)$. Therefore

$$P(A_1) = 4p^4(1 - p).$$

(iii) Likewise for A_2 we must count the number of ways in which Fred can win 2 points. This is just the number of ways of choosing where he can win 2 points, namely

$$\binom{5}{2} = 10.$$

Hence

$$P(A_2) = 10p^4(1 - p)^2.$$

(iv) Finally, Rod can win having been at deuce; we denote the event deuce by D. For D to occur Fred must win 3 points, and so by the argument above

$$P(D) = \binom{6}{3} p^3(1 - p)^3.$$

The probability that Rod wins from deuce is found in example 2.11.5, so combining that result with the above gives

$$P(A_d) = P(A_d|D)P(D)$$

$$= \frac{p^2}{1 - 2p(1 - p)} \binom{6}{3} p^3(1 - p)^3.$$

Thus

$$g = p^4 + 4p^4(1 - p) + 10p^4(1 - p)^2 + \frac{20p^5(1 - p)^3}{1 - 2p(1 - p)}. \qquad \bigcirc$$

Remark.   The first probabilistic analysis of tennis was carried out by James Bernoulli, and included as an appendix to his book published in 1713. Of course he was writing about real tennis (the Jeu de Paume), not lawn tennis, but the scoring system is essentially the same. The play is extremely different.

Exercises for section 3.5

1.   What is the probability that your PIN has exactly one pair of digits the same?

2.   ***Poker dice.***   You roll 5 poker dice. Show that the probability of 2 pairs is

$$\frac{1}{2!} \binom{6}{3} \binom{5}{2, 2, 1} 6^{-5} \simeq 0.23$$

Explain the presence of $1/2!$ in this expression.

3.   ***Bridge.***   Show that the probability that you have x spades and your partner has y spades is

$$\binom{13}{x} \binom{39}{13 - x} \binom{13 - x}{y} \binom{26 + x}{13 - y} \bigg/ \left\{ \binom{52}{13} \binom{39}{13} \right\}.$$

What is the conditional probability that your partner has y spades given that you have x spades?

4.   ***Tennis.***   Check that, in example 3.5.7, when $p = \frac{1}{2}$ we have $g = \frac{1}{2}$ (which we know directly in this case by symmetry).

5.   Suppose Rod and Fred play n independent points. Rod wins each point with probability p, or loses it to Fred with probability $1 - p$. Show that the probability that Rod wins exactly k points is

$$\binom{n}{k} p^k(1 - p)^{n-k}.$$

3.6 APPLICATIONS TO LOTTERIES

Now in the way of Lottery men do also tax themselves in the general, though out of hopes of Advantage in particular: A Lottery therefore is properly a Tax upon unfortunate self-conceited fools; men that have good opinion of their own luckiness, or that have believed some Fortune-teller or Astrologer, who had promised them great success about the time and place of the Lottery, lying Southwest perhaps from the place where the destiny was read.

Now because the world abounds with this kinde of fools, it is not fit that every man that will, may cheat every man that would be cheated; but it is rather ordained, that the Sovereign should have the Guardianship of these fools, or that some Favourite should beg the Sovereign's right of taking advantage of such men's folly, even as in the case of Lunaticks and Idiots.

Wherefore a Lottery is not tollerated without authority, assigning the proportion in which the people shall pay for their errours, and taking care that they be not so much and so often couzened, as they themselves would be.

William Petty (1662)

Lotto's a taxation
On all fools in the nation
But heaven be praised
It's so easily raised.

Traditional

In spite of the above remarks, lotteries are becoming ever more widespread. The usual form of the modern lottery is as follows. There are n numbers available; you choose r of them and the organizers also choose r (without repetition). If the choices are the same, you are a winner.

Sometimes the organizers choose one extra number (or more), called a bonus number. If your choice includes this number and $r-1$ of the other r chosen by the organizers, then you win a consolation prize.

Lotteries in this form seem to have originated in Genoa in the 17th century; for that reason they are often known as Genoese lotteries. The version currently operated in England has $n=49$ and $r=6$, with one bonus number. Just as in the 17th century, the natural question is, what are the chances of winning? This is an easy problem: there are

$$\binom{n}{r}$$

ways of choosing r different numbers from n numbers, and these are equally likely. The probability that your single selection of r numbers wins is therefore

(1) $$p_w = 1 \bigg/ \binom{n}{r}.$$

In this case, when $(n, r) = (49, 6)$, this gives

$$p_w = 1 \bigg/ \binom{49}{6} = \frac{1 \times 2 \times 3 \times 4 \times 5 \times 6}{49 \times 48 \times 47 \times 46 \times 45 \times 44}$$

$$= \frac{1}{13\,983\,816}.$$

It is also straightforward to calculate the chance of winning a consolation prize using the bonus number. The bonus number can replace any one of the r winning numbers to yield your selection of r numbers, so

(2)
$$p_c = r \bigg/ \binom{n}{r}.$$

When $(n, r) = (49, 6)$, this gives

$$p_c = \frac{1}{2330\,636}.$$

An alternative way of seeing the truth of (2) runs as follows. There are r winning numbers and one bonus ball. To win a consolation prize you can choose the bonus ball in just one way, and the remaining $r - 1$ numbers in

$$\binom{r}{r-1}$$

ways. Hence, as before,

$$p_c = \binom{r}{r-1} \times 1 \bigg/ \binom{n}{r}.$$

The numbers drawn in any national or state lottery attract much more attention than most other random events. Occasionally this gives rise to controversy because our intuitive feelings about randomness are not sufficiently well developed to estimate the chances of more complicated outcomes.

For example, whenever the draw yields runs of consecutive numbers, (such as $\{2, 3, 4, 8, 38, 42\}$, which contains a run of length three), it strikes us as somehow less random than an outcome with no runs. Indeed it is not infrequently asserted that there are 'too many' runs in the winning draws, and that this is evidence of bias. (Similar assertions are sometimes made by those who enter football 'pools'.) In fact calculation shows that intuition is misleading in this case. We give some examples.

Example 3.6.1: chance of no runs. Suppose you pick r numbers at random from a sequence of n numbers. What is the probability that no two of them are adjacent, that is to say, the selection contains no runs? We just need to count the number of ways s of choosing r objects from n objects in a line, so that there is at least one unselected object as a *spacer* between each pair of selected objects. The crucial observation is that if we strike out or ignore the $r - 1$ necessary spacers then we have an unconstrained selection of r from $n - (r - 1)$ objects. Here are examples with $n = 4$ and $r = 2$; unselected objects are denoted by $\bigcirc$, selected objects by $\otimes$, and the unselected object used as a spacer is $\bullet$:

$$\otimes\bullet\bigcirc\otimes \equiv \otimes\bigcirc\otimes,$$

$$\bigcirc\otimes\bullet\otimes \equiv \bigcirc\otimes\otimes,$$

and so on. Conversely any selection of r objects from $n - (r - 1)$ objects can be turned into a selection of r objects from n objects with no runs, simply by adding $r - 1$ spacers. Therefore the number we seek is

$$s = \binom{n - (r - 1)}{r}.$$

Hence the probability that the r winning lottery numbers contain no runs at all is

(3) $$p_s = \binom{n+1-r}{r} \Big/ \binom{n}{r}$$

For example, if $n = 49$ and $r = 6$ then

$$p_s = \binom{44}{6} \Big/ \binom{49}{6}$$

$$\simeq 0.505.$$

So in the first six draws you are about as likely to see at least one run as not. This is perhaps more likely than intuition suggests.

When the bonus ball is drawn, the chance of no runs at all is now

$$\binom{43}{7} \Big/ \binom{49}{7} \simeq 0.375.$$

The chance of at least one run is not far short of $\frac{2}{3}$.   ○

Similar arguments will find the probability of any patterns of interest.

Example 3.6.2: a run of 3.   Suppose we have 47 objects in a row. We can choose 4 of these with at least one spacer between each in

$$\binom{47+1-4}{4} = \binom{44}{4}$$

ways. Now we can choose one of these 4 and add two consecutive objects to follow it, in 4 ways. Hence the probability that 6 winning lottery numbers contain exactly one run of length 3, such as $\{2, 3, 4, 8, 38, 42\}$, is

$$4\binom{44}{4} \Big/ \binom{49}{6}.$$ ○

Example 3.6.3: two runs of 2.   Choose 4 non-adjacent objects from 47 in

$$\binom{47+1-4}{4} = \binom{44}{4}$$

ways. Now choose two of them to be pairs in

$$\binom{4}{2}$$

ways. Hence the chance that 6 lottery numbers include just two runs of length 2, such as $\{1, 2, 20, 30, 41, 42\}$, is

$$\binom{4}{2}\binom{44}{4} \Big/ \binom{49}{6}.$$ ○

Exercises for section 3.6

1. A lottery selects 6 numbers from $\{1, 2, \ldots, 49\}$. Show that the probability of exactly one consecutive pair of numbers in the 6 is

$$5\binom{44}{5}\bigg/\binom{49}{6}\simeq 0.065.$$

2. A lottery selects r numbers from the first n integers. Show that the probability that all r numbers have at least k spacers between each pair of them is

$$\binom{n-(r-1)k}{r}\bigg/\binom{n}{r}, \quad (r-1)k \le n.$$

3. A lottery selects r numbers from n. Show that the probability that exactly k of your r selected numbers match k of the winning r numbers is

$$\binom{r}{k}\binom{n-r}{r-k}\bigg/\binom{n}{r}.$$

4. **Example 3.6.1 revisited: no runs.** You pick r numbers at random from a sequence of n numbers (without replacement). Let $s(n, r)$ be the number of ways of doing this such that no two of the r selected are adjacent. Show that

$$s(n, r) = s(n-2, r-1) + s(n-1, r).$$

Now set $s(n, r) = c(n - r + 1, r) = c(m, k)$, where $m = n - r + 1$. Show that $c(m, k)$ satisfies the same recurrence relation, (9) of section 3.3, as the binomial coefficients. Deduce that

$$s(n, r) = \binom{n-r+1}{r}.$$

3.7 THE PROBLEM OF THE POINTS

Prolonged gambling differentiates people into two groups; those playing with the odds, who are following a trade or profession; and those playing against the odds, who are indulging a hobby or pastime, and if this involves a regular annual outlay, this is no more than what has to be said of most other amusements.

John Venn

In this section we consider just one problem, which is of particular importance in the history and development of probability. In previous sections we have looked at several problems involving dice, cards, and other simple gambling devices. The application of the theory is so natural and useful that it might be supposed that the creation of probability parallelled the creation of dice and cards. In fact this is far from being the case. The greatest single initial step in constructing a theory of probability was made in response to a more recondite question, the problem of the points.

Roughly speaking the essential question is this.

Two players, traditionally called A and B, are competing for a prize. The contest takes the form of a sequence of independent similar trials; as a result of each trial one of the contestants is awarded a point. The first player to accumulate n points is the winner; in colloquial parlance A and B are playing the best of $2n - 1$ points. Tennis matches are usually the best of five sets; $n = 3$.

The problem arises when the contest has to be stopped or abandoned before either has won n points; in fact A still needs a points (having $n - a$ already) and B still needs b points (having $n - b$ already). How should the prize be fairly divided? (Typically the 'prize' consisted of stakes put up by A and B, and held by the stakeholder.)

For example, in tennis, sets correspond to points and men play the best of five sets. If the players were just beginning the fourth set when the court was swallowed up by an earthquake, say, what would be a fair division of the prize? (assuming a natural reluctance to continue the game on some other nearby court).

This is a problem of great antiquity; it first appeared in print in 1494 in a book by Luca Pacioli, but was almost certainly an old problem even then. In his example, A and B were playing the best of 11 games for a prize of ten ducats, and are forced to abandon the game when A has 5 points (needing 1 more) and B has 2 points (needing 4 more). How should the prize be divided?

Though Pacioli was a man of great talent (among many other things his book includes the first printed account of double-entry book-keeping), he could not solve this problem.

Nor could Tartaglia (who is best known for showing how to find the roots of a cubic equation), nor could Forestani, Peverone, or Cardano, who all made attempts during the 16th century.

In fact the problem was finally solved by Blaise Pascal in 1654, who, with Fermat, thereby officially inaugurated the theory of probability. In that year, probably sometime around Pascal's birthday (19 June; he was 31), the problem of the points was brought to his attention. The enquiry was made by the Chevalier de Méré (Antoine Gombaud) who, as a man-about-town and gambler, had a strong and direct interest in the answer. Within a very short time Pascal had solved the problem in two different ways. In the course of a correspondence with Fermat, a third method of solution was found by Fermat.

Two of these methods use ideas that were well known at that time, and are familiar to you now from the previous section. That is, they relied on counting a number of equally likely outcomes.

Pascal's great step forward was to create a method that did not rely on having equally likely outcomes. This breakthrough came about as a result of his explicit formulation of the idea of the value of a bet or lottery, which we discussed in chapters 1 and 2. That is, if you have a probability p of winning \$1 then the game is worth \$$p$ to you.

It naturally follows that, in the problem of the points, the prize should be divided in proportion to the players' respective probabilities of winning if the game were to be continued. The problem is therefore more precisely stated thus.

Precise problem of the points.   A sequence of fair coins is flipped; A gets a point for every head, B a point for every tail. Player A wins if there are a heads before b tails, otherwise B wins. Find the probability that A wins.

Solution.   Let $\alpha(a, b)$ be the probability that A wins and $\beta(a, b)$ the probability that B wins. If the first flip is a head, then A now needs only $a - 1$ further heads to win, so the conditional probability that A wins, given a head, is $\alpha(a - 1, b)$. Likewise the conditional probability that A wins, given a tail, is $\alpha(a, b - 1)$. Hence, by the partition rule,

(1) $\alpha(a, b) = \frac{1}{2}\alpha(a - 1, b) + \frac{1}{2}\alpha(a, b - 1).$

Thus if we know $\alpha(a, b)$ for small values of a and b, we can find the solution for any a and b by this simple recursion. And of course we do know such values of $\alpha(a, b)$, because if $a = 0$ and $b > 0$, then A has won and takes the whole prize: that is to say

(2) $\alpha(0, b) = 1.$

Likewise if $b = 0$ and $a > 0$ then B has won, and so

(3) $$\alpha(a, 0) = 0.$$

How do we solve (1) in general, with (2) and (3)? Recall the fundamental property of Pascal's triangle: the entries $c(j + k, k) = d(j, k)$ satisfy

(4) $$d(j, k) = d(j - 1, k) + d(j, k - 1).$$

You don't need to be a genius to suspect that the solution $\alpha(a, b)$ of (1) is going to be connected with the solutions

$$c(j + k, k) = d(j, k) = \binom{j + k}{k}$$

of (4). We can make the connection even more transparent by writing

$$\alpha(a, b) = \frac{1}{2^{a+b}} u(a, b).$$

Then (1) becomes

(5) $$u(a, b) = u(a - 1, b) + u(a, b - 1)$$

with

(6) $$u(0, b) = 2^b \quad \text{and} \quad u(a, 0) = 0.$$

There are various ways of solving (5) with the conditions (6), but Pascal had the inestimable advantage of having already obtained the solution by another method. Thus he had simply to check that the answer is indeed

(7) $$u(a, b) = 2 \sum_{k=0}^{b-1} \binom{a + b - 1}{k},$$

and

(8) $$\alpha(a, b) = \frac{1}{2^{a+b-1}} \sum_{k=0}^{b-1} \binom{a + b - 1}{k}.$$

At long last there was a solution to this classic problem. We may reasonably ask why Pascal was able to solve it in a matter of weeks, when all previous attempts had failed for at least 150 years. As usual the answer lies in a combination of circumstances: mathematicians had become better at counting things; the binomial coefficients were better understood; notation and the techniques of algebra had improved immeasurably; and Pascal had a couple of very good ideas.

Pascal immediately realized the power of these ideas and techniques and quickly invented new problems on which to use them. We discuss the best known of them in the next section.

Exercises for section 3.7

1. Check that the solution given by (8) does satisfy the recurrence (1) and the boundary conditions (2) and (3).

2. Suppose the game is not fair, that is, A wins any point with probability p or B wins with probability q, where $p \neq q$. Show that
$$\alpha(a, b) = p\alpha(a - 1, b) + q\alpha(a, b - 1)$$
   with solution

(9)
$$\alpha(a, b) = p^{a+b-1} \sum_{k=0}^{b-1} \binom{a+b-1}{k} \left(\frac{q}{p}\right)^k.$$

3. Calculate the answer to Pacioli's original problem when $a = 1$, $b = 4$, the prize is ten ducats, and the players are of equal skill.

3.8 THE GAMBLER'S RUIN PROBLEM

> In writing on these matters I had in mind the enjoyment of mathematicians, not the benefit of the gamblers; those who waste time on games of chance fully deserve to lose their money as well.
>
> P. de Montmort

Following the contributions of Pascal and Fermat, the next advances were made by Christiaan Huygens, who was Newton's closest rival for top scientist of the 17th century. Born in the Netherlands, he visited Paris in 1655 and heard about the problems Pascal had solved. Returning to Holland, he wrote a short book *Calculations in Games of Chance (van Rekeningh in Speelen van Geluck)*. Meanwhile, Pascal had proposed and solved another famous problem.

Pascal's problem of the gambler's ruin.   Two gamblers, A and B, play with three dice. At each throw, if the total is 11 then B gives a counter to A; if the total is 14 then A gives a counter to B. They start with 12 counters each, and the first to possess all 24 is the winner. What are their chances of winning?

Pascal gives the correct solution. The ratio of their respective chances of winning, $p_A : p_B$, is

$$150\ 094\ 635\ 296\ 999\ 122 : 129\ 746\ 337\ 890\ 625,$$

which is the same as

$$282\ 429\ 536\ 481 : 244\ 140\ 625$$

on dividing by 3^{12}.

Unfortunately it is not certain what method Pascal used to get this result. However, Huygens soon heard about this new problem, and solved it in a few days (sometime between 28 September 1656 and 12 October 1656). He used a version of Pascal's idea of value, which we have discussed several times above:

Huygens' definition of value.   If you are offered $\$x$ with probability p, or $\$y$ with probability q $(p + q = 1)$, then the value of this offer to you is $\$(px + qy)$. $\triangle$

Now of course we do not know for sure if this was Pascal's method, but Pascal was certainly at least as capable of extending his own ideas as Huygens was. The balance of probabilities is that he did use this method. By long-standing tradition this problem is always solved in books on elementary probability, and so we now give a modern version of the solution. Here is a general statement of the problem.

Gambler's ruin.   Two players, A and B again, play a series of independent games. Each game is won by A with probability α, or by B with probability β; the winner of each game gets one counter from the loser. Initially A has m counters and B has n. The victor of the contest is the first to have all $m + n$ counters; the loser is said to be 'ruined', which explains the name of this problem. What are the respective chances of A and B to be the victor?

Note that $\alpha + \beta = 1$, and for the moment we assume $\alpha \neq \beta$.

Just as in the problem of the points, suppose that at some stage A has a counters (so B has $m + n - a$ counters), and let A's chances of victory at that point be $v(a)$. If A wins the next game his chance of victory is now $v(a + 1)$; if A loses the next game his chance of victory is $v(a - 1)$. Hence, by the partition rule,

(1) $$v(a) = \alpha v(a + 1) + \beta v(a - 1), \quad 1 \leqslant a \leqslant m + n - 1.$$

Furthermore we know that

(2) $$v(m + n) = 1$$

because in this case A has all the counters, and

(3) $$v(0) = 0$$

because A then has no counters.

From section 2.15, we know that the solution of (1) takes the form

$$v(a) = c_1 \lambda^a + c_2 \mu^a,$$

where c_1 and c_2 are constants, and λ and μ are the roots of

(4) $$\alpha x^2 - x + \beta = 0.$$

Trivially, the roots of (4) are $\lambda = 1$, and $\mu = \beta/\alpha \neq 1$ (since we assumed $\alpha \neq \beta$). Hence, using (2) and (3), we find that

(5) $$v(a) = \frac{1 - (\beta/\alpha)^a}{1 - (\beta/\alpha)^{m+n}}.$$

In particular, when A starts with m counters,

$$p_A = v(m) = \frac{1 - (\beta/\alpha)^m}{1 - (\beta/\alpha)^{m+n}}.$$

This method of solution of difference equations was unknown in 1656, so other approaches were employed. In obtaining the answer to the gambler's ruin problem, Huygens (and later workers) used intuitive induction with the proof omitted. Pascal probably did use (1) but solved it by a different route. (See the exercises at the end of the section.)

Finally we consider the case when $\alpha = \beta$. Now (1) is

(6) $$v(a) = \tfrac{1}{2}v(a + 1) + \tfrac{1}{2}v(a - 1)$$

and it is easy to check that, for arbitrary constants c_1 and c_2,

$$v(a) = c_1 + c_2 a$$

satisfies (6). Now using (2) and (3) gives

(7) $$v(a) = \frac{a}{m + n}.$$

Exercises for section 3.8

1. **Gambler's ruin.**   Find p_B, the probability that B wins, and show that
$$p_A + p_B = 1.$$

So somebody does win; the probability that the game is unresolved is zero.

2. Solve the equation (1) as follows.
   (a) Rearrange (1) as
   $$\alpha\{v(a+1) - v(a)\} = \beta\{v(a) - v(a-1)\}.$$
   (b) Sum and use successive cancellation to get
   $$\alpha\{v(a+1) - v(1)\} = \beta\{v(a) - v(0)\} = \beta v(a).$$
   (c) Deduce that
   $$v(a) = \frac{1 - (\beta/\alpha)^m}{1 - \beta/\alpha} v(1).$$
   (d) Finally derive (5).
   Every step of this method would have been familiar to Pascal in 1656.

3. Adapt the method of the last exercise to deal with the case when $\alpha = \beta$ in the gambler's ruin problem.

4. Suppose a gambler plays a sequence of fair games, at each of which he is equally likely to lose a point or gain a point. Show that the chance of being a points ahead before first being d points down is $a/(a + d)$.

3.9 SOME CLASSIC PROBLEMS

I have made this letter longer than usual, because I lack the time to make it shorter.

Pascal in a letter to Fermat.

Pascal and Fermat corresponded on the problem of the points in 1654, and on the gambler's ruin problem in 1656. Their exchanges mark the official inauguration of probability theory. (Pascal's memorial in the Church of St Étienne-du-Mont in Paris warrants a visit by any passing probabilist.) These ideas quickly circulated in intellectual circles, and in 1657 Huygens published a book on probability, *On Games of Chance* (in Latin and Dutch editions); an English translation by Arbuthnot appeared in 1692.

This pioneering text was followed in remarkably quick succession by several books on probability. A brief list would include the books of de Montmort (1708), J. Bernoulli (1713), and de Moivre (1718), in French, Latin, and English respectively.

It is notable that the development of probability in its early stages was so extensively motivated by simple games of chance and lotteries. Of course, the subject now extends far beyond these original boundaries, but even today most people's first brush with probability will involve rolling a die in a simple board game, wondering about lottery odds, or deciding which way to finesse the missing queen. Over the years a huge amount of analysis has been done on these simple but naturally appealing problems. We therefore give a brief random selection of some of the better-known classical problems tackled by these early pioneers and their later descendants. (We have seen some of the easier classical problems already in chapter 2, such as Pepys' problem, de Méré's problem, Galileo's problem, Waldegrave's problem, and Huygens' problem.)

Example 3.9.1: problem of the points revisited. As we have noted above, Pascal was probably assisted in his elegant and epoch-making solution of this problem by the fact that he could also solve it another way. A typical argument runs as follows.

Solution.   Recall that A needs a points and B needs b points; A wins any game with probability p. Now let A_k be the event that when A has first won a points, B has won k points at that stage. Then

(1)
$$A_k \cap A_j = \emptyset, \quad j \neq k$$

and

$$P(A_k) = P(A \text{ wins the } (a+k)\text{th game and } a-1 \text{ of the preceding } a+k-1 \text{ games})$$

$$= pP(A \text{ wins } a-1 \text{ of } a+k-1 \text{ games})$$

$$= p^a(1-p)^k \binom{a+k-1}{a-1},$$

by exercise 5 of section 3.5. Now the event that A wins is $\bigcup_{k=0}^{b-1} A_k$, and the solution $\alpha(a, b) = \sum_k P(A_k)$ follows, using (1) above. (See problem 21 also.) ○

Example 3.9.2: problem of the points extended.   It is natural to extend the problem of the points to a group of n players $P_1, \ldots, P_n$, where P_1 needs a_1 games to win, P_2 needs a_2, and so on, and the probability that P_r wins any game is p_r. Naturally $\sum p_r = 1$. The same argument as that used in the previous example shows that if P_1 wins the contest when P_r has won x_r games ($2 \leqslant r \leqslant n$, $x_r < a_r$), this has probability

(2)
$$p_1^{a_1} p_2^{x_2} \cdots p_n^{x_n} \frac{(a_1 + x_1 + \cdots + x_n - 1)!}{(a_1 - 1)! x_2! \cdots x_n!}.$$

Thus the total probability that P_1 wins the contest is the sum of all such terms as each x_r runs over $0, 1, \ldots, a_r - 1$. ○

Example 3.9.3: Banach's matchboxes.   The celebrated mathematician Stefan Banach used to meet other mathematicians in the Scottish Coffee House in Lwów. He arranged for a notebook to be kept there to record mathematical problems and answers; this was the Scottish Book. The last problem in the book, dated 31 May 1941, concerns a certain mathematician who has two boxes of n matches. One is in his right pocket, one is in his left pocket, and he removes matches at random until he finds a box empty. What is the probability p_k that k matches remain in the other box?

Solution.   The mathematician must have removed the boxes from their pockets $n + 1 + n - k$ times. If the last $(n+1)$th (unsuccessful) removal of some box is the right-hand box, then the previous n right-hand removals may be chosen from any of the previous $2n - k$. This has probability

$$2^{-(2n-k+1)} \binom{2n-k}{n}.$$

The same is true for the left pocket, so

$$p_k = 2^{-(2n-k)} \binom{2n-k}{n}. \qquad ○$$

Example 3.9.4: occupancy problem.   Suppose a fair die with s faces (or sides) is rolled r times. What is the probability a that every side has turned up at least once?

Solution.   Let A_j be the event that the jth side has not been shown. Then

(3) $$a = 1 - P(A_1 \cup A_2 \cup \cdots \cup A_s)$$

$$= 1 - \sum_{j=1}^{s} P(A_j) + \sum_{j<k} P(A_j \cap A_k) - \cdots$$

$$+ (-1)^s P(A_1 \cap \cdots \cap A_s)$$

on using problem 18 of section 2.16. Now by symmetry $P(A_j) = P(A_k)$, $P(A_j \cap A_k) = P(A_m \cap A_n)$, and so on. Hence

$$a = 1 - sP(A_1) + \binom{s}{2} P(A_1 \cap A_2) - \cdots + (-1)^s P\left(\bigcap_j A_j\right).$$

Now, by the independence of rolls, for any set of k sides

(4) $$P(A_1 \cap A_2 \cap \cdots \cap A_k) = \left(1 - \frac{k}{s}\right)^r,$$

and hence

(5) $$a = 1 - s\left(1 - \frac{1}{s}\right)^r + \binom{s}{2}\left(1 - \frac{2}{s}\right)^r - \binom{s}{3}\left(1 - \frac{3}{s}\right)^r + \cdots$$

$$+ (-1)^{s-1} \binom{s}{s-1}\left(1 - \frac{s-1}{s}\right)^r$$

$$= \sum_{k=0}^{s} (-1)^k \binom{s}{k}\left(1 - \frac{k}{s}\right)^r. \qquad \bigcirc$$

Remark.   This example may look a little artificial, but in fact it has many practical applications. For example, if you capture, tag (if not already tagged), and release r animals successively in some restricted habitat, what is the probability that you have tagged all the s present? Think of some more such examples yourself.

Example 3.9.5: derangements and coincidences.   Suppose the lottery machine were not stopped after the winning draw, but allowed to go on drawing numbers until all n were removed. What is the probability d that no number r is the rth to be drawn by the machine?

Solution.   Let A_r be the event that the rth number drawn is in fact r; that is to say, the rth ball that rolls out bears the number r. Then

(6)
$$d = 1 - \mathrm{P}\left(\bigcup_{r=1}^{n} A_r\right)$$

$$= 1 - \sum_{r=1}^{n} \mathrm{P}(A_r) + \cdots + (-1)^n \mathrm{P}(A_1 \cap \cdots \cap A_n)$$

$$= 1 - n\mathrm{P}(A_1) + \binom{n}{2}\mathrm{P}(A_1 \cap A_2) - \cdots$$

$$+ (-1)^n \mathrm{P}\left(\bigcap_{r=1}^{n} A_r\right)$$

by problem 18 of section 2.16 and symmetry, as usual. Now for any set of k numbers

(7)
$$\mathrm{P}(A_1 \cap \cdots \cap A_k) = \frac{1}{n}\frac{1}{n-1}\cdots\frac{1}{n-k+1} = \frac{(n-k)!}{n!}.$$

Hence

(8)
$$d = 1 - n\frac{1}{n} + \binom{n}{2}\frac{(n-2)!}{n!} - \cdots$$

$$+ (-1)^k \binom{n}{k}\frac{(n-k)!}{n!} + \cdots + (-1)^n \frac{1}{n!}$$

$$= \frac{1}{2!} - \frac{1}{3!} + \cdots + (-1)^n \frac{1}{n!}$$

It is remarkable that as $n \to \infty$ we have $d \to e^{-1}$. ◯

Exercises for section 3.9

1. **Derangements revisited.**   Suppose n competitors in a tournament organize a sweepstake on the result of the tournament. Their names are placed in an urn, and each player pays a dollar to withdraw one name from the urn. The player holding the name that wins the tournament is awarded the pot of $\$n$.
    (a)   Show that the probability that exactly r players draw their own name is
    $$\frac{1}{r!}\left\{\frac{1}{2!} - \frac{1}{3!} + \cdots + \frac{(-1)^{n-r}}{(n-r)!}\right\}.$$
    (b)   Given that exactly r such matches occur, what is the probability that Fred draws his own name? (Fred is a competitor.)

2. **Derangements once again.**   Let d_n be the number of derangements of the first n integers. Show that $d_{n+1} = nd_n + nd_{n-1}$, by considering which number is in the first place in each derangement.

3.10 STIRLING'S FORMULA

At a very early stage probabilists encountered the fundamental problem of turning theoretical expressions into numerical answers, especially when the solutions to a problem involved large numbers of large factorials. We have seen many examples of this above, especially (for example) in even the simplest problems involving poker hands or suit distributions in bridge hands.

For another example, consider the basic problem of proportions in flipping coins.

Example 3.10.1.   A fair coin is flipped repeatedly. Routine calculations show that

(1) $$P(\text{exactly 6 heads in 10 flips}) = \binom{10}{6} 2^{-10} \simeq 0.2,$$

(2) $$P(\text{exactly 30 heads in 50 flips}) = \binom{50}{30} 2^{-50} \simeq 0.04,$$

(3) $$P(\text{exactly 600 heads in 1000 flips}) = \binom{1000}{600} 2^{-1000} \simeq 10^{-8}.$$

These are simple but not straightforward. The problem is that $n!$ is impossibly large for large n. (Try 1000! on your pocket calculator.)   ○

Furthermore, an obvious next question in flipping coins is to ask for the probability that the proportion of heads lies between 0.4 and 0.6, say, or any other range of interest. Even today, summing the relevant probabilities including factorials would be an exceedingly tedious task, and for 18th century mathematicians it was clearly impossible. de Moivre and others therefore set about finding useful approximations to the value of $n!$, especially for large n. That is, they tried to find a sequence $(a(n); n \geqslant 1)$ such that as n increases

$$\frac{a(n)}{n!} \to 1,$$

and of course, such that $a(n)$ can be relatively easily calculated. For such a sequence we use the notation $n! \sim a(n)$. In 1730 de Moivre showed that a suitable sequence is given by

(4) $$a(n) = Bn^{n+1/2}e^{-n}$$

where

(5) $$\log B \simeq 1 - \frac{1}{12} + \frac{1}{360} - \frac{1}{1260} + \frac{1}{1680} - \cdots.$$

Inspired by this, Stirling showed that in fact

(6) $$B = (2\pi)^{1/2}.$$

We therefore write:

Stirling's formula

(7) $$n! \sim (2\pi n)^{1/2} n^n e^{-n}.$$

This enabled de Moivre to prove the first central limit theorem in 1733. We meet this important result later.

Remark.   Research by psychologists has shown that, before the actual calculations, many people (probabilistically unsophisticated) estimate that the probabilities defined in (1), (2), and (3) are roughly similar, or even the same. This may be called the fallacy of proportion, because it is a strong, but wrongly applied, intuitive feeling for proportionality that leads people into this error. Typically they are also very reluctant to believe the truth, even when it is demonstrated as above.

Exercises for section 3.10

1. Show that the number of ways of dealing the four hands for a game of bridge is

$$M(13, 13, 13, 13) = \frac{52!}{(13!)^4}.$$

   Use Stirling's formula to obtain an approximate value for this. (Then compare your answer with the exact result, 53 644 737 765 488 792 839 237 440 000.)

2. Use Stirling's formula to approximate the number of ways of being dealt one hand at bridge,

$$\binom{52}{13} = 635\,013\,559\,600.$$

3.11 REVIEW

As promised above we have surveyed the preliminaries to probability, and observed its foundation by Pascal, Fermat, and Huygens. This has, no doubt, been informative and entertaining, but are we any better off as a result? The answer is yes, for a number of reasons: principally

(i)   We have found that a large class of interesting problems can be solved simply by counting things. This is good news, because we are all quite confident about counting.

(ii)   We have gained experience in solving simple classical problems which will be very useful in tackling more complicated problems.

(iii)   We have established the following combinatorial results.
   - The number of possible sequences of length r using elements from a set of size n is n^r. (Repetition permitted.)
   - The number of permutations of length r using elements from a set of size n is $n(n-1) \cdots (n-r+1)$. (Repetition not permitted.)
   - The number of combinations (choices) of r elements from a set of size n is
$$\binom{n}{r} = \frac{n(n-1) \cdots (n-r+1)}{r(r-1) \cdots 1}$$
   - The number of subsets of a set of size n is 2^n.
   - The number of derangements of a set of size n is
$$n!\left\{ 1 - \frac{1}{1!} + \frac{1}{2!} - \frac{1}{3!} + \frac{1}{4!} - \cdots + (-1)^n \frac{1}{n!} \right\}.$$

(iv)   We can record the following useful approximations.
   - Stirling's formula says that as n increases
$$\frac{\sqrt{2\pi} n^{n+1/2} e^{-n}}{n!} \to 1.$$

- Robbins' improved formula says that

$$\exp\left(\frac{-1}{12n}\right) < \frac{\sqrt{2\pi}n^{n+1/2}e^{-n}}{n!} < \exp\left(\frac{-1}{12n+1}\right).$$

3.12 APPENDIX. SERIES AND SUMS

Another method I have made use of, is that of Infinite Series, which in many cases will solve the Problems of Chance more naturally than Combinations.

A. de Moivre, *Doctrine of Chances*, 1717

What was true for de Moivre is equally true today, and this is therefore a convenient moment to remind the reader of some general and particular properties of series.

I Finite series

Consider the series

$$s_n = \sum_{r=1}^{n} a_r = a_1 + a_2 + \cdots + a_n.$$

The variable r is a *dummy variable* or *index of summation*, so any symbol will suffice:

$$\sum_{r=1}^{n} a_r \equiv \sum_{i=1}^{n} a_i.$$

In general

$$\sum_{r=1}^{n} (ax_r + by_r) = a\sum_{r=1}^{n} x_r + b\sum_{r=1}^{n} y_r.$$

In particular

$$\sum_{r=1}^{n} 1 = n;$$

$$\sum_{r=1}^{n} r = \tfrac{1}{2}n(n+1), \quad \text{the arithmetic sum;}$$

$$\sum_{r=1}^{n} r^2 = \tfrac{1}{6}n(n+1)(2n+1) = 2\binom{n+1}{3} + \binom{n+1}{2};$$

$$\sum_{r=1}^{n} r^3 = \left(\sum_{r=1}^{n} r\right)^2 = \tfrac{1}{4}n^2(n+1)^2;$$

$$\sum_{r=0}^{n} \binom{n}{r} x^r y^{n-r} = (x+y)^n, \quad \text{the binomial theorem;}$$

$$\sum_{\substack{a+b+c=n \\ a,b,c\geq 0}} M(a, b, c)x^a y^b z^c = \sum_{\substack{a+b+c=n \\ a,b,c\geq 0}} \binom{a+b+c}{a+b}\binom{a+b}{a} x^a y^b z^c$$

$$= (x+y+z)^n, \quad \text{the multinomial theorem;}$$

$$\sum_{r=0}^{n} x^r = \frac{1-x^{n+1}}{1-x}, \quad \text{the geometric sum.}$$

II Limits

Very often we have to deal with infinite series. A fundamental and extremely useful concept in this context is that of the limit of a sequence.

Definition. Let $(s_n; n \geq 1)$ be a sequence of real numbers. If there is a number s such that $|s_n - s|$ may ultimately always be as small as we please then s is said to be the limit of the sequence s_n. Formally we write

$$\lim_{n \to \infty} s_n = s$$

if and only if for any $\varepsilon > 0$, there is a finite n_0 such that

$$|s_n - s| < \varepsilon$$

for all $n > n_0$. $\triangle$

Notice that s_n need never actually take the value s, it must just get closer to it in the long run. (For example, let $s_n = n^{-1}$.)

III Infinite series

Let $(a_r; r \geq 1)$ be a sequence of terms, with partial sums

$$s_n = \sum_{r=1}^{n} a_r, \quad n \geq 1.$$

If s_n has a finite limit s as $n \to \infty$, then the sum $\sum_{r=1}^{\infty} a_r$ is said to *converge* with sum s. Otherwise it *diverges*. If $\sum_{r=1}^{\infty} |a_r|$ converges, then $\sum_{r=1}^{\infty} a_r$ is said to be *absolutely convergent*.

For example, in the geometric sum in I above, if $|x| < 1$ then $|x|^n \to 0$ as $n \to \infty$. Hence

$$\sum_{r=0}^{\infty} x^r = \frac{1}{1-x}, \quad |x| < 1,$$

and the series is absolutely convergent for $|x| < 1$. In particular we have the negative binomial theorem:

$$\sum_{r=0}^{\infty} \binom{n+r-1}{r} x^r = (1-x)^{-n}.$$

This is true even when n is not an integer, so for example

$$(1-x)^{-1/2} = \sum_{r=0}^{\infty} \binom{r-\frac{1}{2}}{r} x^r = \sum_{r=0}^{\infty} \frac{(r-\frac{1}{2})^r}{r!} x^r$$

$$= 1 + \frac{1}{2}x + \frac{3}{2} \times \frac{1}{2} \times \frac{x^2}{2!} + \frac{5}{2} \times \frac{3}{2} \times \frac{1}{2} \times \frac{x^3}{3!} + \cdots$$

$$= \sum_{r=0}^{\infty} \binom{2r}{r} \left(\frac{x}{4}\right)^r.$$

In particular, we often use the case $n = 2$:

$$\sum_{r=0}^{\infty} (r+1)x^r = (1-x)^{-2}.$$

Also, by definition, for all x,

$$\exp x = e^x = \sum_{r=0}^{\infty} \frac{x^r}{r!}.$$

and, for $|x| < 1$,

$$-\log(1 - x) = \sum_{r=1}^{\infty} \frac{x^r}{r}.$$

An important property of e^x is the exponential limit theorem:

$$\text{as } n \to \infty, \quad \left(1 + \frac{x}{n}\right)^n \to e^x.$$

This has a very useful generalization: let $r(n, x)$ be any function such that $nr(n, x) \to 0$ as $n \to \infty$; then

$$\left\{1 + \frac{x}{n} + r(n, x)\right\}^n \to e^x, \quad \text{as } n \to \infty.$$

Finally, note that we occasionally use special identities such as

$$\sum_{r=1}^{\infty} \frac{1}{r^2} = \frac{\pi^2}{6} \quad \text{and} \quad \sum_{r=1}^{\infty} \frac{1}{r^4} = \frac{\pi^4}{90}.$$

3.13 PROBLEMS

1. Assume people are independently equally likely to have any sign of the Zodiac.
    (a) What is the probability that four people have different signs?
    (b) How many people are needed to give a better than evens chance that at least two of them share a sign?
    (There are 12 signs of the Zodiac.)

2. Five digits are selected independently at random (repetition permitted), each from the ten possibilities $\{0, 1, \ldots, 9\}$. Show that the probability that they are all different is 0.3 approximately.
       What is the probability that six such random digits are all different?

3. Four digits are selected independently at random (without repetition) from $\{0, 1, \ldots, 9\}$. What is the probability that
    (a) the four digits form a run? (e.g. 2, 3, 4, 5)
    (b) they are all greater than 5?
    (c) they include the digit 0?
    (d) at least one is greater than 7?
    (e) all the numbers are odd?

4. You roll 6 fair dice. You win a small prize if at least 2 of the dice show the same, and you win a big prize if there are at least 4 sixes. What is the probability that you
    (a) get exactly 2 sixes?
    (b) win a small prize?
    (c) win a large prize?
    (d) win a large prize given that you have won a small prize?

5. Show that the probability that your poker hand contains two pairs is approximately 0.048, and that the probability of three of a kind is approximately 0.021.

6. Show that $n^{\underline{r}}$, the number of permutations of r from n things, satisfies the recurrence relation

$$n^{\underline{r}} = (n - 1)^{\underline{r}} + r(n - 1)^{\underline{r-1}}.$$

7. Show that

$$\binom{2n}{n} = \sum_{k=0}^{n} \binom{n}{k}^2.$$

8. A construction toy comprises n bricks, which can each be any one of c different colours. Let $w(n, c)$ be the number of different ways of making up such a box. Show that
$$w(n, c) = w(n-1, c) + w(n, c-1)$$
and that
$$w(n, c) = \binom{n+c-1}{n}.$$

9. *Pizza problem.* Let R_n be the largest number of bits of a circular pizza which you can produce with n straight cuts. Show that
$$R_n = R_{n-1} + n$$
and that
$$R_n = \binom{n+1}{2} + 1.$$

10. If n people, including Algernon and Zebedee, are randomly placed in a line (queue), what is the probability that there are exactly k people in line between Algernon and Zebedee? What if they were randomly arranged in a circle?

11. A combination lock has n buttons. It opens if k different buttons are depressed in the correct order. What is the chance of opening a lock if you press k different random buttons in random order?

12. In poker a *straight* is a hand such as $\{3, 4, 5, 6, 7\}$, where the cards are not all of the same suit (for that would be a straight flush), and aces may rank high or low. Show that
$$P(\text{straight}) = \left\{ 10 \binom{4}{1}^5 - 10 \binom{4}{1} \right\} \bigg/ \binom{52}{5} \simeq 0.004.$$
Show also that P(straight flush) $\simeq 0.000015$.

13. The Earl of Yarborough is said to have offered the following bet to anyone about to be dealt a hand at whist: if you paid him one guinea, and your hand then contained no card higher than a nine, he would pay you one thousand guineas. Show that the probability y of being dealt such a hand is
$$y = \frac{5394}{9860\,459}.$$
What do you think of the bet?

14. (a) Adonis has k cents and Bubear has $n - k$ cents. They repeatedly roll a fair die. If it is even, Adonis gets a cent from Bubear; otherwise, Bubear gets a cent from Adonis. Show that the probability that Adonis first has all n cents is k/n.
    (b) There are $n + 1$ beer glasses $\{g_0, g_1, \ldots, g_n\}$, in a circle. A wasp is on g_0. At each flight the wasp is equally likely to fly to either of the two neighbouring glasses. Let L_k be the event that the glass g_k is the last one to be visited by the wasp $(k \neq 0)$. Show that $P(L_k) = n^{-1}$.

15. Consider the standard 6 out of 49 lottery.
    (a) Show that the probability that 4 of your 6 numbers match those drawn is
$$\frac{13\,545}{13\,983\,816}.$$
    (b) Find the probability that all 6 numbers drawn are odd.
    (c) What is the probability that at least one number fails to be drawn in 52 consecutive drawings?

16. ***Matching.*** The first n integers are placed in a row at random. If the integer k is in the kth place in the row, that is a match. What is the probability that '1' is first, given that there are exactly m matches?

17. You have n sovereigns and r friends, $n \geqslant r$. Show that the number of ways of dividing the coins among your friends so that each has at least one is

$$\binom{n-1}{r-1}.$$

18. A biased coin is flipped $2n$ times. Show that the probability that the number of heads is the same as the number of tails is

$$\binom{2n}{n}(pq)^n.$$

Use Stirling's formula to show how this behaves as $n \to \infty$.

19. Suppose n objects are placed in a row. The operation S_k is defined thus: 'Pick one of the first k objects at random, and swap it with the object in the kth place'. Now perform S_n, S_{n-1}, $\ldots$, S_1. Show that the final order is equally likely to be any one fo the $n!$ permutations of the objects.

20. Your computer requires you to choose a password comprising a sequence of m characters drawn from an alphabet of a possibilities, with the constraint that not more than two consecutive characters may be the same. Let $t(m)$ be the total number of passwords, for $m > 2$. Show that

$$t(m) = (a-1)\{t(m-1) + t(m-2)\}.$$

Hence find an expression for $t(m)$.

21. Suppose A and B play a series of $a + b - 1$ independent games, each won by A with probability p, or by B with probability $1 - p$. Find the probability that A wins at least a games, and hence obtain the solution (9) in exercise 2 of section 3.7, the problem of the points.

4

Distributions: trials, samples, and approximation

Men that hazard all
Do it in hope of fair advantage.

Shakespeare

4.1 PREVIEW

This chapter deals with one of the most useful and important ideas in probability, that is, the concept of a probability distribution. We have seen in chapter 2 how the probability function P assigns or distributes probability to the events in Ω. We have seen in chapter 3 how the outcomes in Ω are often numbers or can be indexed by numbers. In these, and many other cases, P naturally distributes probability to the relevant numbers, which we may regard as points on the real line. This all leads naturally to the idea of a probability distribution on the real line, which often can be easily and obviously represented by simple and familiar functions.

We shall look at the most important special distributions in detail: Bernoulli, geometric, binomial, negative binomial, and hypergeometric. Then we consider some important and very useful approximations, especially the Poisson, exponential, and normal distributions.

In particular, we shall need to deal with problems in which probability is assigned to intervals in the real line, or even to the whole real line. In such cases we talk of a probability density, using a rather obvious analogy with the distribution of matter.

Finally, probability distributions and densities in the plane are briefly considered.

Prerequisites. We use elementary results about sequences and series, and their limits, such as

$$\lim_{n \to \infty} \left(1 + \frac{x}{n} \right)^n = e^x.$$

See the appendix to chapter 3 for a brief account of these notions.

4.2 INTRODUCTION; SIMPLE EXAMPLES

Very often all the outcomes of some experiment are just numbers. We give some examples.

129

Darts.   You throw a dart, obtaining a score between 0 and 60.

Temperature.   You observe a thermometer and record the temperature to the nearest degree. The outcome is an integer.

Counter.   You turn on your Geiger counter, and note the time when it has counted 10^6 particles. The outcome is a positive real number.

Lottery.   The lottery draw yields seven numbers between 1 and 49.

Obviously we could produce yet another endless list of experiments with random numerical outcomes here: you weigh yourself; you sell your car; you roll a die with numbered faces, and so on. Write some down yourself. In such cases it is customary and convenient to denote the outcome of the experiment before it occurs by some appropriate capital letter, such as X.

We do this in the interests of clarity. Outcomes in general (denoted by ω) can be anything: rain, or heads, or an ace, for example. Outcomes that are denoted by X (or any other capital) can only be numerical. Thus, in the second example above we could say

'Let T be the temperature observed'.

In the third example we might say

'Let X be the time needed to count 10^6 particles'.

In all examples of this kind, events are of course just described by suitable sets of numbers. It is natural and helpful to specify these events by using the previous notation; thus
$$\{a \leqslant T \leqslant b\}$$
means that the temperature recorded lies between a and b degrees, inclusive. Likewise
$$\{T = 0\}$$
is the event that the temperature is zero. In the same way
$$\{X > x\}$$
means that the time needed to count 10^6 particles is greater than x. In all these cases X and T are being used in the same way as we used ω in earlier chapters, e.g. rainy days in example 2.3.2, random numbers in example 2.4.11, and so on.

Finally, because these are events, we can discuss their probabilities. For the events given above, these would be denoted by
$$\mathrm{P}(0 \leqslant T \leqslant b), \quad \mathrm{P}(T = 0), \quad \mathrm{P}(X > x),$$
respectively.

The above discussion has been fairly general; we now focus on a particularly important special case. That is, the case when X can take only integer values.

Definition.   Let X denote the outcome of an experiment in which X can take only integer values. Then the function $p(x)$ given by
$$p(x) = \mathrm{P}(X = x), \quad x \in \mathbb{Z},$$

is called the *probability distribution* of X. Obviously $p(x) \geq 0$, and we shall show that $\sum_x p(x) = 1$. △

Note that we need only discuss this function for integer values of x, but it is convenient (and possible) to imagine that $p(x) = 0$ when x is not an integer. When x is an integer, $p(x)$ then supplies the probability that the event $\{X = x\}$ occurs. Or, more briefly, the probability that $X = x$.

Example 4.2.1: die. Let X be the number shown when a fair die is rolled. As always
$$X \in \{1, 2, 3, 4, 5, 6\},$$
and of course
$$P(X = x) = \tfrac{1}{6}, \quad x \in \{1, 2, 3, 4, 5, 6\}. \qquad \bigcirc$$

Example 4.2.2: Bernoulli trial. Suppose you engage in some activity that entails that you either win or lose, for example, a game of tennis or a bet. All such activities are given the general name of a Bernoulli trial. Suppose that the probability that you win the trial is p.

Let X be the number of times you win. Putting it in what might seem a rather stilted way, we write
$$X \in \{0, 1\}$$
and
$$P(X = 1) = p.$$
Obviously $X = 0$ and $X = 1$ are complementary, and so by the complement rule
$$P(X = 0) = 1 - p$$
$$= q,$$
where $p + q = 1$. The event $X = 1$ is traditionally known as 'success', and $X = 0$ is known as 'failure'. $\bigcirc$

The Bernoulli trial is the simplest, but nevertheless an important, random experiment, and an enormous number of examples are of this type. For illustration consider the following.

(i) Flip a coin; we may let $\{\text{head}\} = \{\text{success}\} = S$.
(ii) Each computer chip produced is tested; $S = \{\text{the chip passes the test}\}$.
(iii) You attempt to start your car one cold morning; $S = \{\text{it starts}\}$.
(iv) A patient is prescribed some remedy; $S = \{\text{he is thereby cured}\}$.

In each case the interpretation of failure is obvious; $F = S^c$.

Inherent in most of these examples is the possibility of repetition. This leads to another important

Definition. By a *sequence of Bernoulli trials*, we understand a sequence of independent repetitions of an experiment in which the probability of success is the same at each trial. △

The above assumptions enable us to calculate the probability of any given sequence of successes and failures very easily, by independence. Thus, with an obvious notation,

$$P(SFS) = pqp = p^2 q,$$
$$P(FFFS) = q^3 p,$$

and so on.

The choice of examples and vocabulary makes it clear in which kind of questions we are interested. For example:

(i)   How long do we wait for the first success?
(ii) How many failures are there in any n trials?
(iii) How long do we wait for the rth success?

The answers to these questions take the form of a collection of probabilities, as we see in the next few sections.

Further natural sources of distributions arise from measurement and counting. For example, suppose n randomly chosen children are each measured to the nearest inch, and N_r is the number of children whose height is recorded as r inches. Then we have argued often above that $\eta_r = N_r/n$ is (or should be) a reasonable approximation to the probability p_r that a randomly selected child in this population is r inches tall. Of course $\eta_r \geq 0$ and

$$\sum_r \eta_r = n^{-1} \sum_r N_r = 1.$$

Thus η_r satisfies the rules for a probability distribution, as well as representing an approximation to p_r. Such a collection is called an *empirical distribution*.

Example 4.2.3: Benford's distribution revisited.   Let us recall this classic problem, stated as follows. Take any large collection of numbers, such as the Cambridge statistical tables, or a report on the Census, or an almanac. Offer to bet, at evens, that a number picked at random from the book will have first significant digit less than 5. The more people you can find to accept this bet, the more you will win.

The untutored instinct expects intuitively that all nine possible numbers should be equally likely. This is not so. Actual experiment shows that empirically the distribution of probability is close to

(1) $$p(k) = \log_{10}\left(1 + \frac{1}{k}\right), \quad 1 \leq k \leq 9.$$

This is Benford's distribution, and the actual values are approximately

$$p(1) = 0.301, \quad p(2) = 0.176, \quad p(3) = 0.125,$$
$$p(4) = 0.097, \quad p(5) = 0.079, \quad p(6) = 0.067,$$
$$p(7) = 0.058, \quad p(8) = 0.051, \quad p(9) = 0.046.$$

You will notice that $p(1) + p(2) + p(3) + p(4) \simeq 0.7$; the odds on your winning are better than two to one. This is perhaps even more flagrant than a lottery.

It turns out that the same rule applies if you look at a larger number of significant digits. For example, if you look at the first two significant digits, then these pairs lie in the set $\{10, 11, \ldots, 99\}$. It is found that they have the probability distribution

$$p(k) = \log_{10}\left(1 + \frac{1}{k}\right), \quad 10 \leqslant k \leqslant 99.$$

Why should the distribution of first significant digits be given by (1)? Superficially it seems rather odd and unnatural. It becomes less unnatural when you recall that the choice of base 10 in such tables is completely arbitrary. On another planet these tables might be in base 8, or base 12, or indeed any base. It would be extremely strange if the first digit distribution was uniform (say) in base 10 but not in the other bases.

We conclude that any such distribution must be in some sense base-invariant. And, recently, T. P. Hill has shown that Benford's distribution is the only one which satisfies this condition. ○

In these and all the other examples we consider, a probability distribution is just a collection of numbers $p(x)$ satisfying the conditions noted above,

$$\sum_x p(x) = 1, \quad p(x) \geqslant 0.$$

This is fine as far as it goes, but it often helps our intuition to represent the collection $p(x)$ as a histogram. This makes it obvious at a glance what is going on. For example, figure 4.1 displays $p(0)$ and $p(1)$ for Bernoulli trials with various values of $p(0)$.

For another example, consider the distribution of probabilities for the sum Z of the scores of two fair dice. We know that

$$p(2) = \tfrac{1}{36}, \quad p(3) = \tfrac{2}{36}, \quad \ldots, \quad p(7) = \tfrac{6}{36},$$
$$p(8) = \tfrac{5}{36}, \quad \ldots, \quad p(12) = \tfrac{1}{36}.$$

where $p(2) = P(Z = 2)$, and so on. This distribution is illustrated in figure 4.2, and is known as a triangular distribution.

Before we turn to more examples let us list the principal properties of a distribution $p(x)$. First, and obviously by the definition,

(2) $$0 \leqslant p(x) \leqslant 1.$$

Second, note that if $x_1 \neq x_2$ then the events $\{X = x_1\}$ and $\{X = x_2\}$ are disjoint. Hence, by the addition rule (3) of section 2.5,

(3) $$P(X \in \{x_1, x_2\}) = p(x_1) + p(x_2).$$

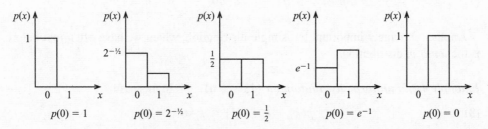

Figure 4.1. Some Bernoulli distributions.

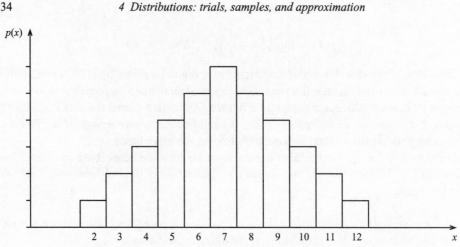

Figure 4.2. Probability distribution of the sum of the scores shown by two fair dice.

More generally, by the extended addition rule (5) of section 2.5, we have the result:

Key rule for distributions

$$(4) \qquad P(X \in C) = \sum_{x \in C} p(x).$$

That is to say, we obtain the probability of any event by adding the probabilities of the outcomes in it. In particular, as we claimed above,

$$(5) \qquad P(X \in \Omega) = \sum_{x \in \mathbb{Z}} p(x) = 1.$$

We make one further important definition.

Definition.   Let X have distribution $p(x)$. Then the function

$$(6) \qquad F(x) = \sum_{t \le x} p(t) = P(X \le x)$$

is called the *distribution function* of X. $\triangle$

Note that this way of thinking about probability distributions suggests a neat way of writing down the probability distribution of a Bernoulli trial.

Example 4.2.4: Bernoulli trial.   The distribution of the number of successes is

$$(7) \qquad p(k) = p^k q^{1-k}, \quad k = 0, 1, \quad p + q = 1. \qquad \bigcirc$$

Another extremely important but simple distribution, which we have often met before, is the *uniform* distribution.

Example 4.2.5: uniform distribution on $\{1, \dots, n\}$.   In this case

$$(8) \qquad p(k) = n^{-1}, \quad 1 \le k \le n.$$

In particular $n = 6$ corresponds to a conventional fair die and $n = 2$ to a fair coin. $\bigcirc$

Returning to the histograms discussed above, we see that in these examples each bar of the histogram is of unit width, and the bar at x is of height $p(x)$; it therefore has area $p(x)$. Thus the algebraic rules laid out from (2) to (5) can be interpreted in terms of areas. Most importantly, we can see that the probability that X lies between a and b, $P(a \leqslant X \leqslant b)$, is just the area of the histogram lying between a and b. This of course is why such diagrams are so appealing. Note that the value of the distribution function $F(x)$ at x is just the area of the histogram to the left of x. Figure 4.3 gives an example.

This idea becomes even more appealing and attractive when we recall that not all experiments have outcomes confined to the integers, or even to a countable set. Weathercocks may point in any direction, isotopes may decay at any time, ropes may snap at any point. In these cases it is natural to replace the discrete bars of the histogram by a smooth curve, so that it is still true that $P(a < X < b)$ is represented by the area under the curve between a and b. Such a curve is called a *density*. The curve has the property that the shaded area yields $P(a < X \leqslant b)$; we denote this by

$$(9) \qquad P(a < X \leqslant b) = \int_a^b f(x)\, dx.$$

We return to this idea in much more detail later.

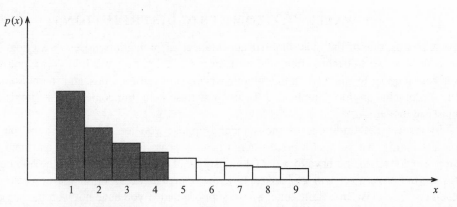

Figure 4.3. Benford's distribution, (1) in section 4.2. The shaded area is $F(4) = P(X \leqslant 4)$, the probability that you win the bet described in example 4.2.3; it equals 0.7.

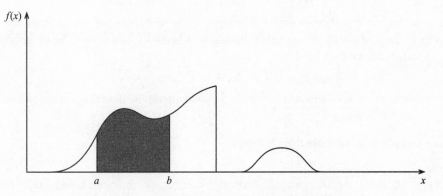

Figure 4.4. A density $f(x)$; probability is represented by area.

Exercises for section 4.2

1. For digital transmission of any signal, it is encoded as a sequence of zeros and ones. Owing to imperfections in the channel (noise and so on) any digit is independently wrongly received with probability p.
    (a) What is the probability of at least one such error in a sequence of n digits?
    (b) To reduce the chance of error each digit is transmitted in triplicate. Each digit of the triple may be wrongly received with probability p, but in each triple the correct symbol is taken to be the one which occurs more often. (Thus 101 would be taken to be 1.) What is the probability that any digit is wrongly received?

2. Aeroplane engines fail with probability q. Assuming that failures occur independently, find the probability that:
    (a) at least two of the engines on a four-engined plane do not fail;
    (b) at least two of the engines on a three-engined plane do not fail.
    Compare these probabilities for all values of q.

3. Explain why it is considered advantageous to own the orange set of properties when playing Monopoly.

4.3 WAITING; GEOMETRIC DISTRIBUTIONS

For all of us, one of the most familiar appearances of probability arises in waiting. You wait for a server to become free, you wait for a traffic light to switch to green, you wait for your number to come up on the roulette wheel, you wait for a bus, and so on. Some of these problems are too complicated for us to analyse here, but some yield a simple and classical model.

To respect the traditions of the subject, suppose you are flipping a coin, on the understanding that you get a prize when a head appears for the first time and then you stop. How long do you have to wait? Obviously there is a physical limit to the number of tosses; the prize-giver will go bankrupt, or the coin will wear out, or the universe may even cease to exist, in a finite time. So we suppose that if you have not won the prize on or before the Kth flip, you quit. Let the probability that you stop on the kth flip be $p(k)$. Clearly you have to flip at least once to win, so

$$p(k) = 0, \quad \text{for } k = 0, -1, -2, \ldots.$$

Then the probability of heads on the first flip is $\frac{1}{2}$; the probability of one tail followed by a head is $\frac{1}{4}$, the probability of two tails followed by a head is $\frac{1}{8}$, and so on. The probability of $k - 1$ tails followed by a head is 2^{-k}, and

$$p(k) = 2^{-k}, \quad \text{for } k = 1, 2, \ldots, K - 1.$$

The probability of $K - 1$ tails is $2^{-(K-1)}$, and you stop on the next flip, whatever it is, so

$$p(K) = 2^{-(K-1)}.$$

Since you never make more than K flips,

$$p(k) = 0, \quad k = K + 1, K + 2, \ldots.$$

Putting all these together we see that the number of flips until you stop has the distribution

$$(1) \qquad p(k) = \begin{cases} 2^{-k}, & 1 \leqslant k \leqslant K - 1 \\ 2^{-(K-1)}, & k = K \\ 0 & \text{otherwise.} \end{cases}$$

Now, the sequence $2^{-1}, 2^{-2}, 2^{-3}, \ldots$ is a geometric series (see the appendix to chapter 3) with ratio $\frac{1}{2}$. It is therefore called the *geometric distribution truncated at K, with parameter* $\frac{1}{2}$.

Suppose we now imagine that the coin can be flipped indefinitely. Then the distribution is

$$(2) \qquad p(k) = \begin{cases} 2^{-k}, & k \geqslant 1 \\ 0 & \text{otherwise.} \end{cases}$$

This is called the *geometric distribution with parameter* $\frac{1}{2}$.

The assumption that the coin can be tossed indefinitely is not as unrealistic as it sounds. After all, however many times it has been flipped, you should be able to toss it once more. And no one objects to the idea of a line being prolonged indefinitely. In both cases we allow continuation indefinitely because it is almost always harmless, and often very convenient.

Example 4.3.1: die. If you roll a die and wait for a six, then the same argument as that used for (2) shows that the number of rolls required has the distribution

$$(3) \qquad p(j) = \frac{1}{6}\left(\frac{5}{6}\right)^{j-1}, \quad j \geqslant 1. \qquad \bigcirc$$

Example 4.3.2: trials. More generally, suppose you have a sequence of independent Bernoulli trials in which you win with probability p, or lose with probability $1 - p$. Then the number of trials you perform until your first win has the distribution

$$(4) \qquad p(i) = p(1 - p)^{i-1}, \quad i \geqslant 1.$$

This is the geometric distribution with parameter p. $\qquad \bigcirc$

A word of warning is appropriate here; you must be quite clear what you are counting. Let $p(i)$ be the distribution of the number of trials *before* you win. This number can be zero if you win on the first trial, so we should write

$$(5) \qquad p(i) = p(1 - p)^{i}, \quad i \geqslant 0.$$

This is not *the* geometric distribution (which is on the positive integers). It is *a* geometric distribution.

In any case it is easy to see that $p(i)$ is indeed a probability distribution, as defined in (4) of section 4.2, because

$$(6) \qquad \sum_{i} p(i) = \sum_{i=0}^{\infty} pq^{i} = \frac{p}{1 - q} = 1.$$

Example 4.3.3: unlucky numbers. Suppose a certain lottery takes place each week. Let p be the probability that some given number d is drawn on any given week. After n successive draws, let $p(k)$ be the probability that d last appeared k weeks ago. What is $p(k)$?

Solution.   Note first that the probability that d does not appear in any given draw is $1 - p$. Now the last occurrence of d is k weeks ago only if it is drawn that week and then is not drawn on k occasions. This yields

(7) $$p(k) = p(1 - p)^k, \quad 0 \leqslant k \leqslant n - 1.$$

Obviously d fails to appear at all with probability $(1 - p)^n$, and in accordance with the principal property of a distribution, (5) in section 4.2, we do indeed have

$$\sum_{k=0}^{n-1} p(k) + (1 - p)^n = 1.$$

Comparison of (7) with data from real lotteries shows it to be an excellent description of reality.   ○

Remark.   Lotteries and roulette wheels publish and keep records of their results. This is for two incompatible reasons. The first is that they wish to demonstrate that the numbers that turn up are indeed completely random. The second is that some gamblers choose to bet on numbers that have not appeared for a long time. The implicit assumption, that such numbers are more likely to appear next time, is the gambler's fallacy. Other gamblers choose to bet on the numbers that have appeared most often. Do you think this is more rational?

Example 4.3.4: 'sudden death'.   Suppose two players A and B undertake a series of trials such that each trial independently yields one of the following:

(a)   a win for A with probability p;
(b)   a win for B with probability q;
(c)   a draw (or no result, or a void trial), with probability $1 - p - q$.

The game stops at the first win by either A or B. This is essentially the format of the game of craps, and such contests are also often used to resolve golf and other tournaments in which players are tied for the lead at the end of normal play. In this context, they are called sudden-death playoffs. We may ask:

(i)   What is the probability a_n that A wins at the nth trial?
(ii)   What is the probability α that A wins overall?
(iii) What is the probability $\lambda(n)$ that the game lasts for n trials?

Solution.   For (i): First we note that A wins at the nth trial if and only if the first $n - 1$ trials are drawn, and A wins the nth. Hence, using independence,

$$a_n = (1 - p - q)^{n-1} p.$$

For (ii): By the addition rule for probabilities,

(8) $$\alpha = \sum_{n=1}^{\infty} a_n = p \sum_{n=1}^{\infty} (1 - p - q)^{n-1} = \frac{p}{p + q}.$$

Of course we already know an alternative method for this. Let A_w be the event that A is the overall winner, and denote the possible results of the first trial by A, B, and D. Then

$$\alpha = P(A_w|A)p + P(A_w|B)q + P(A_w|D)(1 - p - q)$$

$$= p + 0 + \alpha(1 - p - q),$$

which gives (8).

For (iii): Here we just have Bernoulli trials with $P(S) = p + q$, and $P(F) = 1 - p - q$. Hence

$$\lambda(n) = (1 - p - q)^{n-1}(p + q). \qquad \bigcirc$$

Exercises for section 4.3

1. A die is rolled repeatedly until it shows a six. Let A_n be the event that the first six appears on the nth roll, and let E be the event that the number of rolls required for the first six is even.

   Find $P(E)$ and $p_n = P(A_n|E)$. Is $(p_n; n \geqslant 2)$ a geometric distribution?

2. *'Sudden death' continued.* Let D_n be the event that the duration of the game is n trials, and let A_w be the event that A is the overall winner. Show that A_w and D_n are independent.

4.4 THE BINOMIAL DISTRIBUTION AND SOME RELATIVES

As we have remarked above, in many practical applications it is necessary to perform some fixed number, n, of Bernoulli trials. Naturally we would very much like to know the probability of r successes, for various values of r. Here are some obvious examples, some familiar and some new.

(i) A coin is flipped n times. What is the chance of exactly r heads?
(ii) You have n chips. What is the chance that r are defective?
(iii) You treat n patients with the same drug. What is the chance that r respond well?
(iv) You buy n lottery scratch cards. What is the chance of r wins?
(v) You type a page of n symbols. What is the chance of r errors?
(vi) You call n telephone numbers. What is the chance of making r sales?

This is obviously yet another list that could be extended indefinitely, but in every case the underlying problem is the same. It is convenient to standardize our names and notation around Bernoulli trials so we ask the following: in a sequence of n independent Bernoulli trials with $P(S) = p$, what is the probability $p(k)$ of k successes?

For variety, and in deference to tradition, we often speak in terms of coins: if you flip a biased coin n times, what is the probability $p(k)$ of k heads, where $P(H) = p$?

These problems are the same, and the answer is given by the

Binomial distribution. For n Bernoulli trials with $P(S) = p = 1 - q$, the probability $p(k)$ of obtaining exactly k successes is

$$(1) \qquad p(k) = P(k \text{ successes}) = \binom{n}{k} p^k q^{n-k}, \qquad 0 \leqslant k \leqslant n.$$

We refer to this as $B(n, p)$ or 'the $B(n, p)$ distribution'. We can see that this is indeed a probability distribution as defined in section 4.2, because

$$\sum_k p(k) = \sum_{k=0}^n \binom{n}{k} p^k q^{n-k} = (p+q)^n, \quad \text{by the binomial theorem,}$$

$$= 1, \text{ since } p + q = 1.$$

The first serious task is to prove (1).

Proof of (1).   When we perform n Bernoulli trials there are 2^n possible outcomes, because each yields either S or F. How many of these outcomes comprise exactly k successes and $n - k$ failures? The answer is

$$\binom{n}{k},$$

because this is the number of distinct ways of ordering k successes and $n - k$ failures. (We proved this in section 3.3; see especially the lines before (8)). Now we observe that, by independence, any given outcome with k successes and $n - k$ failures has probability $p^k q^{n-k}$. Hence

$$p(k) = \binom{n}{k} p^k q^{n-k}, \quad 0 \leqslant k \leqslant n. \qquad \square$$

It is interesting, and a useful exercise, to obtain this result in a different way by using conditional probability. It also provides an illuminating connection with many earlier ideas, and furthermore illustrates a useful technique for tackling harder problems. In this case the solution is very simple and runs as follows.

Another proof of (1).   Let $A(n, k)$ be the event that n flips show k heads, and let
$$p(n, k) = P(A(n, k)).$$
The first flip gives H or T, so by the partition rule (6) of section 2.8
(2) $$p(n, k) = P(A(n, k)|H)P(H) + P(A(n, k)|T)P(T).$$
But given H on the first flip, $A(n, k)$ occurs if there are exactly $k - 1$ heads in the next $n - 1$ flips. Hence
$$P(A(n, k)|H) = p(n - 1, k - 1).$$
Likewise
$$P(A(n, k)|T) = p(n - 1, k).$$
Hence substituting in (2) yields
(3) $$p(n, k) = pp(n - 1, k - 1) + qp(n - 1, k).$$
Of course we know that $p(n, 0) = q^n$ and $p(n, n) = p^n$, so equation (3) successively supplies values of $p(n, k)$ just as in Pascal's triangle and the problem of the points.

It is now a very simple matter to show that the solution of (3) is indeed given by the binomial distribution

$$p(n, k) = \binom{n}{k} p^k q^{n-k}, \quad 0 \leqslant k \leqslant n. \qquad \square$$

The connection with Pascal's triangle is made completely obvious if the binomial probabilities are displayed as a diagram (or graph) as in figure 4.5. This is very similar to

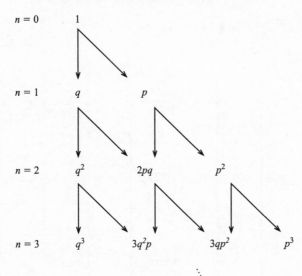

Figure 4.5. Triangle of binomial probabilities.

a tree diagram (though it is not in fact a tree). The process starts at the top where no trials have yet been performed. Each trial yields S or F, with probabilities p and q, and corresponds to a step down to the row beneath. Hence any path of n steps downwards corresponds to a possible outcome of the first n trials. The kth entry in the nth row is the sum of the probabilities of all possible paths to that vertex, which is just $p(k)$. The first entry at the top corresponds to the obvious fact that the probability of no successes in no trials is unity.

The binomial distribution is one of the most useful, and we take a moment to look at some of its more important properties. First we record the simple relationship between $p(k + 1)$ and $p(k)$, namely

$$(4) \qquad p(k+1) = \binom{n}{k+1} p^{k+1}(1 - p)^{n-k+1}$$

$$= \frac{n-k}{k+1} \left\{ \frac{n!}{k!(n-k)!} \right\} \left(\frac{p}{1-p} \right) p^k (1 - p)^{n-k}$$

$$= \frac{n-k}{k+1} \left(\frac{p}{1-p} \right) p(k).$$

This recursion, starting either with $p(0) = (1 - p)^n$ or with $p(n) = p^n$, is very useful in carrying out explicit calculations in practical cases.

It is also very useful in telling us about the shape of the distribution in general. Note that

$$(5) \qquad \frac{p(k)}{p(k+1)} = \frac{k+1}{n-k} \left(\frac{1-p}{p} \right),$$

which is less than 1 whenever $k < (n + 1)p - 1$. Thus the probabilities $p(k)$ increase up to this point. Otherwise, the ratio in (5) is greater than 1 whenever $k > (n + 1)p - 1$;

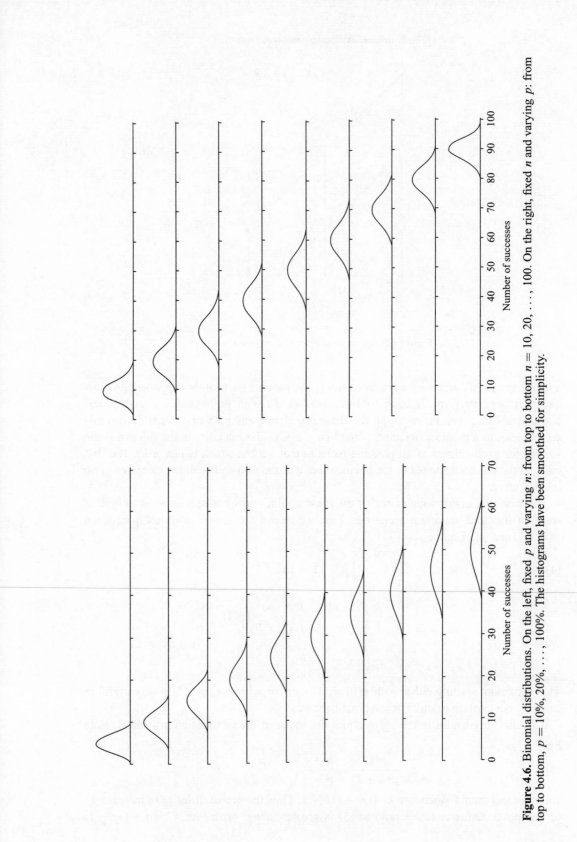

Figure 4.6. Binomial distributions. On the left, fixed p and varying n: from top to bottom $n = 10, 20, \ldots, 100$. On the right, fixed n and varying p: from top to bottom, $p = 10\%, 20\%, \ldots, 100\%$. The histograms have been smoothed for simplicity.

the probabilities decrease past this point. The largest term is $p([(n+1)p])$, where $[(n+1)p]$ is the largest integer not greater than $(n+1)p$. If $(n+1)p$ happens to be exactly an integer then

(6)
$$\frac{p([(n+1)p-1])}{p([(n+1)p])} = 1$$

and both these terms are maximal.

The shape of the distribution becomes even more obvious if we draw it; see figure 4.6, which displays the shape of binomial histograms for various values of n and p.

We shall return to the binomial distribution later; for the moment we continue looking at the simple but important distributions arising from a sequence of Bernoulli trials. So far we have considered the geometric distribution and the binomial distribution. Next we have the

Negative binomial distribution.   A close relative of the binomial distribution arises when we ask the opposite question. The question above is 'Given n flips, what is the chance of k heads?' Suppose we ask instead 'Given we must have k heads, what chance that we need n flips?'

We can rephrase this in a more dignified way as follows. A coin is flipped repeatedly until the first flip at which the total number of heads it has shown is k. Let $p(n)$ be the probability that the total number of flips is n (including the tails). What is $p(n)$? We shall show that the answer to this is

(7)
$$p(n) = p^k q^{n-k} \binom{n-1}{k-1}, \quad n = k, k+1, k+2, \ldots .$$

This is called the *negative binomial* distribution. It is quite easy to find $p(n)$; first we notice that to say 'The total number of flips is n' is the same as saying 'The first $n-1$ flips include $k-1$ heads and the nth is a head'. But this last event is just $A(n-1, k-1) \cap H$. We showed above that

$$P(A(n-1, k-1)) = \binom{n-1}{k-1} p^{k-1} q^{n-k},$$

and $P(H) = p$. Hence by the independence of A and H,

$$p(n) = P(A(n-1, k-1) \cap H)$$

$$= P(A(n-1, k-1))P(H) = p^k q^{n-k} \binom{n-1}{k-1}.$$

By its construction you can see that the negative binomial distribution tends to crop up when you are waiting for a collection of things. For instance

Example 4.4.1: krakens.   Each time you lower your nets you bring up a kraken with probability p. What is the chance that you need n fishing trips to catch k krakens? The answer is given by (7).   ○

We can derive this distribution by conditioning also. Let $p(n, k)$ be the probability that you require n flips to obtain k heads; let $F(n, k)$ be the event that the kth head occurs at the nth flip. Then, noting that the first flip yields either H or T, we have

$$p(n, k) = P(F(n, k))$$
$$= P(F(n, k)|H)p + P(F(n, k)|T)q$$

But reasoning as above shows that

$$P(F(n, k)|H) = p(n - 1, k - 1)$$

and

$$P(F(n, k)|T) = p(n - 1, k).$$

Hence

(8) $$p(n, k) = pp(n - 1, k - 1) + qp(n - 1, k).$$

Since this equation is the same as (3) it is not surprising that the answer involves the same binomial coefficients.

Exercises for section 4.4

1. Use the recursion given in (4) to calculate the 11 terms in the binomial distribution for parameters 10 and $\frac{1}{2}$, namely B$(10, \frac{1}{2})$.

2. Let $(p(k); 0 \leq k \leq n)$ be the binomial B(n, p) distribution. Show that
$$\{p(k)\}^2 \geq p(k + 1)p(k - 1) \quad \text{for all } k.$$

3. Let $(p(k); 0 \leq k \leq n)$ be the binomial B(n, p) distribution and let $(\hat{p}(k); 0 \leq k \leq n)$ be the binomial B$(n, 1 - p)$ distribution. Show that
$$p(k) = \hat{p}(n - k).$$

   Interpret this result.

4. Check that the distribution in (7) satisfies (8).

4.5 SAMPLING

A problem that arises in just about every division of science and industry is that of counting or assessing a divided population. This is a bit vague, but a few examples should make it clear.

Votes.   The population is divided into those who are going to vote for the Progressive party and those who are going to vote for the Liberal party. The politician would like to know the proportions of each.

Soap.   There are those who like 'Soapo' detergent and those who do not. The manufacturers would like to know how many of each there are.

Potatoes.   Some plants are developing scab, and others are not. The farmer would like to know the rate of scab in his crop.

Chips.   These are perfect or defective. The manufacturer would like to know the failure rate.

Fish.   These are normal, or androgynous due to polluted water. What proportion are deformed?

Turkeys. A film has been made. The producers would like to know whether viewers will like it or hate it.

It should now be quite obvious that in all these cases we have a population or collection divided into two distinct non-overlapping classes, and we want to know how many there are in each. The list of similar instances could be prolonged indefinitely; you should think of some yourself.

The examples have another feature in common: it is practically impossible to count all the members of the population to find out the proportion in the two different classes. All we can do is look at a part of the population, and try to extrapolate to the whole of it. Naturally, some thought and care is required here. If the politician canvasses opinion in his own office he is not likely to get a representative answer. The farmer might get a depressing result if he looked only at plants in the damp corner of his field. And so on.

After some thought, you might agree that a sensible procedure in each case would be to take a sample of the population in such a way that each member of the population has an equal chance of being sampled. This ought to give a reasonable snapshot of the situation; the important question is, how reasonable? That is, how do the properties of the sample relate to the composition of the population? To answer this question we build a mathematical model, and use probability.

The classical model is an urn containing balls (or slips of paper). The number of balls (or slips) is the size of the population, the colour of the ball (or slip) denotes which group it is in. Picking a ball at random from the urn corresponds to choosing a member of the population, every member having the same chance to be chosen.

Having removed one ball, we are immediately faced with a problem. Do we put it back or keep it out before the next draw? The answer to this depends on the real population being studied. If a fish has been caught and dissected, it cannot easily be put back in the pool and caught again. But voters can be asked for their political opinions any number of times. In the first case balls are not replaced in the urn, so this is *sampling without replacement*. In the second case they are; so that is *sampling with replacement*. Let us consider an example of the latter.

Example 4.5.1: replacement. A park contains $b + g$ animals of a large and dangerous species. There are b of the first type and g of the second type. Each time a ranger observes one of these animals he notes its type. There is no question of tagging such a dangerous beast, so this is sampling with replacement. The probability that any given observation is of the first type is $b/(b + g)$, and of the second $g/(b + g)$. If there are n such observations, assumed independent, then this amounts to n Bernoulli trials, and the distribution of the numbers of each type seen is binomial:

$$(1) \qquad \text{P(observe } k \text{ of the first type)} = \left(\frac{b}{b + g}\right)^k \left(\frac{g}{b + g}\right)^{n-k} \binom{n}{k}. \qquad \bigcirc$$

Next we consider sampling without replacement.

Example 4.5.2: no replacement; hypergeometry. Our next example of this kind of sampling distribution arises when there are two types of individual and sampling is without replacement. To be explicit, we suppose an urn contains m mauve balls and w

white balls. A sample of r balls is removed at random; what is the probability $p(k)$ that it includes exactly k mauve balls and $r - k$ white balls?

Solution.   There are

$$\binom{m + w}{r}$$

ways of choosing the r balls to be removed; these are equally likely since they are removed at random. To find $p(k)$ we need to know the number of ways of choosing the k mauve balls and the $r - k$ white balls. But this is just

$$\binom{m}{k}\binom{w}{r - k},$$

using the product rule for counting. Hence

$$(2) \qquad p(k) = \frac{\binom{m}{k}\binom{w}{r - k}}{\binom{m + w}{r}}, \qquad 0 \le k \le r.$$

It is easy to show, by expanding all the binomial coefficients as factorials, that this can be written as

$$(3) \qquad p(k) = \frac{\binom{w}{r}}{\binom{m + w}{r}} \frac{\binom{r}{k}\binom{m}{k}}{\binom{w - r + k}{k}}.$$

This may not seem to be a very obvious step but, as it happens, the series

$$(4) \qquad H = \sum_{k=0}^{\infty} \frac{\binom{r}{k}\binom{m}{k}}{\binom{w - r + k}{k}} x^k$$

defines a very famous and well-known function. It is the *hypergeometric function*, which was extensively studied by Gauss in 1812, and before him by Euler and Pfaff. It has important applications in mathematical physics and engineering.

For this reason the distribution (3) is called the *hypergeometric distribution*. We conclude with a typical application of this.   ○

Example 4.5.3: wildlife sampling.   Naturalists and others often wish to estimate the size N of a population of more or less elusive creatures. (They may be nocturnal, or burrowing, or simply shy.) A simple and popular method is capture–recapture, which is executed thus:

(i)   capture a animals and tag (or mark) them;
(ii)   release the a tagged creatures and wait for them to mix with the remaining $N - a$;
(iii) capture n animals and count how many are already tagged (these are recaptures).

Clearly the probability of finding r recaptures in your second group of n animals is hypergeometric:

$$(5) \qquad p(r) = \binom{a}{r}\binom{N-a}{n-r} \Big/ \binom{N}{n}.$$

Now it can be shown (*exercise*) that the value of N for which $p(r)$ is greatest is the integer nearest to an/r. Hence this is a plausible estimate of the unknown population size N, where r is what you actually observe.

This technique has also been used to estimate the number of vagrants in large cities, and to investigate career decisions among doctors. ○

Exercises for section 4.5

1. **Capture recapture.**   Show that in (5), $p(r)$ is greatest when N is the integer nearest to na/r.

2. **Acceptance sampling.**   A shipment of components (called a lot) arrives at your factory. You test their reliability as follows. For each lot of 100 components you take 10 at random, and test these. If no more than one is defective you accept the lot. What is the probability that you accept a lot of size 100 which contains 7 defectives?

3.   In (5), show that $p(r+1)p(r-1) \leq \{p(r)\}^2$.

4.6 LOCATION AND DISPERSION

Suppose we have some experiment, or other random procedure, that yields outcomes in some finite set of numbers D, with probability distribution $p(x)$, $x \in D$. The following property of a distribution turns out to be of great practical and theoretical importance.

Definition.   The *mean* of the distribution $p(x)$ is denoted by μ, where

$$(1) \qquad \mu = \sum_{x \in D} xp(x).$$

The mean is simply a weighted average of the possible outcomes in D; it is also known as the *expectation*. △

Natural questions are, why this number, and why is it useful? We answer these queries shortly; first of all let us look at some simple examples.

Example 4.6.1: coin.   Flip a fair coin and count the number of heads. Trivially $\Omega = \{0, 1\}$, and $p(0) = \frac{1}{2} = p(1)$. Hence the mean is

$$\mu = \tfrac{1}{2} \times 0 + \tfrac{1}{2} \times 1 = \tfrac{1}{2}.$$

This example is truly trivial, but it does illustrate that the mean is not necessarily one of the possible outcomes of the experiment. In this case the mean is half a head. (Journalists and others with an impaired sense of humour sometimes seek to find amusement in this; the average family size will often achieve the same effect, as it involves fractional children. Of course real children are fractious not fractions ...) ○

Example 4.6.2: die.   If you roll a die once, the outcome has distribution $p(x) = \frac{1}{6}$, $1 \leq x \leq 6$. Then

$$\mu = \frac{1}{6} \times 1 + \frac{1}{6} \times 2 + \frac{1}{6} \times 3 + \frac{1}{6} \times 4 + \frac{1}{6} \times 5 + \frac{1}{6} \times 6 = \frac{7}{2}. \qquad \bigcirc$$

Example 4.6.3: craps.   If you roll two dice, then the distribution of the sum of their scores is given in Figure 4.6. After a simple but tedious calculation you will find that $\mu = 7$. $\hfill \bigcirc$

At first sight the mean, μ, may not seem very useful or fascinating, but there are in fact many excellent reasons for our interest in it. Here are some of them.

Mean as value.   In our discussions of probability in chapter 1, we considered the value of an offer of \$$d$ with probability p or nothing with probability $1 - p$. It is clear that the fair value of what you expect to get is \$$dp$, which is just the mean of this distribution.

Likewise if you have a number of disjoint offers (or bets) such that you receive \$$x$ with probability $p(x)$, as x ranges over some finite set, then the fair value of this is just \$$\mu$, where

$$\mu = \sum_x xp(x),$$

is the mean of the distribution.

Sample mean and relative frequency.   Suppose you have a number n of similar objects, n potatoes, say, or n hedgehogs. You could then measure any numerical attribute (such as spines, or weight, or length), and obtain a collection of observations $\{x_1, x_2, \ldots, x_n\}$. It is widely accepted that the average

$$\bar{x} = \frac{1}{n} \sum_{r=1}^{n} x_r$$

is a reasonable candidate for a single number to represent or typify this collection of measurements. Now suppose that some of these numbers are the same, as they often will be in a large set of data. Let the number of times you obtain the value x be $N(x)$; thus the proportion yielding x is

$$P(x) = \frac{N(x)}{n}.$$

We have argued above that, in the long run, $P(x)$ is close to the probability $p(x)$ that x occurs. Now the average $\bar{x}$ satisfies

$$\bar{x} = n^{-1}(x_1 + \cdots + x_n) = n^{-1} \sum_x xN(x) = \sum_x xP(x) \simeq \sum_x xp(x) = \mu,$$

approximately, in the long run. It is important to remark that we can give this informal observation plenty of formal support, later on.

Mean as centre of gravity.   We have several times made the point that probability is analogous to mass; we have a unit lump of probability, which is then split up among the outcomes in D to indicate their respective probabilities. Indeed this is often *called* a probability mass distribution.

We may represent this physically by the usual histogram, with bars of uniform unit density. Where then is the centre of gravity of this distribution? Of course, it is at the point μ given by

$$\mu = \sum xp(x).$$

The histogram, or distribution, is in equilibrium if placed on a fulcrum at μ. See figure 4.7.

Here are some further examples of means.

Example 4.6.4: sample mean. Suppose you have n lottery tickets bearing the numbers $x_1, x_2, \ldots, x_n$ (or perhaps you have n swedes weighing $x_1, x_2, \ldots, x_n$); one of these is picked at random. What is the mean of the resulting distribution?

Of course we have the probability distribution

$$p(x_1) = p(x_2) = \cdots = n^{-1}$$

and so

$$\mu = \sum xp(x) = n^{-1} \sum_{r=1}^{n} x_r = \bar{x}.$$

The sample mean is equal to the average. $\circ$

Example 4.6.5: binomial mean. By definition (1), the mean of the binomial distribution is given by

$$\mu = \sum_{k=0}^{n} kp(k) = \sum_{k=1}^{n} k \frac{n!}{k!(n-k)!} p^k q^{n-k}$$

$$= np \sum_{k=1}^{n} \frac{(n-1)!}{(k-1)!(n-k)!} p^{k-1} q^{n-k} = np \sum_{x=0}^{n-1} \binom{n-1}{x} p^x q^{n-1-x}$$

$$= np(p+q)^{n-1}$$

$$= np.$$

We shall find neater ways of deriving this important result in the next chapter. $\circ$

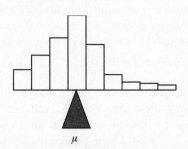

μ

Figure 4.7. Mean as centre of gravity.

Example 4.6.6: geometric mean.   As usual, from (1) the mean of the geometric distribution is

$$\mu = \sum_{k=1}^{\infty} kq^{k-1}p = \frac{p}{(1-q)^2}$$
$$= p^{-1}.$$

Note that we summed the series by looking in appendix 3.12.III.   ○

These examples, and our discussion, make it clear that the mean is useful as a guide to the location of a probability distribution. This is convenient for simple-minded folk such as journalists (and the media in general); if you are replying to a request for information about accident rates, or defective goods, or lottery winnings, it is pointless to supply the press with a distribution; it will be rejected. You will be allowed to use at most one number; the mean is a simple and reasonably informative candidate.

Furthermore, we shall find many more theoretical uses for it later on. But it does have drawbacks, as we now discover; the keen-eyed reader will have noticed already that while the mean tells you where the centre of probability mass is, it does not tell you how spread out or *dispersed* the probability distribution is.

Example 4.6.7.   In a casino the following bets are available for the same price (a price *greater* than $1000).

(i)   You get $1000 for sure.
(ii)   You get $2000 with probability $\frac{1}{2}$, or nothing.
(iii) You get 10^6 with probability 10^{-3}, or nothing.

Calculating the mean of these three distributions we find

For (i), $\mu = \$1000$.
For (ii), $\mu = \frac{1}{2} \times \$2000 + \frac{1}{2} \times \$0 = \$1000$.
For (iii), $\mu = 10^{-3} \times \$10^6 + (1 - 10^{-3}) \times \$0 = \$1000$.

Thus all these three distributions have the same mean, namely $1000. But obviously they are very different bets! Would you be happy to pay the same amount to play each of these games? Probably not; most people would prefer one or another of these wagers, and your preference will depend on how rich you are and whether you are risk-averse or risk-seeking. There is much matter for speculation and analysis here, but we note merely the trivial point that these three distributions vary in how *spread out* they are about their mean. That is to say, (i) is not spread out at all, (ii) is symmetrically disposed not too far from its mean and (iii) is very spread out indeed.   ○

There are various ways of measuring such a dispersion, but it seems natural to begin by ignoring the sign of deviations from the mean μ, and just looking at their absolute magnitude, weighted of course by their probability. It turns out that the algebra is much simplified in general if we use the following measure of dispersion in a probability distribution.

Definition.   The *variance* of the probability distribution $p(x)$ is denoted by σ^2, where

(2)
$$\sigma^2 = \sum_{x \in D} (x - \mu)^2 p(x). \qquad \triangle$$

The variance is a weighted average of the squared distance of outcomes from the mean; it is sometimes called the *second moment* about the mean because of the analogy with mass mentioned often above.

Example 4.6.7 revisited.   For the three bets on offer it is easy to find the variance in each case:

For (i), $\sigma^2 = 0$.
For (ii), $\sigma^2 = \frac{1}{2}(0 - 1000)^2 + \frac{1}{2}(2000 - 1000)^2 = 10^6$.
For (iii), $\sigma^2 = (1 - 10^{-3})(0 - 10^3)^2 + 10^{-3}(10^6 - 10^3)^2 \simeq 10^9$.

Clearly, as the distribution becomes more spread out σ^2 increases dramatically.   ○

In order to keep the same scale, it is often convenient to use σ rather than σ^2.

Definition.   The positive square root σ of the variance σ^2 is known as the *standard deviation* of the distribution. $\triangle$

(3)
$$\sigma = \sqrt{\sum_{x \in D} (x - \mu)^2 p(x)}.$$

Let us consider a few simple examples.

Example 4.6.8: Bernoulli trial.   Here,
$$p(k) = p^k (1 - p)^{1-k}, \quad k = 0, 1,$$
$$\mu = p \times 1 + (1 - p) \times 0 = p,$$

and

$$\sigma^2 = (1 - p)^2 p + (0 - p)^2 (1 - p) = p(1 - p).$$

Thus

$$\sigma = \{p(1 - p)\}^{1/2}. \qquad ○$$

Example 4.6.9: dice.   Here
$$p(k) = \tfrac{1}{6}, \quad 1 \leqslant k \leqslant 6,$$
$$\mu = \sum_{k=1}^{6} \tfrac{1}{6}k = \tfrac{7}{2},$$

and

$$\sigma^2 = \sum_{k=1}^{6} \tfrac{1}{6}\left(k - \tfrac{7}{6}\right)^2 = \tfrac{35}{12},$$

after some arithmetic. Hence $\sigma \simeq 1.71$.   ○

Example 4.6.10.   Show that for any distribution $p(x)$ with mean μ and variance σ^2 we have

$$\sigma^2 = \sum_{x \in D} x^2 p(x) - \mu^2.$$

Solution.   From the definition,

$$\sigma^2 = \sum_{x \in D} (x^2 - 2x\mu + \mu^2) p(x)$$

$$= \sum_{x \in D} x^2 p(x) - 2\mu \sum_{x \in D} x p(x) + \mu^2 \sum_{x \in D} p(x)$$

$$= \sum_{x \in D} x^2 p(x) - 2\mu^2 + \mu^2$$

as required.   ○

We end this section with a number of remarks.

Remark: good behaviour.   When a probability distribution is assigned to a finite collection of real numbers, the mean and variance are always well behaved. However, for distributions on an unbounded set (the integers for example), good behaviour is not guaranteed. The mean may be infinite, or may even not exist. Here are some examples to show what can happen.

Example 4.6.11: distribution with no mean.   Let

$$p(x) = \frac{c}{x^2}, \quad x = \pm 1, \pm 2, \ldots .$$

Since

$$\sum_{x=1}^{\infty} \frac{1}{x^2} = \frac{\pi^2}{6},$$

it follows that $c = 3/\pi^2$, because $\sum p(x) = 1$. Now $\sum_{x=1}^{\infty} x p(x) = \infty$ and $-\sum_{x=-1}^{-\infty} x p(x) = \infty$, so the mean μ does not exist.   ○

Example 4.6.12: distribution with infinite mean.   Let

$$p(x) = \frac{2c}{x^2}, \quad x \geqslant 1.$$

Then as in example 4.6.11 we have

$$\mu = \sum_{x=1}^{\infty} \frac{2c}{x} = \infty.$$ ○

Example 4.6.13: distribution with finite mean but infinite variance.   Let

$$p(x) = \frac{c}{x^3}, \quad x = \pm 1, \pm 2, \ldots ,$$

where $c^{-1} = \sum_{x \neq 0} x^{-3}$. Then $-\sum_{x=-1}^{-\infty} xp(x) = \sum_{x=1}^{\infty} xp(x) = \frac{1}{6} c\pi^2$. Hence $\mu = 0$. However,

$$\sigma^2 = \sum x^2 p(x) = \sum_{x=1}^{\infty} \frac{2c}{x} = \infty. \qquad \bigcirc$$

Remark: median and mode. We have seen in examples 4.6.11 and 4.6.12 above that the mean may not be finite, or even exist. Nevertheless in these examples (and many similar cases) we would like a rough indication of location. Luckily, some fairly obvious candidates offer themselves. If we look at example 4.6.11 we note that the distribution is symmetrical about zero, and the values ± 1 are considerably more likely than any others.

These two observations suggest the following two ideas.

Definition: median. Let $p(x)$, $x \in D$, be a distribution. If m is any number such that

$$\sum_{x \leq m} p(x) \geq \frac{1}{2} \quad \text{and} \quad \sum_{x \geq m} p(x) \geq \frac{1}{2}$$

then m is a *median* of the distribution. $\qquad \triangle$

Definition: mode. Let $p(x)$, $x \in D$, be a distribution. If $\lambda \in D$ is such that
$$p(\lambda) \geq p(x) \quad \text{for all } x \text{ in } D$$
then λ is said to be a *mode* of the distribution. $\qquad \triangle$

Roughly speaking, outcomes are equally likely to be on either side of a median, and the most likely outcomes are modes.

Example 4.6.11 revisited. Here any number in $[-1, 1]$ is a median, and ± 1 are both modes. (Remember there is no mean for this distribution.) $\qquad \bigcirc$

Example 4.6.12 revisited. Here $+1$ is the only mode, and it is also the only median because $6/\pi^2 > 1/2$. (Remember that $\mu = \infty$ in this case.) $\qquad \bigcirc$

Example 4.6.14. Let $p(x)$ be the geometric distribution
$$p(x) = (1 - p)p^{x-1}, \quad x \geq 1, 0 < p < 1;$$
then $\lambda = 1$. Further, let
$$m = \min\{x: 1 - p^x \geq \tfrac{1}{2}\}.$$
If $1 - p^m > \frac{1}{2}$, then m is the unique median. If $1 - p^m = \frac{1}{2}$, then the interval $[m, m + 1]$ is the set of medians. We have shown already that the mean μ is $(1 - p)^{-1}$. $\qquad \bigcirc$

Remark: mean and median. It is important to stress that the mean is only a crude summary measure of the distribution. It tells you something about the distribution of probability, but not much. In particular it does *not* tell you that

$$\sum_{x > \mu} p(x) = \tfrac{1}{2}.$$

This statement is false in general, but is nevertheless widely believed in a vague unfocused way. For example, research has shown that many people will agree with the following statement:

> If the average lifespan is 75 years, then it is an evens chance that any newborn infant will live for more than 75 years.

This is not true, because the mean is not in general equal to the median. It is true that the mean μ and median m are quite close together when the variance is small. In fact it can be shown that

$$(\mu - m)^2 \leqslant \sigma^2$$

where σ^2 is the variance of the distribution.

Exercises for section 4.6

1. **Uniform distribution.**   Let $p(k) = n^{-1}$, $1 \leqslant k \leqslant n$. Show that $\mu = \frac{1}{2}(n+1)$, and $\sigma^2 = \frac{1}{12}(n^2 - 1)$.

2. **Binomial variance.**   Let
$$p(k) = \binom{n}{k} p^k (1-p)^{n-k}, \quad 0 \leqslant k \leqslant n.$$
   Show that $\sigma^2 = np(1-p)$.

3. **Geometric variance.**   When $p(k) = q^{k-1} p$, $k \geqslant 1$, show that $\sigma^2 = qp^{-2}$.

4. **Poisson mean.**   Let $p(k) = \lambda^k e^{-\lambda}/k!$. Show that $\mu = \lambda$.

5. **Benford.**   Show that the expected value of the first significant digit in (for example) census data is 3.44, approximately. (See example 4.2.3 for the distribution.)

4.7 APPROXIMATIONS: A FIRST LOOK

At this point the reader may observe this expanding catalogue of different distributions with some dismay. Not only are they too numerous to remember with enthusiasm, but many comprise a tiresomely messy collection of factorials that promise tedious calculations ahead.

Fortunately, things are not as bad as they seem because, for most practical purposes, many of the distributions we meet can be effectively approximated by much simpler functions. Let us recall an example to illustrate this.

Example 4.7.1: polling voters.   Voters belong either to the red party or the green party. There are r reds, g greens, and $v = r + g$ voters altogether. You take a random sample of size n, without asking any voter twice. Let A_k be the event that your sample includes k greens. This is sampling without replacement, and so of course from (2) of section 4.5 you know that

(1)
$$P(A_k) = \binom{g}{k}\binom{r}{n-k} \bigg/ \binom{g+r}{n},$$

a hypergeometric distribution. This formula is rather disappointing, as calculating it for many values of the parameters is going to be dull and tedious at best. And results are unlikely to appear in a simple form.

However, it is often the case that v, g, and r are very large compared with k and n. (Typically n might be 1000, while r and g are in the millions.) In this case if we set
$$p = g/v, \quad q = 1 - p = r/v$$

and remember that k/v and n/v are very small, we can argue as follows. For fixed n and k, as v, g and r becomes increasingly large,
$$\frac{g-1}{v} \to p, \quad \ldots, \quad \frac{g-k+1}{v} \to p$$
$$\frac{v-k+1}{v} \to 1, \quad \frac{r-n+k+1}{v} \to q,$$

and so on. Hence

(2)
$$P(A_k) = \frac{g!}{k!(g-k)!}\frac{r!}{(n-k)!(r-n+k)!}\frac{n!(r+g-n)!}{(r+g)!}$$

$$= \binom{n}{k}\left\{\left(\frac{g}{v}\right)\cdots\left(\frac{g-k+1}{v}\right)\right\}$$

$$\times \left\{\left(\frac{r}{v}\right)\cdots\left(\frac{r-n+k+1}{v}\right)\right\}\left\{\left(\frac{v}{v}\right)\cdots\left(\frac{v-n+1}{v}\right)\right\}$$

$$\simeq \binom{n}{k}p^k q^{n-k}$$

for large r, g, and v. Thus in these circumstances the hypergeometric distribution is very well approximated by the binomial distribution, for many practical purposes.   ○

This is very pleasing, but we can often go further in many cases.

Example 4.7.2: rare greens.   Suppose in the above example that there are actually very few greens; naturally we need to make our sample big enough to have a good chance of registering a reasonable number of them. Now if g, and hence p, are very small, we have

(3)
$$P(A_1) = np(1-p)^{n-1}.$$

For this to be a reasonable size as p decreases we must increase n in such a way that np stays at some desirable constant level, λ say.

In this case, if we set $np = \lambda$, which is fixed as n increases, we have as $n \to \infty$
$$\left(1 - \frac{1}{n}\right) \to 1, \quad \ldots, \quad 1 - \frac{k-1}{n} \to 1,$$
$$(1-p)^k = \left(1 - \frac{\lambda}{n}\right)^k \to 1,$$

and

$$\left(1 - p\right)^n = \left(1 - \frac{\lambda}{n}\right)^n \to e^{-\lambda}.$$

Hence

(4)
$$P(A_k) = \binom{n}{k} p^k (1 - p)^{n-k}$$

$$= \left(1 - \frac{\lambda}{n}\right)^n \binom{n}{k} \left(\frac{\lambda}{n}\right)^k \left(1 - \frac{\lambda}{n}\right)^{-k}$$

$$= \left(1 - \frac{\lambda}{n}\right)^n \frac{\lambda^k}{k!} \left(1 - \frac{1}{n}\right) \cdots \left(1 - \frac{k-1}{n}\right) \left(1 - \frac{\lambda}{n}\right)^k$$

$$\to e^{-\lambda} \frac{\lambda^k}{k!}, \quad \text{as } n \to \infty.$$

This is called the *Poisson distribution*. We should check that it is a distribution; it is, since each term is positive and

$$e^{\lambda} = \sum_{k=0}^{\infty} \lambda^k / k!.$$

It is so important that we devote the next section to it, giving a rather different derivation. ○

Exercise for section 4.7

1. *Mixed sampling.* A lake contains g gudgeon and r roach. You catch a sample of size n, on the understanding that roach are returned to the lake after being recorded, whereas gudgeon are retained in a keep-net. Find the probability that your sample includes k gudgeon. Show that as r and g increase in such a way that $g/(r+g) \to p$, the probability distribution tends to the binomial.

4.8 SPARSE SAMPLING; THE POISSON DISTRIBUTION

Another problem that arises in almost every branch of science is that of counting rare events. Once again, this slightly opaque statement is made clear by examples.

Meteorites. The Earth is bombarded by an endless shower of meteorites. Rarely, they hit the ground. It is natural to count how many meteorite-strikes there are on some patch of ground during a fixed period. (For example: on your house, while you are living there.)

Accidents. Any stretch of road, or road junction, is subject to the occasional accident. How many are there in a given stretch of road? How many are there during a fixed period at some intersection?

Misprints. An unusually good typesetter makes a mistake very rarely. How many are there on one page of a broadsheet newspaper? How many does she make in a year?

Currants. A frugal baker adds a small packet of currants to his batch of dough. How many currants are in each bun? How many in each slice of currant loaf?

Clearly this is another list which could be extended indefinitely. You have to think only for a moment of the applications to counting: colonies of bacteria on a dish; flaws in a carpet; bugs in a program; earwigs in your dahlias; daisy plants in your lawn; photons in your telescope; lightning strikes on your steeple; wasps in your beer; mosquitoes on your neck; and so on.

Once again we need a canonical example that represents or acts as a model for all the rest. Tradition is not so inflexible in this case (we are not bound to coins and urns as we were above). For a change, we choose to count the meteorites striking Bristol during a time period of length t, $[0, t]$, say.

The period is divided up into n equal intervals; as we make the intervals smaller (weeks, days, seconds, ...), the number n becomes larger. We assume that the intervals are so small that the chance of two or more strikes in the same interval is negligible.

Furthermore meteorites take no account of our calendar, so it is reasonable to suppose that strikes in different intervals are independent, and that the chance of a strike is the same for each of the n intervals, p say. (A more advanced model would take into account the fact that meteorites sometimes arrive in showers.) Thus the total number of strikes in the n intervals is the same as the number of successes in n Bernoulli trials, with distribution

$$p(k) = \binom{n}{k} p^k (1 - p)^{n-k}, \quad 0 \leqslant k \leqslant n,$$

which is binomial. These assumptions are in fact well supported by observation.

Now obviously p depends on the size of the interval; there must be more chance of a strike during a month than during a second. Also it seems reasonable that if p is the chance of a strike in one minute, then the chance of a strike in two minutes should be about $2p$, and so on. This amounts to the assumption that np/t is a constant, which we call λ. So

$$np = \lambda t.$$

Thus as we increase n and decrease p so that λt is fixed, we have exactly the situation considered in example 4.7.2, with λ replaced by λt. Hence, as $n \to \infty$,

$$P(k \text{ strikes in } [0, t]) \to \frac{e^{-\lambda t}(\lambda t)^k}{k!},$$

the Poisson distribution of (4) in section 4.7.

The important point about the above derivation is that it is generally applicable to many other similar circumstances. Thus, for example, we could replace 'meteorites' by 'currants' and 'the interval $[0, t]$' by 'the cake'; the 'n divisions of the interval' then become the 'n slices of the cake', and we find that a fruit cake made from a large batch of well-mixed dough will contain a number of currants with a Poisson distribution, approximately.

The same argument has yielded approximate Poisson distributions observed for flying bomb hits on London in 1939–45, soldiers disabled by horse-kicks in the Prussian Cavalry, accidents along a stretch of road, and so on. In general, rare events that occur independently but consistently in some region of time or space, or both, will often follow a Poisson distribution. For this reason it is sometimes called the law of rare events.

Notice that we have to count events that are isolated, that is to say occur singly, because we have assumed that only one event is possible in a short enough interval. Therefore we do not expect the number of *people* involved in accidents at a junction to have a simple Poisson distribution, because there may be several in each vehicle. Likewise the number of daisy flowers in your lawn may not be Poisson, because each plant has a cluster of flowers. And the number of bacteria on a Petri dish may not be Poisson, because the separate colonies form tightly packed groups. The colonies, however, may well have an approximately Poisson distribution.

We have come a long way from the hypergeometric distribution but, surprisingly, we can go further still. It will turn out that for large values of the parameter λ, the Poisson distribution can be usefully approximated by an even more important distribution, the normal distribution. But this lies some way ahead.

Exercises for section 4.8

1. A cook adds 200 chocolate chips to a batch of dough, and makes 40 biscuits. What is the approximate value of the probability that a random biscuit has
   (a) at least 4 chips?
   (b) no chips?

2. A jumbo jet carries 400 passengers. Any passenger independently fails to show up with probability 10^{-2}. If the airline makes 404 reservations, what is the probability that it has to bump at least one passenger?

3. Find the mode of the Poisson distribution
$$p(x) = \lambda^x e^{-\lambda}/x!, \quad x \geq 0.$$
   Is it always unique?

4.9 CONTINUOUS APPROXIMATIONS

We have seen above that, in many practical situations, complicated and unwieldy distributions can be usefully replaced by simpler approximations; for example, sometimes the hypergeometric distribution can be approximated by a binomial distribution; this in turn can sometimes approximated by a Poisson distribution. We are now going to extend this idea even further.

First of all consider an easy example. Let X be the number shown by a fair n-faced die. Thus X has a uniform distribution on $\{1, \ldots, n\}$, and its distribution function is shown in Figure 4.8, for some large unspecified value of n.

Now if you were considering this distribution for large values of n, and sketched it many times everyday, you would in general be content with the picture in Figure 4.9.

The line

(1) $$y = \frac{x}{n}, \quad 0 \leq x \leq n$$

is a very good approximation to the function

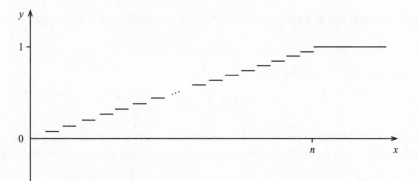

Figure 4.8. The uniform distribution function $y = P(X \leqslant x) = [x]/n$.

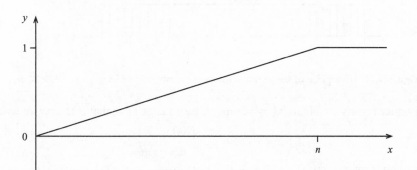

Figure 4.9. The line $y = x/n$ provides a reasonable approximation to figure 4.8.

(2)
$$F(x) = \frac{[x]}{n},$$

(recall that $[x]$ means 'the integer part of x') as is obvious from the figures. Indeed, for all x

$$|y - F(x)| = \left| \frac{x}{n} - \frac{[x]}{n} \right| \leqslant \frac{1}{n},$$

so if we use the approximation to calculate probabilities, we find that we have

$$P(a < X \leqslant b) \simeq \frac{b}{n} - \frac{a}{n},$$

where the exact result is

$$P(a < X \leqslant b) = \frac{[b]}{n} - \frac{[a]}{n}.$$

The difference between the exact and approximate answers is always less than $2/n$, which may be negligible for large n. The function $y = x/n$ is an excellent continuous approximation to $F(x)$ for large n.

We can also consider a natural continuous approximation to the actual discrete distribution

$$p(k) = \frac{1}{n}, \quad 1 \leqslant k \leqslant n,$$

when $p(k)$ is displayed as a histogram; see figure 4.10. Clearly the constant function

$$y' = \frac{1}{n}, \quad 0 \leqslant x \leqslant n,$$

fits $p(k)$ exactly. Now remember that

$$F(x) = \sum_{k \leqslant x} p(k),$$

so that $F(x)$ is just the area under the histogram to the left of x. It is therefore important to note also that $y = x/n$ is the area under $y' = 1/n$ to the left of x, as we would wish.

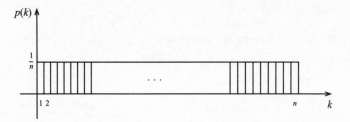

Figure 4.10. Histogram of the uniform discrete distribution $p(k) = 1/n$, $1 \leqslant k \leqslant n$.

Even neater results are obtained by scaling X by a factor of n; that is to say, we write

$$P\left(a < \frac{X}{n} \leqslant b\right) = \frac{[nb]}{n} - \frac{[na]}{n}$$

exactly, and then define $U(x) = x$ and $u(x) = 1$, for $0 \leqslant x \leqslant 1$. Then for large n

(3) $$P\left(a < \frac{X}{n} \leqslant b\right) \simeq b - a$$

$$= U(b) - U(a).$$

In particular note that when h is small this gives

$$P\left(x < \frac{X}{n} \leqslant x + h\right) \simeq h = u(x)h.$$

The uniform distribution is so simply dealt with as to be almost dull. Next let us consider a much more important and interesting case. The geometric distribution

(4) $$p(k) = p(1 - p)^{k-1}, \quad k \geqslant 1,$$

arose when we looked at the waiting time X for a success in a sequence of Bernoulli trials. Once again we consider two figures. Figure 4.11 shows the distribution function

$$P(X \leqslant x) = F(x) = \sum_{k=1}^{x} p(k) = 1 - (1 - p)^x, \quad x \geqslant 1,$$

for a reasonably small value of p. Figure 4.12 shows what you would be content to sketch in general, to gain a good idea of how the distribution behaves. We denote this curve by $E(x)$.

It is not quite so obvious this time what $E(x)$ actually is, so we make a simple calculation.

Let $p = \lambda/n$, where λ is fixed and n may be as large as we please. Now for any fixed x

we can proceed as follows:

(5)
$$P(X > nx) = P(X > [nx])$$

$$= 1 - F([nx]) = (1 - p)^{[nx]}$$

$$= \left(1 - \frac{\lambda}{n}\right)^{[nx]}$$

$$\simeq e^{-\lambda x}$$

for large values of n, corresponding to small values of p.

Thus in this case the function

(6)
$$E(x) = 1 - e^{-\lambda x}, \quad x \geqslant 0,$$

provides a good fit to the discrete distribution

(7)
$$F([nx]) = 1 - \left(1 - \frac{\lambda}{n}\right)^{[nx]}.$$

Once again we can use this to calculate good simple approximations to probabilities.

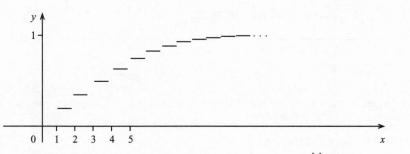

Figure 4.11. The geometric distribution function $y(x) = P(X \leqslant x) = \sum_{k=1}^{[x]} p(k) = 1 - (1 - p)^{[x]}$.

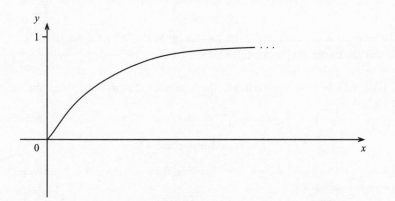

Figure 4.12. The function $y = E(x)$ provides a reasonable approximation to figure 4.11.

Thus

(8)
$$P\left(a < \frac{X}{n} \leqslant b\right) = (1 - p)^{[na]} - (1 - p)^{[nb]}$$

$$= \left(1 - \frac{\lambda}{n}\right)^{[na]} - \left(1 - \frac{\lambda}{n}\right)^{[nb]}$$

$$\simeq e^{-\lambda a} - e^{-\lambda b}$$

$$= E(b) - E(a).$$

It can be shown that, for some constant c,

$$\left|\left(1 - \frac{\lambda}{n}\right)^{[na]} - e^{-\lambda a}\right| \leqslant \frac{c}{n}$$

so this approximation is not only simple, it is close to the correct expression for large n.

Just as for the uniform distribution, we can obtain a natural continuous approximation to the actual discrete distribution (3), when expressed as a histogram.

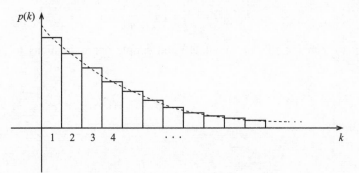

Figure 4.13. Histogram of the geometric distribution $p(k) = q^{k-1} p$, together with the continuous approximation $y = \lambda e^{-\lambda x}$ (broken line).

From (8) we have, for small h,

$$P\left(a < \frac{X}{n} \leqslant a + h\right) \simeq e^{-\lambda a} - e^{-\lambda(a-h)} \simeq \lambda e^{-\lambda a} h.$$

Thus, as Figure 4.13 and (8) suggest, the distribution (3) is well fitted by the curve

$$e(x) = \lambda e^{-\lambda x}.$$

Again, just as $F(x)$ is the area under the histogram to the left of x, so also does $E(x)$ give the area under the curve $e(x)$ to the left of x.

These results, though interesting, are supplied mainly as an introduction to our principal task, which is to approximate the binomial distribution. That we do next, in section 4.10.

Exercise for section 4.9

1. You roll two fair dice each having n sides. Let X be the absolute value of the difference between their two scores. Show that

$$p(k) = P(X = k) = \frac{2(n - k)}{n^2}, \quad 1 \leqslant k \leqslant n.$$

Find functions $T(x)$ and $t(x)$ such that for large n

$$P\left(a < \frac{X}{n} \leqslant b\right) \simeq T(b) - T(a),$$

and for small h

$$P\left(x < \frac{X}{n} \leqslant x + h\right) \simeq t(x)h.$$

4.10 BINOMIAL DISTRIBUTIONS AND THE NORMAL APPROXIMATION

Let us summarize what we did in section 4.9. If X is uniform on $\{1, 2, \ldots, n\}$, then we have functions $U(x)$ and $u(x)$ such that for large n

$$P\left(\frac{X}{n} \leqslant x\right) \simeq U(x) = x$$

and for small h

$$P\left(x < \frac{X}{n} \leqslant x + h\right) \simeq u(x)h = h, \quad 0 \leqslant x \leqslant 1.$$

Likewise if X is geometric with parameter p, then we have functions $E(x)$ and $e(x)$ such that for large n, and $np = \lambda$,

$$P\left(\frac{X}{n} \leqslant x\right) \simeq E(x) = 1 - e^{-\lambda x},$$

and for small h

$$P\left(x < \frac{X}{n} \leqslant x + h\right) \simeq e(x)h = \lambda e^{-\lambda x}h, \quad x > 0.$$

Of course the uniform and geometric distributions are not very complicated, so this seems like hard work for little reward. The rewards come, though, when we apply the same ideas to the binomial distribution

(1) $$P(X = k) = \binom{n}{k} p^k q^{n-k}, \quad p + q = 1,$$

with mean $\mu = np$ and variance $\sigma^2 = npq$ (you showed this in exercise 2 of section 4.6).

In fact we shall see that, when X has the binomial distribution $B(n, p)$ (see (1) of section 4.4), there are functions $\Phi(x)$ and $\phi(x)$ such that for large n

(2) $$P\left(\frac{X - \mu}{\sigma} \leqslant x\right) \simeq \Phi(x)$$

and for small h

(3) $$P\left(x < \frac{X - \mu}{\sigma} \leqslant x + h\right) \simeq \phi(x)h$$

where

(4) $$\phi(x) = (2\pi)^{-1/2} e^{-x^2/2}, \quad -\infty < x < \infty.$$

That is to say, the functions $\Phi(x)$ and $\phi(x)$ play the same role for the binomial distribution as $U(x)$, $u(x)$, $E(x)$, and $e(x)$ did for the uniform and geometric distributions respectively. And $\Phi(x)$ gives the area under the curve $\phi(x)$ to the left of x; see figure 4.14.

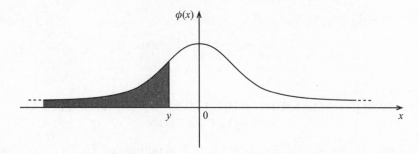

Figure 4.14. The normal function $\phi(x) = (2\pi)^{-\frac{1}{2}}\exp\left(-\frac{1}{2}x^2\right)$. The shaded area is $\Phi(y) = \int_{-\infty}^{y} \phi(x)\,dx$. Note that $\phi(-x) = \phi(x)$ and $\Phi(-x) = 1 - \Phi(x)$.

This is one of the most remarkable and important results in the theory of probability; it was first shown by de Moivre. A natural question is, why is the result so important that de Moivre expended much effort proving it, when so many easier problems could have occupied him?

The most obvious motivating problem is typified by the following. Suppose you perform 10^6 Bernoulli trials with $P(S) = \frac{1}{2}$, and for some reason you want to know the probability α that the number of successes lies between $a = 500\,000$ and $b = 501\,000$. This probability is given by

$$(5) \qquad\qquad \alpha = \sum_{k=a}^{b} \binom{10^6}{k} 2^{-10^6}.$$

Calculating α is a very unattractive prospect indeed; it is natural to ask if there is any hope for a useful approximation. Now, a glance at the binomial diagrams in section 4.4 shows that there is some hope. As n increases, the binomial histograms are beginning to get closer and closer to a bell-shaped curve. To a good approximation, therefore, we might hope that adding up the huge number of small but horrible terms in (5) could be replaced by finding the appropriate area under this bell-shaped curve; if the equation of the curve were not too difficult, this might be an easier task. It turns out that our hope is justified, and there is such a function. The bell-shaped curve is the very well-known function

$$(6) \qquad\qquad f(x) = \frac{1}{(2\pi)^{1/2}\sigma}\exp\left\{-\frac{1}{2}\left(\frac{x-\mu}{\sigma}\right)^2\right\}.$$

This was first realized and proved by de Moivre in 1733. He did not state his results in this form, but his conclusions are equivalent to the following celebrated theorem.

Normal approximation to the binomial distribution.   Let the number of successes in n Bernoulli trials be X. Thus X has a binomial distribution with mean $\mu = np$ and variance $\sigma^2 = npq$, where $p + q = 1$, as usual. Then there are functions $\Phi(x)$ and $\phi(x)$, where $\phi(x)$ is given in (4), such that, for large n,

Table 4.1. *The normal functions $\phi(x)$ and $\Phi(x)$*

$\Phi(x)$	0.500	0.691	0.841	0.933	0.977	0.994	0.999	0.9998
$\phi(x)$	0.399	0.352	0.242	0.13	0.054	0.018	0.004	0.0009
x	0	0.5	1	1.5	2	2.5	3	3.5

(7)
$$P(a < X \leq b) \simeq \Phi\left(\frac{b-\mu}{\sigma}\right) - \Phi\left(\frac{a-\mu}{\sigma}\right)$$

and

(8)
$$P(a < X \leq a+1) \simeq \frac{1}{\sigma}\phi\left(\frac{a-\mu}{\sigma}\right).$$

Alternatively, as we did in (4.9), we can scale X and write (7) and (8) in the equivalent forms

(9)
$$P\left(a < \frac{X-\mu}{\sigma} \leq b\right) \simeq \Phi(b) - \Phi(a)$$

and, for small h,

(10)
$$P\left(a < \frac{X-\mu}{\sigma} \leq a+h\right) \simeq h\phi(a).$$

As in our previous examples, $\Phi(x)$ supplies the area under $\phi(x)$, to the left of x; there are tables of this function in many books on probability and statistics (and elsewhere), so that we can use the theorem in practical applications. Table 4.1 lists $\Phi(x)$ and $\phi(x)$ for some half-integer values of x.

We give a sketch proof of de Moivre's theorem later on, in section 4.15; for the moment let us concentrate on showing how useful it is. For example, consider the expression (5) above for α. By (7), to use the normal approximation we need to calculate

$$\mu = np = 500\,000$$

and

$$\sigma = (npq)^{1/2} = 500.$$

Now de Moivre's theorem says that, approximately,

(11)
$$\alpha = \Phi(2) - \Phi(0)$$

$$\simeq 0.997 - 0.5$$

$$= 0.497,$$

from Table 4.1. This is so remarkably easy that you might suspect a catch; however, there is no catch, this is indeed our answer. The natural question is, how good is the approximation? We answer this by comparing the exact and approximate results in a number of cases. From the discussion above, it is obvious already that the approximation should be good for large enough n, for it is in this case that the binomial histograms can be best fitted by a smooth curve.

For example, suppose $n = 100$ and $p = \frac{1}{2}$, so that $\mu = 50$ and $\sigma = 5$. Then

$$p(55) = P(X = 55) = \binom{100}{55} 2^{-100} \simeq 0.0485$$

and

$$p(50) = P(X = 50) = \binom{100}{50} 2^{-100} \simeq 0.0796.$$

The normal approximation given by (8) yields for the first

$$p(55) \simeq \frac{1}{\sigma} \phi\left(\frac{55 - \mu}{\sigma}\right) = \frac{1}{5}\phi(1) \simeq 0.0485$$

and for the second

$$p(50) \simeq \frac{1}{\sigma} \phi\left(\frac{50 - \mu}{\sigma}\right) = \frac{1}{5}\phi(0) \simeq 0.0798.$$

This seems very satisfactory. However, for smaller values of n we cannot expect so much; for example, suppose $n = 4$ and $p = \frac{1}{2}$, so that $\mu = 2$ and $\sigma = 1$. Then

$$p(3) = \binom{4}{3} 2^{-4} = 0.25 \quad \text{and} \quad p(2) = \binom{4}{2} 2^{-4} = 0.375.$$

The normal approximation (8) now gives for the first

$$p(3) \simeq \frac{1}{\sigma} \phi\left(\frac{3 - \mu}{\sigma}\right) = \phi(1) = 0.242$$

and for the second

$$p(2) \simeq \frac{1}{\sigma} \phi\left(\frac{2 - \mu}{\sigma}\right) = \phi(0) = 0.399.$$

This is not so good, but is still surprisingly accurate for such a small value of n.

We conclude this section by recording an improved and more accurate version of the normal approximation theorem. First, let us ask just how (9) and (10) could be improved? The point lies in the fact that X actually takes discrete values and, when a is an integer,

$$(12) \qquad\qquad P(X = a) = \binom{n}{a} p^a q^{n-a} > 0.$$

However, if we let $a \to b$ in (9), or $h \downarrow 0$ in (10), in both cases the limit is zero.

The problem arises because we have failed to allow for the discontinuous nature of the histogram of the binomial distribution. If we reconsider our estimates using the mid-points of the histogram bars instead of the end-points, then we can improve the approximation in the theorem. This gives the so-called *continuity correction* for the normal approximation.

In its corrected form, de Moivre's result (7) becomes

$$(13) \qquad P(a < X \leqslant b) \simeq \Phi\left(\frac{b + \frac{1}{2} - \mu}{\sigma}\right) - \Phi\left(\frac{a - \frac{1}{2} - \mu}{\sigma}\right)$$

and for the actual distribution we can indeed now allow $a \to b$, to obtain

(14)
$$P(X = a) \simeq \Phi\left(\frac{a + \frac{1}{2} - \mu}{\sigma}\right) - \Phi\left(\frac{a - \frac{1}{2} + \mu}{\sigma}\right)$$

$$\simeq \frac{1}{\sigma}\phi\left(\frac{a - \mu}{\sigma}\right).$$

We omit any detailed proof of this, but it is intuitively clear, if you just remember that $\Phi(x)$ measures the area under $\phi(x)$. (Draw a diagram.)

The result (14) is sometimes called the *local limit theorem*.

One further approximate relationship that is occasionally useful is *Mills' ratio*: for large positive x

(15)
$$1 - \Phi(x) \simeq \frac{1}{x}\phi(x).$$

We offer no proof of this either.

It is intuitively clear from all these results that the normal approximation is *better*

the larger n is,

the nearer p is to q,

the nearer k is to np.

The approximation is *worse*

the smaller n is,

the smaller p (or q) is,

the further k is from np.

It can be shown with much calculation, which we omit, that for $p = \frac{1}{2}$ and $n \geqslant 10$, the error in the approximation (13) is always less than 0.01, when you use the continuity correction. For $n \geqslant 20$, the maximum error is halved again.

If $p \neq \frac{1}{2}$, then larger values of n are required to keep the error small. In fact the worst error is given approximately by the following rough and ready formula when $npq \geqslant 10$:

(16)
$$\text{worst error} \simeq \frac{|p - q|}{10(npq)^{1/2}}.$$

If you do not use the continuity correction the errors may be larger, especially when $|a - b|$ is small.

Here are some examples to illustrate the use of the normal approximation. In each case you should spend a few moments appreciating just how tiresome it would be to answer the question using the binomial distribution as it stands.

Example 4.10.1. In the course of a year a fanatical gambler makes 10 000 fair wagers. (That is, winning and losing are equally likely.) The gambler wins 4850 of these and loses the rest. Was this very unlucky? (Hint: $\Phi(-3) = 0.0013$.)

Solution.   The number X of wins (before the year begins) is binomial, with $\mu = 5000$ and $\sigma = 50$. Now

$$P(X \leqslant 4850) = P\left(\frac{X - 5000}{50} \leqslant -3\right)$$

$$\simeq \Phi(-3)$$

$$\simeq 0.0013.$$

The chance of winning 4850 games or fewer was only 0.0013, so one could regard the actual outcome, losing 4850 games, as unlucky.   ○

Example 4.10.2: rivets.   A large steel plate is fixed with 1000 rivets. Any rivet is flawed with probability 10^{-2}. If the plate contains more than 100 flawed rivets it will spring in heavy seas. What is the probability of this?

Solution.   The number X of flawed rivets is binomial $B(10^3, 10^{-2})$, with $\mu = 10$ and $\sigma^2 = 9.9$. Hence

$$P(X > 100) = 1 - P(X \leqslant 100) \simeq 1 - P\left(\frac{X - 10}{3.2} \leqslant \frac{90}{3.2}\right)$$

$$\simeq 1 - \Phi(28) \simeq \tfrac{1}{28}\phi(28), \text{ by (15)}$$

$$\simeq \tfrac{1}{28}\exp(-392).$$

This number is so small that it can be ignored for all practical purposes. The ship would have rusted to nothing while you were waiting. Its seaworthiness could depend on how many plates like this were used, but we do not investigate further here.   ○

Example 4.10.3: cheat or not?   You suspect that a die is crooked, i.e. that it has been weighted to show a six more often than it should. You decide to roll it 180 times and count the number of sixes. For a fair die the expected number of sixes is $\frac{1}{6} \times 180 = 30$, and you therefore contemplate adopting the following rule. If the number of sixes is between 25 and 35 inclusive then you will accept it as fair. Otherwise you will call it crooked. This is a serious allegation, so naturally you want to know the chance that you will call a fair die crooked. The probability that a fair die will give a result in your 'crooked' region is

$$p(c) = 1 - \sum_{k=25}^{35} \binom{180}{k}\left(\frac{1}{6}\right)^k\left(\frac{5}{6}\right)^{180-k}.$$

Calculating this is fairly intimidating. However, the normal approximation easily and quickly gives

$$p(c) = 1 - P\left(-1 \leqslant \frac{X - 30}{5} \leqslant 1\right) \simeq \Phi(1) + \Phi(-1)$$

$$\simeq 0.32.$$

This value is rather greater than you would like: there is a chance of about a third that you accuse an honest player of cheating. You therefore decide to weaken the test, and accept that the die is honest if the number of sixes in 180 rolls lies between 20 and 40.

The normal approximation now tells you that the chance of calling the die crooked when it is actually fair is

$$1 - P\left(-2 \leqslant \frac{X - 30}{5} \leqslant 2\right) \simeq 1 - \Phi(2) + \Phi(-2)$$

$$\simeq 0.04.$$

Whether this is a safe level for false accusations depends on whose die it is. ○

Example 4.10.4: airline overbooking. Acme Airways has discovered by long experience that there is a $\frac{1}{10}$ chance that any passenger with a reservation fails to show up for the flight. If AA accepts 441 reservations for a 420 seat flight, what is the probability that they will need to bump at least one passenger?

Solution. We assume that passengers show up or not independently of each other. The number that shows up is binomial $B\left(441, \frac{9}{10}\right)$, and we want the probability that this number exceeds 420. The normal approximation to the binomial shows that this probability is very close to

$$1 - \Phi\left(\frac{420 - \frac{1}{2} - 396.9}{(441 \times \frac{1}{10} \times \frac{9}{10})^{1/2}}\right) \simeq 1 - \Phi(0.36)$$

$$\simeq 1 - 0.64$$

$$\simeq 0.36. \qquad ○$$

Exercises for section 4.10

1. Let X be binomial with parameters 16 and $\frac{1}{2}$, that is, $B(16, \frac{1}{2})$. Compare the normal approximation with the true value of the distribution for $P(X = 12)$ and $P(X = 14)$. (Note: $\phi(2) \simeq 0.054$ and $\phi(3) \simeq 0.0044$.)

2. Show that the mode (= most likely value) of a binomial distribution $B(n, p)$ has probability given approximately by
$$p(m) = (2\pi npq)^{-1/2}.$$

3. ***Trials.*** A new drug is given to 1600 patients and a rival old drug is given to 1600 matched controls. Let X be the number of pairs in which the new drug performs better than the old (so that it performs worse in $1600 - X$ pairs; ties are impossible). As it happens, they are equally effective, so the chance of performing better in each pair is $\frac{1}{2}$. Find the probability that it does better in at least 880 of the pairs. What do you think the experimenters would conclude if they got this result?

4.11 DENSITY

We have devoted much attention to discrete probability distributions, particularly those with integer outcomes. But, as we have remarked above, many experiments have outcomes that may be anywhere on some interval of the line; a rope may break anywhere, a meteorite may strike at any time. How do we deal with such cases? The answer is suggested by the previous sections, in which we approximated probabilities by expressing them as areas under some curve. And this idea was mentioned even earlier in example 4.2.4, in which we pointed out that areas under a curve can represent probabilities.

We therefore make the following definition.

Definition.   Let X denote the outcome of an experiment such that $X \in \mathbb{R}$. If there is a function $f(x)$ such that* for all $a < b$

(1) $$P(a < X \leqslant b) = \int_a^b f(x)\, dx$$

then $f(x)$ is said to be the *density* of X. $\triangle$

We already know of one density.

Example 4.11.1: uniform density.   Suppose a rope of length l under tension is equally likely to fail at any point. Let X be the point at which it *does* fail, supposing one end to be at the origin. Then, for $0 \leqslant a \leqslant b < l$,

$$P(a < X \leqslant b) = (b - a)l^{-1}$$

$$= \int_a^b l^{-1}\, dx.$$

Hence X has density

$$f(x) = l^{-1}, \quad 0 \leqslant x \leqslant l.$$ $\bigcirc$

Remark.   Note that $f(x)$ is not itself a probability; only the area under $f(x)$ can be a probability. This is obvious from the above example, because if the rope is short, and $l < 1$, then $f(x) > 1$. This is not possible for a probability.

When h is small and $f(x)$ is smooth, we can write from (1)

$$P(x < X \leqslant x + h) = \int_x^{x+h} f(x)\, dx$$

$$\simeq hf(x).$$

Thus $hf(x)$ is the approximate probability that X lies within the small interval $(x, x + h)$; this idea replaces discrete probabilities.

Obviously from (1) we have that

(2) $$f(x) \geqslant 0,$$

and

(3) $$\int_{-\infty}^{\infty} f(x)\, dx = 1.$$

Furthermore we have, as in the discrete case, the following

---

*See appendix 4.14 for a discussion of the integral. For now, just read $\int_a^b f(x)\, dx$ as the area under $f(x)$ between a and b.

Key rule for densities.   Let C be any subset of $\mathbb{R}$ such that $P(X \in C)$ exists. Then

$$P(X \in C) = \int_{x \in C} f(x)\, dx.$$

We shall find this very useful later on. For the moment here are some simple examples.

First, from the above definition of density we see that *any* function $f(x)$ satisfying (2) and (3) can be regarded as a density. In particular, and most importantly, we see that the functions used to approximate discrete distributions in sections 4.9 and 4.10 are densities.

Example 4.11.2: exponential density.   This most important density arose as an approximation to the geometric distribution with

(4) $f(x) = \lambda e^{-\lambda x}; \quad x \geq 0, \lambda > 0.$

If we wind a thread onto a spool or bobbin and let X be the position of the first flaw, then in practice it is found that X has approximately the density given by (4). The reasons for this should be clear from section 4.9. $\bigcirc$

Even more important is our next density.

Example 4.11.3: normal density.   This arose as an approximation to the binomial distribution, with

(5) $f(x) = \phi(x) = \dfrac{1}{\sqrt{2\pi}} \exp\left(-\dfrac{x^2}{2}\right).$

More generally the function

$$f(x) = \frac{1}{\sqrt{2\pi\sigma^2}} \exp\left\{ -\frac{1}{2}\left(\frac{x-\mu}{\sigma}\right)^2 \right\}$$

is known as the normal density with parameters μ and σ^2, or $N(\mu, \sigma^2)$ for short. The special case $N(0, 1)$, given by $\phi(x)$ in (5), is called the *standard normal density*. $\bigcirc$

We return to these later. Here is one final complementary example which provides an interpretation of the above remarks.

Example 4.11.4.   Suppose you have a lamina L whose shape is the region lying between $y = 0$ and $y = f(x)$, where $f(x) \geq 0$ and L has area 1. Pick a point P at random in L, with any point equally likely to be chosen. Let X be the x-coordinate of the point P. Then by construction $P(a < X \leq b) = \int_a^b f(x)\, dx$, and so $f(x)$ is the density of X. $\bigcirc$

Exercises for section 4.11

1.   A point P is picked at random within in the unit disc, $x^2 + y^2 \leq 1$. Let X be the x-coordinate of P. Show that the density of X is

$$f(x) = \frac{2}{\pi}(1 - x^2)^{1/2}, \quad -1 \leq x \leq 1.$$

2. Let X have the density

$$f(x) = \begin{cases} a(x+3), & -3 \leqslant x \leqslant 0 \\ 3a(1 - \frac{1}{2}x), & 0 \leqslant x \leqslant 2. \end{cases}$$

What is a? Find $P(|X| > 1)$.

4.12 DISTRIBUTIONS IN THE PLANE

Of course, many experiments have outcomes that are not simply a real number.

Example 4.12.1. You roll two dice. The possible outcomes are the set of ordered pairs $\{(i, j); 1 \leqslant i \leqslant 6, 1 \leqslant j \leqslant 6\}$. ○

Example 4.12.2. Your doctor weighs and measures you. The possible outcomes are of the form (x grams, y millimetres), where x and y are positive and less than 10^6, say. (We assume your doctor's scales and measures round off to whole grams and millimetres respectively.) ○

You can easily think of other examples yourself. The point is that these outcomes are not single numbers, so we cannot usefully identify them with points on a line. But we can usefully identify them with points in the plane, using Cartesian coordinates for example. Just as scalar outcomes yielded distributions on the line, so these outcomes yield distributions on the plane. We give some examples to show what is going on.

The first natural way in which such distributions arise is in the obvious extension of Bernoulli trials to include ties. Thus each trial yields one of

$$\{S, F, T\} \equiv \{\text{success, failure, tie}\}.$$

We shall call these *de Moivre trials*.

Example 4.12.3: trinomial distribution. Suppose n independent de Moivre trials each result in success, failure, or a tie. Let X and Y denote the number of successes and failures respectively. Show that

$$(1) \qquad P(X = x, Y = y) = \frac{n!}{x!y!(n-x-y)!} p^x q^y (1 - p - q)^{n-x-y},$$

where $p = P(S)$ and $q = P(F)$.

Solution. Just as for the binomial distribution of n Bernoulli trials, there are several different ways of showing this. The simplest is to note that, by independence, any sequence of n trials including exactly x successes and y failures has probability $p^x q^y (1 - p - q)^{n-x-y}$, because the remaining $n - x - y$ trials are all ties. Next, by (4) of section 3.3, the number of such sequences is the trinomial coefficient

$$\frac{n!}{x!y!(n-x-y)!},$$

and this proves (1) above. ○

Example 4.12.4: uniform distribution. Suppose you roll two dice, and let X and Y be their respective scores. Then by the independence of the dice

$$P(X = x, \ Y = y) = \tfrac{1}{36}, \quad 0 \leqslant x, \quad y \leqslant 6.$$

This is the uniform distribution on $\{1, 2, 3, 4, 5, 6\}^2$. ○

It should now be clear that, at this simple level, such distributions can be treated using the same ideas and methods as we used for distributions on the line. There is of course a regrettable increase in the complexity of notation and equations, but this is inevitable. All consideration of the more complicated problems that can arise from such distributions is postponed to chapter 6, but we conclude with a brief glance at location and spread.

Given our remarks about probability distributions on the line, it is natural to ask what can be said about the location and spread of distributions in the plane, or in three dimensions. The answer is immediate if we pursue the analogy with mass. Recall that we visualized a discrete probability distribution $p(x)$ on the line as being a unit mass divided up so that a mass $p(x)$ is found at x. Then the mean is just the centre of gravity of this mass distribution, and the variance is its moment of inertia about the mean.

With this in mind it seems natural to regard a distribution in $\mathbb{R}^2$ (or $\mathbb{R}^3$) as being a distribution of masses $p(x, y)$ such that $\sum_{x,y} p(x, y) = 1$. Then the centre of gravity is at $G = (\bar{x}, \bar{y})$ where

$$\bar{x} = \sum_{x,y} x p(x, y) \quad \text{and} \quad \bar{y} = \sum_{x,y} y p(x, y).$$

We define the mean of the distribution $p(j, k)$ to be the point $(\bar{x}, \bar{y})$. By analogy with the spread of mass, the spread of this distribution is indicated by its moments of inertia with respect to the x- and y- axes,

$$\sigma_1^2 = \sum_{x,y} (x - \bar{x})^2 p(x, y)$$

and

$$\sigma_2^2 = \sum_{x,y} (y - \bar{y})^2 p(x, y).$$

Exercises for section 4.12

1. An urn contains three tickets, bearing the numbers 1, 2, and 3 respectively. Two tickets are removed at random, without replacement. Let the numbers they show be X and Y respectively. Find the distribution

$$p(x, y) = P(X = x, \ Y = y), \quad 1 \leqslant x, \ y \leqslant 3.$$

What is the probability that the sum of the two numbers is 3? Find the mean and variance of X and Y.

2. You roll a die, which shows X, and then flip X fair coins, which show Y heads. Find $P(X = x, \ Y = y)$, and hence calculate the mean of Y.

4.13 REVIEW

In this chapter we have looked at the simplest models for random experiments. These give rise to several important probability distributions. We may note in particular the following.

Bernoulli trial

$$P(S) = P(\text{success}) = p = 1 - q$$

$$P(F) = P(\text{failure}) = q.$$

Binomial distribution for n independent Bernoulli trials:

$$p(k) = P(k \text{ successes}) = \binom{n}{k} p^k q^{n-k}, \quad 0 \leqslant k \leqslant n.$$

Geometric distribution for the first success in a sequence of independent Bernoulli trials:

$$p(k) = P(k \text{ trials for 1st success}) = pq^{k-1}, \quad k \geqslant 1.$$

Negative binomial distribution for the number of Bernoulli trials needed to achieve k successes:

$$p(n) = p^k q^{n-k} \binom{n-1}{k-1}, \quad n \geqslant k.$$

Hypergeometric distribution for sampling without replacement:

$$p(k) = \binom{m}{k} \binom{w}{r-k} \Big/ \binom{m+w}{r}, \quad 0 \leqslant k \leqslant r \leqslant m \wedge w.$$

We discussed approximating one distribution by another, showing in particular that the binomial distribution could be a good approximation to the hypergeometric, and that the Poisson could approximate the binomial.

Poisson distribution

$$p(k) = e^{-\lambda} \lambda^k / k!, \quad k \geqslant 0.$$

We introduced the ideas of mean and variance as measures of the location and spread of the probability in a distribution. Table 4.2 shows some important means and variances.

Very importantly, we went on to note that probability distributions could be well approximated by continuous functions, especially as the number of sample points becomes large. We use these approximations in two ways. First, the *local* approximation, which says that if $p(k)$ is a probability distribution, there may be a function $f(x)$ such

Table 4.2. *Means and variances*

Distribution	$p(x)$	Mean	Variance
uniform	$n^{-1}, 1 \leqslant x \leqslant n$	$\frac{1}{2}(n+1)$	$\frac{1}{12}(n^2-1)$
Bernoulli	$p^x(1-p)^{1-x}$, $x \in \{0, 1\}$	p	$p(1-p)$
binomial	$\binom{n}{x} p^x(1-p)^{n-x}$, $0 \leqslant x \leqslant n$	np	$np(1-p)$
geometric	$(1-p)^{x-1}p, x \geqslant 1$	p^{-1}	$(1-p)p^{-2}$
Poisson	$\dfrac{e^{-\lambda}\lambda^x}{x!}, x \geqslant 1$	λ	λ
negative binomial	$\binom{x-1}{k-1} p^k(1-p)^{x-k}$, $x \geqslant k$	kp^{-1}	$k(1-p)p^{-2}$
hypergeometric	$\dfrac{\binom{m}{x}\binom{w}{n-x}}{\binom{m+w}{n}}$, $0 \leqslant x \leqslant n$	$\dfrac{nm}{m+w}$	$\dfrac{nmw(m+w-n)}{(m+w-1)(m+w)^2}$

that $p(k) \simeq f(x)$. Second, the *global* approximation, which follows from the local:

$$\sum_{k=a}^{b} p(k) \simeq \text{the area under } f(x) \text{ between } a \text{ and } b$$

$$= F(b) - F(a) \text{ for some function } F(x).$$

Occasionally it is useful to improve these approximations by making a *continuity correction* to take account of the fact that $f(x)$ is continuous but $p(k)$ is not.

In particular we considered the *normal* approximation to the binomial distribution

$$p(k) = \binom{n}{k} p^k q^{n-k}$$

with mean $\mu = np$ and variance $\sigma^2 = npq$. This approximation is given by

$$p(k) \simeq \frac{1}{\sigma}\phi\left(\frac{k-\mu}{\sigma}\right)$$

where

$$\phi(x) = \frac{1}{(2\pi)^{1/2}} \exp\left(-\tfrac{1}{2}x^2\right)$$

and

$$F(x) = \Phi(x) = \int_{-\infty}^{x} \phi(u)\, du.$$

Using the continuity correction gives

$$\sum_{k=a}^{b} p(k) \simeq \Phi\left(\frac{b+\frac{1}{2}-\mu}{\sigma}\right) - \Phi\left(\frac{a-\frac{1}{2}-\mu}{\sigma}\right).$$

We shall give sketch proofs of the theorems that justify these approximations, especially de Moivre's theorem, in appendix 4.15.

Finally, we considered distributions of probability over the plane and in higher dimensions. In particular we considered sequences of independent de Moivre trials and the multinomial distributions.

4.14 APPENDIX. CALCULUS

Fundamental to calculus is the idea of taking limits of functions. This in turn rests on the idea of convergence.

Convergence.   Let $(x_n; n \geq 1)$ be a sequence of real numbers. Suppose that there is a real number a such that $|x_n - a|$ is always ultimately as small as we please; formally,

$$|x_n - a| < \varepsilon \quad \text{for all } n > n_0,$$

where ε is arbitrarily small and n_0 is finite.   △

In this case the sequence (x_n) is said to converge to the limit a. We write *either*

$$x_n \to a \quad \text{as } n \to \infty,$$

or

$$\lim_{n\to\infty} x_n = a.$$

Now let $f(x)$ be any function defined in some interval (α, β), except possibly at the point $x = a$.

Let (x_n) be a sequence converging to a, such that $x_n \neq a$ for any n. Then $(f(x_n); n \geq 1)$ is also a sequence; it may converge to a limit l.

Limits of functions.   If the sequence $(f(x_n))$ converges to the same limit l for every sequence (x_n) converging to a, $x_n \neq a$, then we say that the limit of $f(x)$ at a is l. We write *either*

$$f(x) \to l \text{ as } x \to a, \quad \text{or } \lim_{x\to a} f(x) = l.$$ △

Suppose now that $f(x)$ is defined in the interval (α, β), and let $\lim_{x\to a} f(x)$ be the limit of $f(x)$ at a. This may or may not be equal to $f(a)$. Accordingly we define:

Continuity.   The function $f(x)$ is continuous in (a, b) if, for all $a \in (\alpha, \beta)$,

$$\lim_{x\to a} f(x) = f(a).$$ △

Now, given a continous function $f(x)$, we are often interested in two principal questions about $f(x)$.

(i)   What is the slope (or gradient) of $f(x)$ at the point $x = a$?
(ii) What is the area under $f(x)$ lying between a and b?

Question (i) is answered by looking at chords of $f(x)$. For any two points a and x, the slope of the chord from $f(a)$ to $f(x)$ is

$$s(x) = \frac{f(x) - f(a)}{x - a}.$$

If $s(x)$ has a limit as $x \to a$, then this is what we regard as the slope of $f(x)$ at a. We call it the *derivative* of $f(x)$, and say that $f(x)$ is *differentiable* at a.

Derivative.   The derivative of $f(x)$ at a is denoted by $f'(a)$, where

$$f'(a) = \lim_{x \to a} \frac{f(x) - f(a)}{x - a}. \qquad\qquad \triangle$$

We also write this as

$$f'(a) = \frac{df}{dx}\bigg|_{x=a}.$$

In this notation $df/dx = df(x)/dx$ is the function of x that takes the value $f'(a)$ when $x = a$.

For question (ii), let $f(a)$ be a function defined on $[a, b]$. Then the area under the curve $f(x)$ in $[a, b]$ is denoted by

$$\int_a^b f(x)\, dx,$$

and is called the integral of $f(x)$ from a to b. In general, areas below the x-axis are counted as negative; for a probability density this case does not arise, because density functions are never negative.

The integral is also defined as a limit, but any general statements would take us too far afield. For well-behaved positive functions you can determine the integral as follows. Plot $f(x)$ on squared graph paper with interval length $1/n$. Let S_n be the number of squares lying entirely between $f(x)$ and the x-axis between a and b. Set

$$I_n = S_n/n^2.$$

Then

$$\lim_{n \to \infty} I_n = \int_a^b f(x)\, dx.$$

The function $f(x)$ is said to be integrable.

Of course we almost never obtain integrals by performing such a limit. We almost always use a method that relies on the following, most important, connexion between differentiation and integration.

Fundamental theorem of calculus

Let $f(x)$ be a continuous function defined on $[a, b]$, and suppose that $f(x)$ is integrable. Define the function $F_a(x)$ by

$$F_a(x) = \int_a^x f(t)\, dt.$$

Then the derivative of $F_a(x)$ is $f(x)$; formally,

$$F_a'(x) = f(x).$$

This may look like sorcery, but actually it is intuitively obvious. The function $F_a'(x)$ is the slope of $F_a(x)$, that is, it measures the rate at which area is appearing under $f(x)$ as x increases. Now just draw a picture of $f(x)$ to see that extra area is obviously appearing at the rate $f(x)$. So $F_a'(x) = f(x)$. We omit any proof.

Summary of elementary properties

(i) $f'(x) = df/dx = f'$. It follows that

If f is constant then $f' = 0$.

$$\frac{d}{dx}(cf + g) = cf' + g'.$$

$$\frac{d}{dx}(fg) = f'g + fg'.$$

$$\frac{d}{dx}f(g) = f'(g)g'.$$

(ii) $F(x) = \int_{-\infty}^{x} f(t)\,dt$. It follows that

If f is constant then $F(b) - F(a) = \int_{a}^{b} f(t)\,dt \propto b - a$.

If $f < g$ then $\int_{a}^{b} f\,dx < \int_{a}^{b} g\,dx$.

$$\int_{a}^{b}(cf + g)\,dx = c\int_{a}^{b} f\,dx + \int_{a}^{b} g\,dx.$$

$$\int_{a}^{b} f'g\,dx + \int_{a}^{b} fg'\,dx = \int_{a}^{b}(fg)'\,dx = f(b)g(b) - f(a)g(a).$$

(iii) $\log x = \int_{1}^{x}(1/y)\,dy$.

Functions of more than one variable

We note briefly that the above ideas can be extended quite routinely to functions of more than one variable. For example, let $f(x, y)$ be a function of x and y. Then

(i) $f(x, y)$ is continuous in x at (a, y) if $\lim_{x \to a} f(x, y) = f(a, y)$.

(ii) $f(x, y)$ is continuous at (a, b) if $\lim_{(x,y) \to (a,b)} f(x, y) = f(a, b)$.

(iii) $f(x, y)$ is differentiable in x at (a, y) if

$$\lim_{x \to a} \frac{f(x, y) - f(a, y)}{x - a} = f_1(a, y)$$

exists. We denote this limit by $\partial f/\partial x$. That is to say $\partial f/\partial x$ is the function of x and y that takes the value $f_1(a, y)$ when $x = a$. Other derivatives, such as $\partial f/\partial y$ and $\partial^2 f/\partial x \partial y$, are defined in exactly the same way.

Finally, we note a small extension of the fundamental theorem of calculus, which is used more often than you might expect:

$$\frac{\partial}{\partial x}\int^{g(x)} f(u, y)\,du = \frac{dg(x)}{dx} f(g(x), y).$$

4.15 APPENDIX. SKETCH PROOF OF THE NORMAL LIMIT THEOREM

In this section we give an outline of a proof that the binomial distribution is well approximated by the bell-shaped curve $\phi(x)$. We will do the symmetric case first, since the algebra is more transparent and enables the reader to appreciate the method unobscured by technical details. To be precise, the symmetric binomial distribution is

$$p(k) = \binom{n}{k} 2^{-n},$$

with mean $\mu = n/2$ and variance $\sigma^2 = n/4$. It is the claim of the normal limit theorem that for large n and moderate $k - \mu$

$$p(k) \simeq \frac{1}{(2\pi)^{1/2}\sigma} \exp\left\{-\frac{1}{2}\frac{(k-\mu)^2}{\sigma^2}\right\}.$$

(It can be shown more precisely that a sufficient condition is that $(k - \mu)^3/n^2$ must be negligibly small as n increases.)

To demonstrate the theorem we need to remember three things.

Stirling's formula $(2\pi)^{-1/2}n! \sim n^{n+1/2}e^{-n}$, which implies that

(1)
$$\binom{n}{n/2} = \frac{n!}{(n/2)!(n/2)!} \simeq \frac{2^n}{(2\pi n)^{1/2}}$$

Arithmetic series

(2)
$$\sum_{k=1}^{n-1} k = \tfrac{1}{2}n(n-1)$$

Exponential approximation

$$e^x = 1 + x + \frac{x^2}{2!} + \cdots,$$

which implies that for small x we can write

(3)
$$\frac{1}{1+x} \simeq 1 - x \simeq e^{-x},$$

and the quadratic and higher terms can be neglected.

Now to sketch the theorem. Assume $k > n/2$ and n even, without much loss of generality. Then

$$p(k) = \frac{n!\,2^{-n}}{k!(n-k)!}$$

$$= \frac{n!\,2^{-n}}{\left(\frac{n}{2}\right)!\left(\frac{n}{2}\right)!}\; \frac{\frac{n}{2}\left(\frac{n}{2}-1\right)\cdots\left\{\frac{n}{2}-\left(k-\frac{n}{2}-1\right)\right\}}{k(k-1)\cdots\left(\frac{n}{2}+1\right)}$$

Now we use Stirling's formula in the first term, and divide top and bottom of the second term by $(n/2)^{k-n/2}$, to give

$$p(k) \simeq \frac{\left(1-\frac{2}{n}\right)\left(1-\frac{4}{n}\right)\cdots\left\{1-\left(k-\frac{n}{2}-1\right)\frac{2}{n}\right\}}{(2\pi)^{1/2}\left(\frac{n}{4}\right)^{1/2}\left(1+\frac{2}{n}\right)\left(1+\frac{4}{n}\right)\cdots\left\{\left(1+k-\frac{n}{2}\right)\frac{2}{n}\right\}}$$

Now we use the exponential approximation, and then sum the arithmetic series as follows. From the above,

$$p(k) \simeq \frac{1}{(2\pi\sigma)^{1/2}} \exp\left\{-\frac{4}{n}-\frac{8}{n}-\cdots-\frac{4}{n}\left(k-\frac{n}{2}-1\right)-\frac{2}{n}\left(k-\frac{n}{2}\right)\right\}$$

$$= \frac{1}{(2\pi\sigma)^{1/2}} \exp\left\{-\frac{2}{n}\left(k-\frac{n}{2}\right)\left(k-\frac{n}{2}-1\right)-\frac{2}{n}\left(k-\frac{n}{2}\right)\right\}$$

$$= \frac{1}{(2\pi\sigma)^{1/2}} \exp\left\{-\frac{1}{2}\left(\frac{k-\mu}{\sigma}\right)^2\right\},$$

as required.

In the asymmetric case, when $p \neq q$, we use exactly the same line of argument, but the algebra becomes a bit more tedious. We have in this case

$$p(k) = \binom{n}{k} p^k q^{n-k}.$$

We shall assume for convenience that $k > np$, and that np and nq are integers. The method can still be forced through when they are not, but with more fiddly details. Remember that

$$\mu = np$$

and

$$\sigma^2 = npq.$$

Recall that we regard terms like $(k - np)^3/n^2$ and $(k - np)/n$, and anything smaller, as negligible as n increases. Off we go:

$$P(k \text{ successes}) = p(k) = \binom{n}{k} p^k q^{n-k}$$

$$= \frac{n! \, p^k q^{n-k}}{(nq)!(np)!} \frac{nq(nq-1)\cdots(nq-k+1+np)}{(np+k-np)(np+k-np-1)\cdots(np+1)}$$

$$\simeq \frac{1}{(2\pi npq)^{1/2}} \frac{\left(1 - \dfrac{1}{nq}\right)\cdots\left(1 - \dfrac{k-1-np}{nq}\right)}{\left(1 + \dfrac{1}{np}\right)\cdots\left(1 + \dfrac{k-np}{np}\right)}, \quad \text{by Stirling}$$

$$\simeq \frac{1}{(2\pi)^{1/2}\sigma} \exp\left\{-\frac{1}{nq} - \frac{1}{np} - \cdots - \frac{k-1-np}{nq} \right.$$
$$\left. - \frac{k-1-np}{np} - \frac{k-np}{np}\right\}, \quad \text{by (3)}$$

$$= \frac{1}{(2\pi)^{1/2}\sigma} \exp\left\{-\frac{1}{npq} - \cdots - \frac{k-np-1}{npq} - \frac{k-np}{np}\right\}$$

$$\simeq \frac{1}{(2\pi)^{1/2}\sigma} \exp\left\{-\frac{(k-np)^2}{2npq}\right\}, \quad \text{ignoring } \frac{(k-np)(p-q)}{npq}\frac{}{2}$$

$$= \frac{1}{(2\pi)^{1/2}\sigma} \exp\left\{-\frac{1}{2}\left(\frac{k-\mu}{\sigma}\right)^2\right\},$$

as required.

4.16 PROBLEMS

1. You roll n dice; all those that show a six are rolled again. Let X be the number of resulting sixes. What is the distribution of X? Find its mean and variance.

2. For what values of c_1 and c_2 are the following two functions probability distributions?
    (a) $p(x) = c_1 x, \quad 1 \leqslant x \leqslant n.$
    (b) $p(x) = c_2/\{x(x+1)\}, \quad x \geqslant 1.$

3. Show that for the Poisson distribution $p(x) = \lambda^x e^{-\lambda}/x!$ the variance is λ. Show also that $\{p(x)\}^2 \geqslant p(x-1)p(x+1).$

4. Let $p(n)$ be the negative binomial distribution

$$p(n) = \binom{n-1}{k-1} p^k q^{n-k}, \quad n \geq k.$$

For what value of n is $p(n)$ largest? Show that $\{p(n)\}^2 \geq p(n-1)p(n+1)$.

5. Show that for the distribution on the positive integers

$$p(x) = \frac{90}{\pi^4} \left(\frac{1}{x^4} \right),$$

we have

$$\{p(x)\}^2 \leq p(x-1)p(x+1).$$

Find a distribution on the positive integers such that $\{p(x)\}^2 = p(x-1)p(x+1)$.

6. You perform a sequence of independent de Moivre trials with $P(S) = p$, $P(F) = q$, $P(T) = r$, where $p + q + r = 1$. Let X be the number of trials up to and including the first trial at which you have recorded at least one success and at least one failure. Find the distribution of X, and its mean.

   Now let Y be the number of trials until you have recorded at least j successes and at least k failures. Find the distribution of Y.

7. A coin shows a head with probability p. It is flipped until it first shows a tail. Let D_n be the event that the number of flips required is divisible by n. Find
   (a) $P(D_2)$, (b) $P(D_r)$, (c) $P(D_r|D_s)$ when r and s are coprime.

8. Two players (Alto and Basso) take turns throwing darts at a bull; their chances of success are α and β respectively at each attempt. They flip a fair coin to decide who goes first. Let X be the number of attempts up to and including the first one to hit the bull. Find the distribution of X, and the probability of the event E that X is even. Let B be the event that Basso is the first to score a bull. Are any of the events B, E, and $\{X = 2n\}$ independent of each other?

9. Let $p(n)$ be the negative binomial distribution,

$$p(n) = \binom{n-1}{k-1} p^k q^{n-k}, \quad n \geq k;$$

let $q \to 0$ and $k \to \infty$ in such a way that $kq = \lambda$ is fixed. Show that

$$p(n+k) \to \lambda^n e^{-\lambda}/n!,$$

the Poisson distribution. Interpret this result.

10. **Tagging.**   A population of n animals has had a number t of its members captured, tagged, and released back into the population. At some later time animals are captured again, without replacement, until the first capture at which m tagged animals have been caught. Let X be the number of captures necessary for this. Show that X has the distribution

$$p(k) = P(X = k) = \frac{t}{n} \binom{t-1}{m-1} \binom{n-t}{k-m} \bigg/ \binom{n-1}{k-1}$$

where $m \leq k \leq n - t + m$.

11. **Runs.**   You flip a coin $h + t$ times, and it shows h heads and t tails. An unbroken sequence of heads, or an unbroken sequence of tails, is called a run. (Thus the outcome *HTTHH* contains 3 runs.) Let X be the number of runs in your sequence. Show that X has the distribution

$$p(x) = P(X = x) = \binom{h-1}{x-1} \binom{t+1}{x} \bigg/ \binom{h+t}{h}.$$

What is the distribution of the number of runs of tails?

12. You roll two fair n-sided dice, each bearing the numbers $\{1, 2, \ldots, n\}$. Let X be the sum of their scores. What is the distribution of X? Find continuous functions $T(x)$ and $t(x)$ such that for large n

$$P\left(\frac{X}{n} \leq x\right) \simeq T(x)$$

and

$$P\left(x < \frac{X}{n} \leq x + h\right) \simeq t(x)h.$$

13. You roll two dice; let X be the score shown by the first die, and let W be the sum of the scores. Find

$$p(x, w) = P(X = x, W = w).$$

14. Consider the standard 6–49 lottery (six numbers are chosen from $\{1, \ldots, 49\}$). Let X be the largest number selected. Show that X has the distribution

$$p(x) = \binom{x-1}{5} \bigg/ \binom{49}{6}, \qquad 6 \leq x \leq 49.$$

What is the distribution of the smallest number selected?

15. When used according to the manufacturer's instructions, a given pesticide is supposed to kill any treated earwig with probability 0.96. If you apply this treatment to 1000 earwigs in your garden, what is the probability that there are more than 20 survivors? (Hint: $\Phi(3.2) \simeq 0.9993$.)

16. Candidates to compete in a quiz show are screened; any candidate passes the screen-test with probability p. Any contestant in the show wins the jackpot with probability t, independently of other competitors. Let X be the number of candidates who apply until one of them wins the jackpot. Find the distribution of X.

17. Find the largest term in the hypergeometric probability distribution, given in (2) of section 4.5. If $m + w = t$, find the value of t for which (2) is largest, when m, r, and k are fixed.

18. You perform n independent de Moivre trials, each with r possible outcomes. Let X_i be the number of trials that yield the ith possible outcome. Prove that

$$P(X_1 = x_1, \ldots, X_r = x_r) = p_1^{x_1} \cdots p_r^{x_r} \frac{n!}{x_1! \cdots x_r!},$$

where p_i is the probability that any given trial yields the ith possible outcome.

19. Consider the standard 6–49 lottery again, and let X be the largest number of the six selected, and Y the smallest number of the six selected.
    (a) Find the distribution $P(X = x, Y = y)$.
    (b) Let Z be the number of balls drawn that have numbers greater than the largest number not drawn. Find the distribution of Z.

20. Two integers are selected at random with replacement from $\{1, 2, \ldots, n\}$. Let X be the absolute difference between them $(X \geq 0)$. Find the probability distribution of X, and its expectation.

21. A coin shows heads with probability p, or tails with probability q. You flip it repeatedly. Let X be the number of flips until at least two heads and at least two tails have appeared. Find the distribution of X, and show that it has expected value $2\{(pq)^{-1} - 1 - pq\}$.

22. Each day a robot manufactures $m + n$ capeks; each capek has probability δ of being defective, independently of the others. A sample of size n (without replacement) is taken from each day's output, and tested $(n \geq 2)$. If two or more capeks are defective, then every one in that day's output is tested and corrected. Otherwise the sample is returned and no action is taken. Let X be the number of defective capeks in the day's output after this procedure. Show that

$$P(X = x) = \frac{\{m + 1 + (n-1)x\}m!}{(m - x + 1)!x!}\,\delta^x(1 - \delta)^{m+n-x}, \quad x \leqslant m + 1.$$

Show that X has expected value

$$\{m + n + \delta m(n - 1)\}\delta(1 - \delta)^{n-1}.$$

23. (a) Let X have a Poisson distribution with parameter λ. Use the Poisson and normal approximations to the binomial distribution to deduce that for large enough λ

$$P\left(\frac{X - \lambda}{\sqrt{\lambda}} \leqslant x\right) \simeq \Phi(x).$$

   (b) In the years 1979–99 in Utopia the average number of deaths per year in traffic accidents is 730. In the year 2000 there are 850 deaths in traffic accidents, and on New Year's Day 2001, there are 5 such deaths, more than twice the daily average for 1979–99.

   The newspaper headlines speak of 'New Year's Day carnage', without mentioning the total figures for the year 2000. Is this rational?

24. **Families.**   A woman is planning her family and considers the following possible schemes.
   (a) Bear children until a girl is born, then stop.
   (b) Bear children until the family first includes children of both sexes, and then stop.
   (c) Bear children until the family first includes two girls and two boys, then stop.
   Assuming that boys and girls are equally likely, and multiple births do not occur, find the mean family size in each case.

25. Three points A, B, and C are chosen independently at random on the perimeter of a circle. Let $p(a)$ be the probability that at least one of the angles of the triangle ABC exceeds $a\pi$. Show that

$$p(a) = \begin{cases} 1 - (3a - 1)^2, & \frac{1}{3} \leqslant a \leqslant \frac{1}{2} \\ 3(1 - a)^2, & \frac{1}{2} \leqslant x \leqslant 1. \end{cases}$$

26. (a) Two players play a game comprising a sequence of points in which the loser of a point serves to the following point. The probability is p that a point is won by the player who serves. Let f_m be the expected number of the first m points that are won by the player who serves first. Show that

$$f_m = pm + (1 - 2p)f_{m-1}.$$

   Find a similar equation for the number that are won by the player who first receives service. Deduce that

$$f_m = \frac{m}{2} - \frac{1 - 2p}{4p}\{1 - (1 - 2p)^m\}.$$

   (b) Now suppose that the winner of a point serves to the following point, things otherwise being as above. Of the first m points, let e_m be the expected number that are won by the player who serves first. Find e_m. Is it larger or smaller than f_m?

Review of Part A, and preview of Part B

I have yet to see a problem, however complicated, which, when you looked at
it in the right way, did not become still more complicated.
P. Anderson, *New Scientist*, 1969

We began by discussing several intuitive and empirical notions of probability, and how
we experience it. Then we defined a mathematical theory of probability using the
framework of experiments, outcomes, and events. This included the ideas of independence and conditioning. Finally, we considered many examples in which the outcomes
were numerical, and this led to the extremely important idea of probability distributions
on the line and in higher dimensions. We also introduced the ideas of mean and variance,
and went on to look at probability density. All this relied essentially on our definition of
probability, which proved extremely effective at tackling these simple problems and
ideas.

Now that we have gained experience and insight at this elementary level, it is time to
turn to more general and perhaps more complicated questions of practical importance.
These often require us to deal with several random quantities together, and in more
technically demanding ways. It is also desirable to have a unified structure, in which
probability distributions and densities can be treated together.

For all these reasons, we now introduce the ideas and methods of random variables,
which greatly aid us in the solution of problems that cannot easily be tackled using the
naive machinery of Part A. This is particularly important, as it enables us to get to grips
with modern probability. Everything in Part A would have been familiar in the 19th
century, and much of it was known to de Moivre in 1750. The idea of a random variable
was finally made precise only in 1933, and this has provided the foundations for all the
development of probability since then. And that growth has been swift and enormous.
Part B provides a first introduction to the wealth of progress in probability in the 20th
century.

Part B

Random Variables

5

Random variables and their distributions

5.1 PREVIEW

It is now clear that for most of the interesting and important problems in probability, the outcomes of the experiment are numerical. And even when this is not so, the outcomes can nevertheless often be represented uniquely by points on the line, or in the plane, or in three or more dimensions. Such representations are called *random variables*. In the preceding chapter we have actually been studying random variables without using that name for them. Now we develop this idea with new notation and background. There are many reasons for this, but the principal justification is that it makes it much easier to solve practical problems, especially when we need to look at the joint behaviour of several quantities arising from some experiment. There are also important theoretical reasons, which appear later.

In this chapter, therefore, we first define random variables, and introduce some new notation that will be extremely helpful and suggestive of new ideas and results. Then we give many examples and explore their connections with ideas we have already met, such as independence, conditioning, and probability distributions. Finally we look at some new tasks that we can perform with these new techniques.

Prerequisites. We shall use some very elementary ideas from calculus; see the appendix to chapter 4.

5.2 INTRODUCTION TO RANDOM VARIABLES

In chapter 4 we looked at experiments in which the outcomes in Ω were numbers; that is to say, $\Omega \subseteq \mathbb{R}$ or, more generally, $\Omega \subseteq \mathbb{R}^n$. This enabled us to develop the useful and attractive properties of probability distributions and densities. Now, experimental outcomes are not always numerical, but we would still like to use the methods and results of chapter 4. Fortunately we can do so, if we just assign a number to any outcome $\omega \in \Omega$ in some natural or convenient way. We denote this number by $X(\omega)$. This procedure simply defines a function on Ω; often there will be more than one such function. In fact it is almost always better to work with such functions than with events in the original sample space.

Of course, the key to our success in chapter 4 was using the probability distribution function

$$F(x) = P(X \leqslant x) = P(B_x),$$

where the event B_x is given by

$$B_x = \{X \leqslant x\} = \{\omega: X(\omega) \leqslant x\}.$$

We therefore make the following definition.

Definition.   A *random variable* X is a real-valued function defined on a sample space Ω, such that B_x as defined above is an event for all x.   △

If ω is an outcome in Ω, then we sometimes write X as $X(\omega)$ to make the function clearer. Looking back to chapter 4, we can now see that there we considered exclusively the special case of random variables for which $X(\omega) = \omega$, $\omega \in \Omega$; this made the analysis particularly simple. The above relation will no longer be the case in general, but at least it has helped us to become familiar with the ideas and methods we now develop. Note that, as in chapter 4, random variables are always denoted by capital letters, such as X, Y, U, W^*, Z_1, and so on. An unspecified numerical value is always denoted by a lower-case letter, such as x, y, u, k, m, n, and so forth.

Remark.   You may well ask, as many students do on first meeting the idea, why we need these new functions. Some of the most important reasons arise in slightly more advanced work, but even at this elementary level you will soon see at least four reasons.

(i)   This approach makes it much easier to deal with two or more random variables;
(ii) dealing with means, variances, and related quantities, is very much simpler when we use random variables;
(iii) this is by far the best machinery for dealing with functions and transformations;
(iv) it unifies and simplifies the notation and treatment for different kinds of random variable.

Here are some simple examples.

Example 5.2.1.   You roll a conventional die, so we can write $\Omega = \{1, 2, 3, 4, 5, 6\}$ as usual. If X is the number shown, then the link between X and Ω is rather obvious:

$$X(j) = j, \quad 1 \leqslant j \leqslant 6.$$

If Y is the number of sixes shown then

$$Y(j) = \begin{cases} 1, & j = 6 \\ 0 & \text{otherwise.} \end{cases}$$ ○

Example 5.2.2.   You flip three coins. As usual

$$\Omega = \{H, T\}^3 = \{HHH, HHT, \ldots, TTT\}.$$

If X is the number of heads, then X takes values in $\{0, 1, 2, 3\}$, and we have

$$X(HTH) = 2, \quad X(TTT) = 0,$$

and so on. If Y is the signed difference between the number of heads and the number of tails then $Y \in \{-3, -2, -1, 0, 1, 2, 3\}$, and $Y(TTH) = -1$, for example.   ○

Example 5.2.3: medical. You go for a check-up. The sample space is rather too large to describe here, but what you are interested in is a collection of numbers comprising your height, weight, and values of whatever other physiological variables your physician measures. ○

Example 5.2.4: breakdowns. You buy a car. Once again, the sample space is large, but you are chiefly interested in the times between breakdowns, and the cost of repairs each time. These are numbers, of course. ○

Example 5.2.5: opinion poll. You ask people whether they approve of the present government. The sample space could be

Ω = {approve strongly, approve, indifferent, disapprove, disapprove strongly}.

You might find it very convenient in analysing your results to represent Ω by the numerical scale

$$S = \{-2, -1, 0, 1, 2\},$$

or if you prefer, you could use the non-negative scale

$$Q = \{0, 1, 2, 3, 4\}.$$

You are then dealing with random variables. ○

In very many examples, obviously, it is natural to consider two or more random variables defined on the same sample space. Furthermore, they may be related to each other in important ways; indeed this is usually the case. The following simple observations are therefore very important.

Corollary to definition of random variable. If X and Y are random variables defined on Ω, then any real-valued function of X and Y, $g(X, Y)$, is also a random variable, if Ω is countable.

Remark. When Ω is not countable, it is possible to define unsavoury functions $g(\cdot)$ such that $g(X)$ is not a random variable. In this text we never meet any of these, but to exclude such cases we may need to add the same condition that we imposed in the definition: that is, that $\{g(X) \leq x\}$ is an event for all x.

Example 5.2.6: craps. You roll two dice, yielding X and Y. You play the game using the combined score $Z = X + Y$, where $2 \leq Z \leq 12$, and Z is a random variable. ○

Example 5.2.7: medical. Your physician may measure your weight and height, yielding the random variables X kilograms and Y metres. It is then customary to find the value V of your *body–mass index*, where

$$V = \frac{X}{Y^2}.$$

It is felt to be desirable that the random variable V should be inside, or not too far outside, the interval [20, 25]. ○

Example 5.2.8: poker.   You are dealt a hand at poker. The sample space comprises

$$\binom{52}{5}$$

possible hands. What you are interested in is the number of pairs, and whether or not you have three of a kind, four of a kind, a flush, and so on. This gives you a short set of numbers telling you how many of these desirable features you have.   ○

Example 5.2.9: election.   In an election, let the number of votes garnered by the ith candidate be X_i. Then in the simple first-past-the-post system the winner is the one with the largest number of votes $Y(= \max_i X_i)$.   ○

Example 5.2.10: coins.   Flip a coin n times. Let X be the number of heads and Y the number of tails. Clearly $X + Y = n$. Now let Z be the remainder on dividing X by 2, i.e.

$$Z = X \text{ modulo } 2.$$

Then Z is a random variable taking values in $\{0, 1\}$. If X is even then $Z = 0$; if X is odd then $Z = 1$.   ○

If we take account of order in flipping coins, we can construct a rich array of interesting random variables with complicated relationships (which we will explore later).

It is important to realize that although all random variables have the above structure, and share many properties, there are significantly different types. The following example shows this.

Example 5.2.11.   You devise an experiment that selects a point P randomly from the interval $[0, 2]$, where any point may be chosen. Then the sample space is $[0, 2]$, or formally

$$\Omega = \{\omega : \omega \in [0, 2]\}.$$

Now define X and Y by

$$X(\omega) = \begin{cases} 0 & \text{if } 0 \leqslant \omega \leqslant 1 \\ 1 & \text{if } 1 < \omega \leqslant 2 \end{cases} \quad \text{and} \quad Y(\omega) = \omega^2.$$

Naturally X and Y are both random variables, as they are both suitable real-valued functions on Ω. But clearly they are very different in kind; X can take one of only two values, and is said to be *discrete*. By contrast Y can take any one of an uncountable number of values in $[0, 4]$; it is said to be *continuous*.   ○

We shall develop the properties of these two kinds of random variable side by side throughout this book. They share many properties, including much of the same notation, but there are some differences, as we shall see.

Furthermore, even within these two classes of random variable there are further subcategories, which it is often useful to distinguish. Here is a short list of some of them.

Constant random variable.   If $X(\omega) = c$ for all ω, where c is a constant, then X is a constant.

Indicator random variable. If X can take only the values 0 or 1, then X is said to be an *indicator*. If we define the event on which $X = 1$,

$$A = \{\omega: X(\omega) = 1\},$$

then X is said to be the indicator of A.

Discrete random variable. If X can take any value in a set D that is countable, then X is said to be discrete. Usually D is some subset of the integers, so we assume in future that any discrete random variable is integer valued unless it is stated otherwise.

Of course, as we saw in example (5.2.11), it is not necessary that Ω be countable, even if X is.

Finally we turn to the most important property of random variables; they all have probability distributions. We show this in the next section, but first recall what we did in chapter 4. In that chapter we were entirely concerned with random variables such that $X(\omega) = \omega$, so it was intuitively obvious that in the discrete case we simply define

$$p(x) = P(X = x) = P(A_x),$$

where A_x is the event that $X = x$, that is, as we now write it,

$$A_x = \{\omega: X(\omega) = x\}.$$

In general things are not so simple as this; we need to be a little more careful in defining the probability distribution of X. Let us sum up what we know so far.

Summary
(i) We have an experiment, a sample space Ω, and associated probabilities given by P. That is, it is the job of the function $P(\cdot)$ to tell us the probability of any event in Ω.
(ii) We have a random variable X defined on Ω. That is, given $\omega \in \Omega$, $X(\omega)$ is some real number, x, say.

Now of course the possible values x of X are more or less likely depending on P and X. What we need is a function to tell us the probability that X takes any value up to x. To find that, we simply define the event

$$B_x = \{\omega: X(\omega) \leq x\}.$$

Then, obviously,

$$P(X \leq x) = P(B_x).$$

This is the reason why (as we claimed above) random variables have probability distributions just like those in chapter 4. We explore the consequences of this in the rest of the chapter.

Exercises for section 5.2

1. Let X be a random variable. Is it true that $X - X = 0$, and $X + X = 2X$? If so, explain why.

2. Let X and Y be random variables. Explain when and why $X + Y$, XY, and $X - Y$ are random variables.

3. Example 5.2.10 continued. Suppose you are flipping a coin that moves you 1 metre east when it shows a head, or 1 metre west when it shows a tail. Describe the random variable W denoting your position after n flips.

4. Give an example in which Ω is uncountable, but the random variable X defined on Ω is discrete.

5.3 DISCRETE RANDOM VARIABLES

Let X be a random variable that takes values in some countable set D. Usually this set is either the integers or some obvious subset of the integers, such as the positive integers. In fact we will take this for granted, unless it is explicitly stated otherwise. In the first part of this book we used the function $P(\cdot)$, which describes how probability is distributed around Ω. Now that we are using random variables, we need a different function to tell us how probability is distributed over the possible values of X.

Definition.   The function $p(x)$ given by

(1) $$p(x) = P(X = x), \quad x \in D,$$

is the *probability distribution* of X. It is also known as the probability distribution function or the probability mass function. (These names may sometimes be abbreviated to p.d.f. or p.m.f.)   △

Remark.   Recall from section 5.2 that $P(X = x)$ denotes $P(A_x)$, where $A_x = \{\omega: X(\omega) = x\}$. Sometimes we use the notation $p_X(x)$, to avoid ambiguity.

Of course $p(x)$ has exactly the same properties as the distributions in chapter 4, namely

(2) $$0 \leq p(x) \leq 1$$

and

(3) $$\sum_{x \in D} p(x) = 1.$$

Here are some simple examples to begin with, several of which are already familiar.

Trivial random variable.   Let X be constant, that is to say $X(\omega) = c$ for all ω. Then
$$p(c) = 1.$$ ○

Indicator random variable.   Let X be an indicator. Then
$$p(1) = p = 1 - p(0).$$ ○

Uniform random variable.   Let X be uniform on $\{1, \ldots, n\}$. Then
$$p(x) = n^{-1}, \quad 1 \leq x \leq n.$$ ○

Triangular random variable.   In this case we have
$$p(x) = \begin{cases} c(n - x), & 0 \leq x \leq n \\ c(n + x), & -n \leq x \leq 0. \end{cases}$$
where $c = n^{-2}$.

We saw an example of a triangular random variable, in a different location, when we looked at the distribution of the sum of the scores of two dice in the game of craps. In that case we found

$$p(x) = 6^{-2}\min\{x-1, 13-x\}, \quad 2 \leqslant x \leqslant 12. \qquad \bigcirc$$

We have remarked several times that the whole point of the distribution $p(x)$ is to tell us how probability is distributed over the possible values of X. The most important demonstration of this follows.

Key rule for the probability distribution. Let X have distribution $p(x)$, and let C be any collection of possible values of X. Then

(4)
$$P(X \in C) = \sum_{x \in C} p(x).$$

This is essentially the same rule as we had in chapter 4, and we prove it in the same way. As usual $A_x = \{\omega: X(\omega) = x\}$. Thus, if $x \neq y$ we have $A_x \cap A_y = \emptyset$. Hence, by the addition rule for probabilities,

$$P(X \in C) = P\left(\bigcup_{x \in C} A_x\right)$$

$$= \sum_{x \in C} P(A_x)$$

$$= \sum_{x \in C} p(x). \qquad \square$$

An important application of the key rule (4) arises when we come to consider functions of random variables. In practice, it is often useful or necessary (or both) to consider such functions, as the following examples show.

Example 5.3.1: indicator function. In many industrial processes it is routine to monitor the levels of damaging or undesirable factors. For example, let X be the number of bacteria in a sample of water taken from a treatment plant. If X exceeds a critical level c, then the process is stopped and the filter beds renewed (say). That is, we define

$$Y(X) = \begin{cases} 1, & X \geqslant c \\ 0, & X < c \end{cases}$$

and $Y(X)$ is the indicator of the event that the process is stopped. From (4) we have

$$P(Y = 1) = P(X \geqslant c)$$

$$= \sum_{x=c}^{\infty} P(X = x). \qquad \bigcirc$$

Example 5.3.2: switch function. Let T denote the temperature in some air-conditioned room. If $T > b$, then the a.c. unit refrigerates; if $T < b$, then the a.c. unit heats. Otherwise it is off. The state of the a.c. unit is therefore given by $S(T)$, where

$$S(T) = \begin{cases} 1 & \text{if } T > b \\ 0 & \text{if } a \leqslant T \leqslant b \\ -1 & \text{if } T < a. \end{cases}$$

Naturally, using (4),

$$P(S = 0) = \sum_{t=a}^{b} P(T = t).$$ ◯

Example 5.3.3.   Let X be the score shown by a fair die. Then
$$P(3X \leqslant 10) = P\left(X \leqslant \tfrac{10}{3}\right) = P(X \leqslant 3) = \tfrac{1}{2};$$
$$P\left(\left|X - \tfrac{7}{2}\right| \leqslant 2\right) = P\left(\left|\tfrac{3}{2} \leqslant X \leqslant \tfrac{11}{2}\right|\right) = \tfrac{2}{3};$$
$$P\left(\left(X - \tfrac{7}{2}\right)^2 \leqslant 2\right) = P\left(\tfrac{7}{2} - \sqrt{2} \leqslant X \leqslant \tfrac{7}{2} + \sqrt{2}\right)$$
$$= P(2.1 \leqslant X \leqslant 4.9) = \tfrac{1}{3}.$$ ◯

These examples all demonstrate applications of a general argument, which runs as follows.

One-to-one functions.   If $Y = g(X)$ is a one-to-one function of X, where X has probability distribution $p(x)$, then for any value y of Y such that $g(x) = y$, we can write
(5) $$P(Y = y) = P(g(X) = y) = P(X = g^{-1}(y))$$

$$= p(g^{-1}(y)).$$

Here is another example.

Example 5.3.4.   Let X be the score shown by a die, and let $Y = 2X$. Then $g(x) = 2x$, so that $g^{-1}(x) = \tfrac{1}{2}y$ and
$$P(Y = 4) = P(X = 2)$$

$$= \tfrac{1}{6},$$

which is obvious anyway, of course.   ◯

If $g(x)$ is not a one-to-one function, then several values of X may give rise to the same value $y = g(x)$. In this case we must sum over all these values to get
(6) $$P(Y = y) = \sum_{x: g(x) = y} P(X = x).$$

We can write this argument out in more detail as follows. Let A_y be the event that $g(X) = y$. That is,
$$A_y = \{\omega: g(X(\omega)) = y\}.$$
Then using the key rule (4) we have

$$P(Y = y) = P(X \in A_y)$$

$$= \sum_{x:\, g(x)=y} p(x),$$

as asserted. Here are some simple examples.

Example 5.3.5. Let X be Poisson with parameter λ, and set
$$Y = X \text{ modulo } 2.$$

Then

$$P(Y = 0) = P(X \text{ is even}) = \sum_{r=0}^{\infty} \frac{e^{-\lambda}\lambda^{2r}}{(2r)!} = e^{-\lambda} \cosh \lambda.$$

Likewise

$$P(Y = 1) = e^{-\lambda} \sinh \lambda. \qquad \bigcirc$$

Example 5.3.6. Let X have a two-sided geometric distribution given by
$$p(x) = \begin{cases} \alpha q^{x-1} p, & x \geq 1 \\ (1-\alpha)q^{-x-1} p, & x \leq -1. \end{cases}$$
Then if $Y = |X|$, it is easy to see that Y is geometric, with $p_Y(y) = q^{y-1} p$. $\qquad \bigcirc$

Finally, we define another function that is very useful in describing the behaviour of random variables. As we have seen in many examples above, we are often interested in quite simple events and properties of X, such as $X \leq x$, or $X > x$. For this reason we introduce the distribution function, as follows.

Definition. Let X have probability distribution $p(x)$. Then the *cumulative distribution function* (or c.d.f.) of X is $F(x)$, where
$$(7) \qquad \qquad F(x) = P(X \leq x)$$

$$= \sum_{y=-\infty}^{x} p(y), \quad \text{by (4).} \qquad \triangle$$

It is usually known simply as the *distribution function* of X. It has a companion.

Definition. If X has distribution function $F(x)$, then
$$(8) \qquad \qquad \overline{F}(x) = 1 - F(x) = P(X > x)$$
is called the *survival function* of X. $\qquad \triangle$

Remark. If we wish to stress the role of X, or avoid ambiguity, we often use the notation $p_X(x)$ and $F_X(x)$ to denote respectively the distribution and the distribution function of X.

It is important to note that knowledge of $F(x)$ also determines $p(x)$, because
$$(9) \qquad \qquad p(x) = P(X \leq x) - P(X \leq x - 1)$$

$$= F(x) - F(x - 1).$$

This is often useful when X is indeed integer valued.

Informally, when $x > 0$ is not too small, $\overline{F}(x)$ and $F(-x)$ are known as the right tail and left tail of X, respectively.

Example 5.3.7: uniform random variable.   Here

$$p(x) = \frac{1}{n}, \quad 1 \leq x \leq n,$$

and

$$F(x) = \sum_{y=1}^{x} \frac{1}{n} = \frac{x}{n}, \quad 0 \leq x \leq n. \qquad \bigcirc$$

Example 5.3.8: geometric random variable.   Here

$$p(x) = q^{x-1}p, \quad x \geq 1,$$

and

$$\overline{F}(x) = \sum_{x+1}^{\infty} q^{y-1}p = q^{x}. \qquad \bigcirc$$

Remark.   If two random variables are the same, then they have the same distribution. That is, if $X(\omega) = Y(\omega)$ for all ω then obviously

$$P(X = x) = P(Y = x).$$

However, the converse is not necessarily true. To see this, flip a fair coin once and let X be the number of heads, and Y the number of tails. Then

$$P(X = 1) = P(Y = 1) = \tfrac{1}{2} = P(X = 0) = P(Y = 0),$$

so X and Y have the same distribution. But

$$X(H) = 1, \ X(T) = 0 \quad \text{and} \quad Y(H) = 0, \ Y(T) = 1.$$

Hence X and Y are *never* equal.

Exercises for section 5.3

1. Find the distribution function $F(x)$ for the triangular random variable.

2. Show that any discrete probability distribution is the probability distribution of some random variable.

3. ***Change of units.***   Let X have distribution $p(x)$, and let $Y = a + bX$, for some a and b. Find the distribution of Y in terms of $p(x)$, and the distribution function of Y in terms of $F_X(x)$, when $b > 0$. What happens if $b \leq 0$?

5.4 CONTINUOUS RANDOM VARIABLES; DENSITY

We discovered in section 5.3 that discrete random variables have discrete distributions, and that any discrete distribution arises from an appropriate discrete random variable. What about random variables that are not discrete? As before, the answer has been foreshadowed in chapter 4.

Let X be a random variable that may take values in an uncountable set C, which is all or part of the real line $\mathbb{R}$. We need a function to tell us how probability is distributed over the possible values of X. It cannot be discrete; we recall the idea of *density*.

Definition. The random variable X is said to be *continuous*, with density $f(x)$, if, for all $a \leq b$,

$$(1) \qquad\qquad P(a \leq X \leq b) = \int_a^b f(x)\,dx. \qquad\qquad \triangle$$

The probability density $f(x)$ is sometimes called the p.d.f. When we need to avoid ambiguity, or stress the role of X, we may use $f_X(x)$ to denote the density.

Of course $f(x)$ has the properties of densities in chapter 4, which we recall as

$$(2) \qquad\qquad f(x) \geq 0$$

and

$$(3) \qquad\qquad \int_{-\infty}^{\infty} f(x)\,dx = 1.$$

We usually specify densities only at those points where $f(x)$ is not zero.

From the definition above it is possible to deduce the following basic identity, which parallels that for discrete random variables, (4) in section 5.3.

Key rule for densities. Let X have density $f(x)$. Then, for $B \subseteq \mathbb{R}$,

$$(4) \qquad\qquad P(X \in B) = \int_{x \in B} f(x)\,dx.$$

Just as in the discrete case, $f(x)$ shows how probability is distributed over the possible values of X. Then the key rule tells us just how likely X is to fall in any subset B of its values (provided of course that $P(X \in B)$ exists).

It is important to remember one basic difference between continuous and discrete random variables: the probability that a continuous random variable takes any particular value x is zero. That is, from (4) we have

$$(5) \qquad\qquad P(X = x) = \int_x^x f(u)\,du = 0.$$

Such densities also arose in chapter 4 as useful approximations to discrete distributions. (Very roughly speaking, the idea is that if probability masses become very small and close together, then for practical purposes we may treat the result as a density.) This led to the continuous uniform density as an approximation to the discrete uniform distribution, and the exponential density as an approximation to the geometric distribution. Most importantly, it led to the normal density as an approximation to the binomial distribution. We can now display these in our new format. Remember that, as we remarked in chapter 4, it is possible that $f(x) > 1$, because $f(x)$ is not a probability. However, informally we can observe that, for small h, $P(x \leq X \leq x + h) \simeq f(x)h$. The probability that X lies in $(x, x + h)$ is approximately $hf(x)$. The smaller h is, the better the approximation.

As in the discrete case, the two properties (2) and (3) characterize all densities; that is to say any nonnegative function $f(x)$, such that the area under $f(x)$ is 1, is a density.

Here are some examples of common densities.

Example 5.4.1: uniform density.   Let X have density

(6)
$$f(x) = \begin{cases} (b-a)^{-1}, & 0 < x < b \\ 0 & \text{otherwise.} \end{cases}$$

Then X is uniform on (a, b); we are very familiar with this density already. In general, $P(X \in B)$ is just $|B|(b-a)^{-1}$, where $|B|$ is the sum of the lengths of the intervals in (a, b) that comprise B.   ○

Example 5.4.2: two-sided exponential density.   Let a, b, α, β be positive, and set

(7)
$$f(x) = \begin{cases} ae^{-\alpha x}, & x > 0, \\ be^{\beta x}, & x < 0, \end{cases}$$

where
$$\frac{a}{\alpha} + \frac{b}{\beta} = 1.$$

Then $f(x)$ is a density, by (3).   ○

Example 5.4.3: an unbounded density.   In contrast to the discrete case, densities not only can exceed 1, they need not even be bounded. Let X have density

(8) $f(x) = \frac{1}{2}x^{-1/2}, \quad 0 < x < 1.$

Then $f(x) > 0$ and, as required,

$$\int_0^1 f(x)\, dx = \left[x^{1/2}\right]_0^1 = 1,$$

but $f(x)$ is not bounded as $x \to 0$.   ○

Our next two examples are perhaps the most important of all densities. Firstly:

Example 5.4.4: normal density.   The standard normal random variable X has density $\phi(x)$, where

(9) $\phi(x) = (2\pi)^{-1/2} \exp\left(-\frac{1}{2}x^2\right), \quad -\infty < x < \infty.$

We met this density as an approximation to a binomial distribution in chapter 4, and we shall meet it again in similar circumstances in chapter 7. For the reasons suggested by that result, it is a distribution that is found empirically in huge areas of science and statistics. It is easy to see that $\phi(x) \geq 0$, but not so easy to see that (3) holds. We postpone the proof of this to chapter 6.   ○

Secondly:

Example 5.4.5: exponential density with parameter λ.   Let X have density function

(10)
$$f(x) = \begin{cases} \lambda e^{-\lambda x}, & x > 0 \\ 0 & \text{otherwise.} \end{cases}$$

Clearly we require $\lambda > 0$, so that we satisfy the requirement that

$$\int_0^\infty f(x)\,dx = 1.$$

By the key rule, for $0 < a < b$,

(11)
$$P(a < X < b) = \int_a^b \lambda e^{-\lambda x}\,dx = e^{-\lambda a} - e^{-\lambda b}. \qquad \bigcirc$$

We met the exponential density in Chapter 4 as an approximation to geometric probability mass functions. These arise in models for waiting times, so it is not surprising that the exponential density is also used as a model in situations where you are waiting for some event which can occur at any nonnegative time. The following example provides some explanation and illustration of this.

Example 5.4.6: the Poisson process and the exponential density.   Recall our derivation of the Poisson distribution. Suppose that events can occur at random anywhere in the interval $[0, t]$, and these events are independent, 'rare', and 'isolated'. We explained the meaning of these terms in section 4.8, in which we also showed that, on these assumptions, the number of events $N(t)$ in $[0, t]$ turns out to have approximately a Poisson distribution,

(12)
$$P(N(t) = n) = \frac{e^{-\lambda t}(\lambda t)^n}{n!}.$$

When t is interpreted as time (or length), then the positions of the events are said to form a *Poisson process*. A natural way to look at this is to start from $t = 0$, and measure the interval X until the first event. Then X is said to be the *waiting time* until the first event, and is a random variable.

Clearly X is greater than t if and only if $N(t) = 0$. From (12) this gives

$$P(X > t) = P(N(t) = 0) = e^{-\lambda t}.$$

Now from (10) we find that, if X is exponential with parameter λ,

(13)
$$P(X > t) = \int_t^\infty \lambda e^{-\lambda u}\,du = e^{-\lambda t}.$$

We see that the waiting time in this Poisson process does have an exponential density. $\bigcirc$

Just as for discrete random variables, one particular probability is so important and useful that it has a special name and notation.

Definition.   Let X have density $f(x)$. Then the *distribution function* $F(x)$ of X is given by

(14)
$$F(x) = \int_{-\infty}^x f(u)\,du = P(X \leqslant x) = P(X < x),$$

since $P(X = x) = 0$. $\triangle$

Sometimes this is called the cumulative distribution function, but not by us. As in the discrete case the survival function is given by

$$(15) \qquad \overline{F}(x) = 1 - F(x) = P(X > x),$$

and we may denote $F(x)$ by $F_X(x)$, to avoid ambiguity. We have seen that the distribution function $F(x)$ is defined in terms of the density $f(x)$ by (14). It is a very important and useful fact that the density can be derived from the distribution function by differentiation:

$$(16) \qquad f(x) = \frac{dF(x)}{dx} = F'(x).$$

This is just the fundamental theorem of calculus, which we discussed in appendix 4.14. This means that in solving problems, we can choose to use either F or f, since one can always be found from the other.

Here are some familiar densities and their distributions.

Uniform distribution. When $f(x) = (b - a)^{-1}$, it is easy to see that

$$(17) \qquad F(x) = \frac{x - a}{b - a}, \quad a \leqslant x \leqslant b.$$

Exponential distribution. When $f(x) = \lambda e^{-\lambda x}$, then

$$(18) \qquad F(x) = 1 - e^{-\lambda x}, \quad x \geqslant 0.$$

Normal distribution. When $f(x) = \phi(x)$, then there is a special notation:

$$(19) \qquad F(x) = \Phi(x) = \int_{-\infty}^{x} \phi(y)\, dy.$$

We have already used $\Phi(x)$ in chapter 4, of course.

Next, observe that it follows immediately from (14), and the properties of $f(x)$, that $F(x)$ satisfies

$$(20) \qquad \lim_{x \to -\infty} F(x) = 0, \quad \lim_{x \to \infty} F(x) = 1,$$

and

$$(21) \qquad F(y) - F(x) \geqslant 0, \quad \text{for } x \leqslant y.$$

These properties characterize distribution functions just as (2) and (3) do for densities. Here are two examples to show how we use them.

Example 5.4.7: Cauchy density. We know that $\tan^{-1}(-\infty) = -\pi/2$, and $\tan^{-1}(\infty) = \pi/2$, and $\tan^{-1}(x)$ is an increasing function. Hence

$$(22) \qquad F(x) = \frac{1}{2} + \frac{1}{\pi} \tan^{-1} x$$

is a distribution function, and differentiating gives

$$(23) \qquad f(x) = \frac{1}{\pi(1 + x^2)},$$

which is known as the Cauchy density.   ○

Example 5.4.8: doubly exponential density.   By inspection we see that

(24) $F(x) = \exp\{-\exp(-x)\}$

satisfies all the conditions for being a distribution, and differentiating gives the density

(25) $f(x) = e^{-x}\exp(-e^{-x}).$ ○

We can use the distribution function to show that the Poisson process, and hence the exponential density, has intimate links with another important family of densities.

Example 5.4.9: the gamma density.   As in example 5.4.6, let $N(t)$ be the number of events of a Poisson process that occur in $[0, t]$. Let Y_r be the time that elapses from $t = 0$ until the moment when the rth event occurs. Now a few moments' thought show that

$$Y_r > t \quad \text{if and only if} \quad N(t) < r.$$

Hence

(26) $1 - F_Y(t) = P(Y_r > t) = P(N(t) < r)$

$$= \sum_{x=0}^{r-1} e^{-\lambda t}(\lambda t)^x/x!, \quad \text{by (12).}$$

Therefore Y_r has density $f_Y(y)$ obtained by differentiating (26):

(27) $f_Y(y) = (\lambda y)^{r-1}\lambda e^{-\lambda y}/(r-1)!, \quad 0 \leqslant y < \infty.$

This is known as the gamma density, with parameters λ and r.   ○

Remark.   You may perhaps be wondering what happened to the sample space Ω and the probability function $P(\cdot)$, which played a big part in early chapters. The point is that, since random variables take real values, we might as well let Ω be the real line $\mathbb{R}$. Then any event A is a subset of $\mathbb{R}$ with length $|A|$, and

$$P(A) = \int_{x \in A} f(x)\,dx.$$

We do not really need to mention Ω again explicitly. However, it is worth noting that this shows that any non-negative function $f(x)$, such that $\int f(x)\,dx = 1$, is the density function of some random variable X.

Exercises for section 5.4

1.   Let X have density function $f(x) = cx$, $0 \leqslant x \leqslant a$. Find c, and the distribution function of X.

2.   Let $f_1(x)$ and $f_2(x)$ be densities, and λ any number such that $0 \leqslant \lambda \leqslant 1$. Show that $\lambda f_1 + (1-\lambda)f_2$ is a density. Is $f_1 f_2$ a density?

3.   Let $F_1(x)$ and $F_2(x)$ be distributions, and $0 \leqslant \lambda \leqslant 1$. Show that $\lambda F_1 + (1-\lambda)F_2$ is a distribution. Is $F_1 F_2$ a distribution?

4.   (a)   Find c when X_1 has the beta density $\beta(\frac{3}{2}, \frac{3}{2})$, $f_1(x) = c\{x(1-x)\}^{1/2}$.
     (b)   Find c when X_2 has the arcsin density, $f_2(x) = c\{x(1-x)\}^{-1/2}$.
     (c)   Find the distribution function of X_2.

5.5 FUNCTIONS OF A CONTINUOUS RANDOM VARIABLE

Just as for discrete random variables, we are often interested in functions of continuous random variables.

Example 5.5.1.   Many measurements have established that if R is the radius of the trunk, at height one metre, of a randomly selected tree in Siberia, then R has a certain density $f(r)$. The cross-sectional area of such a tree at height one metre is then roughly

$$A = \pi R^2.$$

What is the density of A?   ○

This exemplifies the general problem, which is: given random variables X and Y, such that

$$Y = g(X)$$

for some function g, what is the distribution of Y in terms of that of X? In answering this we find that the distribution function appears much more often in dealing with continuous random variables than it did in the discrete case. The reason for this is rather obvious; it is the fact that $P(X = x) = 0$ for random variables with a density. The elementary lines of argument, which served us well for discrete random variables, sometimes fail here for that reason. Nevertheless, the answer is reasonably straightforward, if $g(X)$ is a one-to-one function. Let us consider the simplest example.

Example 5.5.2: scaling and shifting.   Let X have distribution $F(x)$ and density $f(x)$, and suppose that

(1) $$Y = aX + b, \; a > 0.$$

Then, arguing as we did in the discrete case,

(2) $$F_Y(y) = P(Y \leq y)$$

$$= P(aX + b \leq y)$$

$$= P\left(X \leq \frac{y - b}{a}\right)$$

$$= F\left(\frac{y - b}{a}\right).$$

Thus the distribution of Y is just the distribution F of X, when it has been *shifted* a distance b along the axis and *scaled* by a factor a.

The scaling factor becomes even more apparent when we find the density of Y. This is obtained by differentiating $F_Y(y)$, to give

(3) $$f_Y(y) = \frac{d}{dy} F_Y(y) = \frac{d}{dy} F\left(\frac{y - b}{a}\right) = \frac{1}{a} f\left(\frac{y - b}{a}\right).$$

You may wonder why we imposed the condition $a > 0$. Relaxing it shows the reason, as follows. Let $Y = aX + b$ with no constraints on a. Then we note that if $a = 0$ then Y is just a constant b, which is to say that

(4) $$P(Y = b) = 1, \; a = 0.$$

If $a \neq 0$, we must consider its sign. If $a > 0$ then

(5) $$P(aX \leqslant y - b) = P\left(X \leqslant \frac{y-b}{a}\right) = F\left(\frac{y-b}{a}\right).$$

If $a < 0$ then

(6) $$P(aX \leqslant y - b) = P\left(X \geqslant \frac{y-b}{a}\right) = 1 - F\left(\frac{y-b}{a}\right).$$

In each case, when $a \neq 0$ we obtain the density of Y by differentiating $F_Y(y)$ to get

$$f(y) = \frac{1}{a} f_X\left(\frac{y-b}{a}\right), \quad a > 0,$$

or

$$f(y) = -\frac{1}{a} f_X\left(\frac{y-b}{a}\right), \quad a < 0.$$

We can combine these to give

(7) $$f_Y(y) = \frac{1}{|a|} f_X\left(\frac{y-b}{a}\right), \quad a \neq 0. \qquad \bigcirc$$

The general case, when $Y = g(X)$, can be tackled in much the same way. The basic idea is rather obvious; it runs as follows. Because $Y = g(X)$, we have

(8) $$F_Y(y) = P(Y \leqslant y) = P(g(X) \leqslant y).$$

Next, we differentiate to get the density of Y:

(9) $$f_Y(y) = \frac{d}{dy} F_Y(y) = \frac{d}{dy} P(g(X) \leqslant y).$$

Now if we play about with the right-hand side of (9), we should obtain useful expressions for $f_Y(y)$, when $g(\cdot)$ is a friendly function.

We can clarify this slightly hazy general statement by examples.

Example 5.5.3. Let X be uniform on $(0, 1)$, and suppose $Y = -\lambda^{-1} \log X$. Then, following the above prescription, we have for $\lambda \geqslant 0$,

$$F_Y(y) = P(Y \leqslant y) = P(\log X \geqslant -\lambda y)$$

$$= P(X \geqslant e^{-\lambda y}) = 1 - e^{-\lambda y}.$$

Hence

$$f_Y(y) = \frac{d}{dy} F_Y(y) = \lambda e^{-\lambda y},$$

and Y has an exponential density. $\qquad \bigcirc$

Example 5.5.4. Let X have a continuous distribution function $F(x)$, and let

$$Y = F(X).$$

Then, as above,

$$F_Y(y) = P(Y \leqslant y) = P(F(X) \leqslant y)$$

$$= P(X \leqslant F^{-1}(y)),$$

where F^{-1} is the inverse function of F. Hence

$$F_Y(y) = F(F^{-1}(y)) = y,$$

and Y is uniform on $(0, 1)$.

Example 5.5.5: normal densities.   Let X have the standard normal density

$$f(x) = \phi(x) = \frac{1}{(2\pi)^{1/2}} \exp\left(-\frac{x^2}{2}\right),$$

and suppose $Y = \mu + \sigma X$, where $\sigma \neq 0$. Then, by example 5.5.2,

$$F_Y(y) = \begin{cases} \Phi\left(\dfrac{y - \mu}{\sigma}\right), & \sigma > 0, \\[2ex] 1 - \Phi\left(\dfrac{y - \mu}{\sigma}\right), & \sigma < 0. \end{cases}$$

Differentiating shows that Y has density

$$f_Y(y) = \frac{1}{(2\pi\sigma^2)^{1/2}} \exp\left\{-\frac{1}{2}\left(\frac{y - \mu}{\sigma}\right)^2\right\}.$$

We can write this in terms of $\phi(x)$ as

$$f_Y(y) = \frac{1}{|\sigma|} \phi\left(\frac{y - \mu}{\sigma}\right)$$

and this is known as the N(μ, σ^2) density, or the general normal density.

Conversely, of course, if we know Y to be N(μ, σ^2), then the random variable

$$X = \frac{Y - \mu}{\sigma}$$

is a standard normal random variable. This is a very useful little result.

Example 5.5.6: powers.   Let X have density f and distribution F. What is the density of Y, where $Y = X^2$?

Solution.   Here some care is needed, for the function is not one–one. We write, as usual,

$$F_Y(y) = P(X^2 \leq y)$$
$$= P(-\sqrt{y} \leq X \leq \sqrt{y})$$
$$= F(\sqrt{y}) - F(-\sqrt{y})$$

so that Y has density

$$f_Y(y) = \frac{d}{dy} F_Y(y) = \frac{1}{2\sqrt{y}}\{f(\sqrt{y}) + f(-\sqrt{y})\}.$$

Example 5.5.7: continuous to discrete.   Let X have an exponential density, and let $Y = [X]$, where $[X]$ is the integer part of X. What is the distribution of Y?

Solution.   Trivially for any integer n, we have $[x] \geqslant n$ if and only if $x \geqslant n$. Hence

$$P(Y \geqslant n) = e^{-\lambda n}, \quad n \geqslant 0$$

and so

$$P(Y = n) = P(Y \geqslant n) - P(Y \geqslant n + 1)$$

$$= e^{-\lambda n}(1 - e^{-\lambda n}), \quad n \geqslant 0.$$

Thus Y has a geometric distribution.   ○

Exercises for section 5.5

1.   Let X be a standard normal random variable. Find the density of $Y = X^2$.

2.   Let X be uniform in $[0, m]$ with density $f_X(x) = m^{-1}, 0 \leqslant x \leqslant m$. What is the distribution of $Y = [X]$?

3.   Let X have density f and distribution F. What is the density of $Y = X^3$?

4.   Let X have density $f = 6x(1 - x), 0 \leqslant x \leqslant 1$. What is the density of $Y = 1 - X$?

5.6 EXPECTATION

In chapter 4 we introduced the ideas of mean μ and variance σ^2 for a probability distribution. These were suggested as guides to the location and spread of the distribution, respectively. Recall that for a discrete distribution $(p(x); x \in D)$, we defined

$$(1) \qquad\qquad \mu = \sum_x xp(x)$$

and

$$(2) \qquad\qquad \sigma^2 = \sum_x (x - \mu)^2 p(x).$$

Now, since any discrete random variable has such a probability distribution, it follows that we can calculate its mean using (1). This is such an important and useful attribute that we give it a formal definition.

Definition.   Let X be a discrete random variable. Then the *expectation* of X is denoted by $\mathrm{E}X$, where

$$(3) \qquad\qquad \mathrm{E}X = \sum_x x\mathrm{P}(X = x).$$

Note that this is also known as the *expected value* of X, or the *mean* of X, or the *first moment* of X. Note also that we assume that the summation converges absolutely, that is to say, $\sum_x |x| p(x) < \infty$.   △

Now suppose that X is a continuous random variable with density $f(x)$. We remarked in chapter 4 that such a density has a mean value (just as mass distributed as a density has a centre of gravity). We therefore make a second definition.

Definition.   Let X be a continuous random variable with density $f(x)$. Then the *expectation* of X is denoted by EX, where

(4) $$EX = \int_{-\infty}^{\infty} xf(x)\,dx.$$

This is also known as the *mean* or *expected value*. (Just as in the discrete case, it exists if $\int_{-\infty}^{\infty} |x| f(x)\,dx < \infty$; this is known as the condition of absolute convergence.)   △

Remark.   We note that (3) and (4) immediately demonstrate one of the advantages of using the concept of random variables. That is, EX denotes the mean of the distribution of X, *regardless of its type*; (discrete, continuous, or whatever). The use of the expectation symbol unifies these ideas for all categories of random variable.

Now the definition of expectation in the continuous case may seem a little arbitrary, so we expend a brief moment on explanation. Recall that we introduced probability densities originally as continuous approximations to discrete probability distributions. Very roughly speaking, as the distance h between discrete probability masses decreases, so they merge into what is effectively a probability density. Symbolically, as $h \to 0$, we have for $X \in A$

$$P(X \in A) = \sum_{x \in A} f_X(x), \quad \text{where } f_X(x) \text{ is a discrete distribution}$$

$$\to \int_{x \in A} f(x)\,dx, \quad \text{where } f(x) \text{ is a density function.}$$

Likewise we may appreciate that, as $h \to 0$,

$$EX = \sum x f_X(x)$$

$$\to \int xf(x)\,dx.$$

We omit the details that make this argument a rigorous proof; the basic idea is obvious. Let us consider some examples of expectation.

Example 5.6.1: indicators.   If X is an indicator then it takes the value 1 with probability p, or 0 with probability $1 - p$. In line with the above definition then

(5) $$EX = 1 \times p + 0 \times (1 - p) = p.$$ ○

Though simple, this equation is more important than it looks! We recall from chapter 2 that it was precisely this relationship that enabled Pascal to make the first nontrivial calculations in probability. It was a truly remarkable achievement to combine the notions of probability and expectation in this way. He also used the following.

Example 5.6.2: two possible values. Let X take the value a with probability $p(a)$, or b with probability $p(b)$. Of course $p(a) + p(b) = 1$. Then

$$(6) \qquad EX = ap(a) + bp(b).$$

This corresponds to a wager in which you win a with probability $p(a)$, or b with probability $p(b)$, where your stake is included in the value of the payouts. The wager is said to be fair if $EX = 0$. ○

Example 5.6.3: random sample. Suppose n bears weigh $x_1, x_2, \ldots, x_n$ kilograms respectively; we catch one bear and weigh it, with equal probability of catching any. The recorded weight X is uniform on $\{x_1, \ldots, x_n\}$, with distribution

$$p(x_r) = n^{-1}, \quad 1 \leqslant r \leqslant n.$$

Hence

$$EX = n^{-1} \sum_{r=1}^{n} x_r = \bar{x}.$$

The expectation is the population mean. ○

Example 5.6.4. Let X be uniform on the integers $\{1, 2, \ldots, n\}$. Then

$$(7) \qquad EX = n^{-1} \sum_{r=1}^{n} r = \tfrac{1}{2}(n+1).$$ ○

Example 5.6.5: uniform density. Let X be uniform on (a, b). Then

$$(8) \qquad EX = \int_a^b xf(x)\,dx = \int_a^b \frac{x}{b-a}\,dx = \tfrac{1}{2}\left(\frac{b^2 - a^2}{b-a}\right)$$

$$= \tfrac{1}{2}(a+b),$$

which is what you would anticipate intuitively. ○

Example 5.6.6: exponential density. Let X be exponential with parameter λ. Then

$$(9) \qquad EX = \int_0^\infty x\lambda e^{-\lambda x}\,dx$$

$$= \lambda^{-1}.$$ ○

Example 5.6.7: normal density. If X has a standard normal density then

$$EX = \int_{-\infty}^\infty x\phi(x)\,dx$$

$$= 0, \quad \text{by symmetry.}$$ ○

Let us return to consider discrete random variables for a moment. When X is integer valued, and non-negative, the following result is often useful.

Example 5.6.8: tail sum.   When $X \geqslant 0$, and X is integer valued, show that

$$\text{(10)} \qquad EX = \sum_{r=0}^{\infty} P(X > r) = \sum_{r=0}^{\infty} \{1 - F(r)\}.$$

Solution.   By definition

$$\text{(11)} \qquad EX = \sum_{r=1}^{\infty} rp(r) = p(1)$$

$$+ p(2) + p(2)$$
$$+ p(3) + p(3) + p(3)$$
$$\vdots$$

$$= \sum_{r=1}^{\infty} p(r) + \sum_{r=2}^{\infty} p(r) + \sum_{r=3}^{\infty} p(r) + \cdots$$

on summing the columns on the right-hand side of (11). This is just (10), as required. ○

For an application consider this.

Example 5.6.9: geometric mean.   Let X be geometric with parameter p. Then

$$EX = \sum_{r=0}^{\infty} P(X > r) = \sum_{r=0}^{\infty} q^r = p^{-1}. \qquad\qquad ○$$

It is natural to wonder whether some simple expression similar to (10) holds for continuous random variables. Remarkably, the following example shows that it does.

Example 5.6.10: tail integral.   Let the non-negative continuous random variable X have density $f(x)$ and distribution function $F(x)$. Then

$$\text{(12)} \qquad EX = \int_0^{\infty} \{1 - F(x)\} \, dx = \int_0^{\infty} P(X > x) \, dx.$$

In general, for any continuous random variable X

$$\text{(13)} \qquad EX = \int_0^{\infty} P(X > x) \, dx - \int_0^{\infty} P(X < -x) \, dx.$$

The proof is the second part of problem 25 at the end of the chapter. Here we use this result in considering the exponential density.

Example 5.6.11.   If X is exponential with parameter λ then by (12)

$$EX = \int_0^{\infty} e^{-\lambda x} \, dx = \lambda^{-1}. \qquad\qquad ○$$

Note that expected values need not be finite.

Example 5.6.12. Let X have the density

$$f(x) = x^{-2}, \quad x \geq 1$$

so that

$$F(x) = 1 - x^{-1}, \quad x \geq 1.$$

Hence, by (12),

$$EX = \int_0^\infty \{1 - F(x)\}\, dx = \int_1^\infty x^{-1}\, dx = \infty. \qquad \bigcirc$$

Finally in this section, we note that since the mean gives a measure of location, it is natural in certain circumstances to obtain an idea of the probability in the tails of the distribution by scaling with respect to the mean. This is perhaps a bit vague; here is an example to make things more precise. We see more such examples later.

Example 5.6.13. Let X be exponential with parameter λ, so $EX = \lambda^{-1}$. Then

$$P\left(\frac{X}{EX} > t\right) = P(X > tEX) = 1 - F(tEX)$$

$$= \exp(-\lambda t \lambda^{-1}) = e^{-t};$$

note that this does not depend on λ. In particular, for any exponential random variable X,

$$P(X > 2EX) = e^{-2}. \qquad \bigcirc$$

Example 5.6.14: leading batsmen. In any innings a batsman faces a series of balls. At each ball (independently), he is out with probability r, or scores a run with probability p, or scores no run with probability $q = 1 - p - r$. Let his score in any innings be X. Show that his average score is $a = EX = p/r$ and that, for large a, the probability that his score in any innings exceeds twice his average is approximately e^{-2}.

Solution. First we observe that the only relevant balls are those in which the batsman scores, or is out. Thus, by conditional probability,

$$P(\text{scores}|\text{relevant ball}) = \frac{p}{p+r},$$

$$P(\text{out}|\text{relevant ball}) = \frac{r}{p+r}.$$

Thus X is geometric, with parameter $r/(p+r)$, and we know that

$$P(X > n) = \left(\frac{p}{p+r}\right)^{n+1}, \quad n \geq 0$$

and

$$a = EX = \frac{p+r}{r} - 1 = \frac{p}{r},$$

Hence

$$P(X > 2a) = \left(\frac{p}{p+r}\right)^{2a+1} = \left(1 - \frac{1}{1+p/r}\right)^{2a+1}$$

$$= \left(1 - \frac{1}{a+1}\right)^{2a+1}$$

$$\simeq e^{-2} \quad \text{for large } a. \qquad \bigcirc$$

Remark.   This result is due to Hardy and Littlewood (*Math. Gazette*, 1934), who derived it in connexion with the batting statistics of some exceptionally prolific cricketers in that season.

This is a good moment to stress that despite appearing in different definitions, discrete and continuous random variables are very closely related; you may regard them as two varieties of the same species.

Broadly speaking, continuous random variables serve exactly the same purposes as discrete random variables, and behave in the same way. The similarities make themselves apparent immediately since we use the same notation: X for a random variable, EX for its expectation, and so on.

There are some differences in development, and in the way that problems are approached and solved. These differences tend to be technical rather than conceptual, and lie mainly in the fact that probabilities and expectations may need to be calculated by means of integrals in the continuous case. In a sense this is irrelevant to the probabilistic properties of the questions we want to investigate. This is why we choose to treat them together, in order to emphasize the shared ideas rather than the technical differences.

Exercises for section 5.6

1.   Show that if X is triangular on $(0, 1)$ with density $f(x) = 2x$, $0 \leqslant x \leqslant 1$, then E$X = \frac{2}{3}$.

2.   Let X be triangular on the integers $\{1, \dots, n\}$ with distribution

$$p(x) = \frac{2x}{n(n+1)}, \quad 1 \leqslant x \leqslant n.$$

Find EX.

3.   Let X have the gamma density

$$f(x) = \frac{\lambda^r}{(r-1)!} x^{r-1} e^{-\lambda x}, \quad x \geqslant 0.$$

Find EX.

5.7 FUNCTIONS AND MOMENTS

We have now seen many examples (especially in this chapter) demonstrating that we are very often interested in functions of random variables. For example scientists or statisticians, having observed some random variable X, may very well wish to consider a change of location and scale, defining

$$(1) \qquad Y = aX + b.$$

Sometimes the change of scale is not linear; it is quite likely that you have seen, or even used, logarithmic graph paper, and so your interest may be centred on

$$(2) \qquad Z = \log X.$$

Even more frequently, we need to combine two or more random variables to yield functions like

$$U = X + Y, \quad V = \sum_1^n X_i, \quad W = XY,$$

and so on; we postpone consideration of several random variables to the next chapter.

In any case such new random variables have probability distributions, and it is very often necessary to know the expectation in each case. If we proceed directly, we can argue as follows. Let

$$Y = g(X),$$

where X is discrete with distribution $p(x)$. Then Y has distribution

$$(3) \qquad p_Y(y) = \sum_{x:\,g(x)=y} p(x)$$

and by definition

$$(4) \qquad EY = \sum_y y p_Y(y).$$

Likewise if X and Y are continuous, where

$$Y = g(X),$$

then we have supplied methods for finding the density of Y in section 5.5, and hence its expectation.

However, the prospect of performing the two summations in (3) and (4) to find EY, in the discrete case, is not one that we relish. And the procedure outlined when X and Y are continuous is even less attractive. Fortunately these tedious approaches are rendered unnecessary by the following timely, useful, and attractive result.

Theorem: expectation of functions. (i) Let X and Y be discrete, with $Y = g(X)$. Then

$$(5) \qquad EY = \sum_x g(x)P(X = x)$$

$$= \sum_x g(x)p(x).$$

(ii) Let X be continuous with density $f(x)$, and suppose $Y = g(X)$. Then

$$(6) \qquad EY = \int_{\mathbb{R}} g(x)f(x)\,dx.$$

The point of (5) and (6) is that we do not need to find the distribution of Y in order to find its mean.

Remark.   Some labourers in the field of probability make the mistake of assuming that (5) and (6) are the definitions of EY. This is not so. They are unconscious of the fact that EY is actually defined in terms of its own distribution, as (4) states in the discrete case. For this reason the theorem is occasionally known as the *law of the unconscious statistician.*

Remark.   Of course it is *not true* in general that

$$Eg(X) = g(EX).$$

You need to remember this. A simple example is enough to prove it. Let X have distribution

$$p(1) = \tfrac{1}{2} = p(-1),$$

so that $EX = 0$. Then $P(X^2 = 1) = 1$. Hence $0 = (EX)^2 \neq EX^2 = 1$.

Proof of (i).   Consider the right-hand side of (5), and rearrange the sum so as to group all the terms in which $g(x) = y$, for some fixed y. Then for these terms

$$\sum_x g(x)p(x) = \sum_x yp(x) = y \sum_{x:g(x)=y} p(x)$$

$$= yp_Y(y).$$

Now summing over all y, we obtain the definitive expression for EY, as required.

The proof of (ii) is similar in conception but a good deal more tedious in the exposition, so we omit it.   □

The above theorem is one of the most important properties of expectation, and it has a vital corollary.

Corollary: linearity of expectation.   Let X be any random variable and let the random variable Z satisfy

$$Z = g(X) + h(X)$$

for functions g and h. Then

(7) $$EZ = Eg(X) + Eh(X).$$

Proof for discrete case.   By (5),

$$EZ = \sum_x \{g(x) + h(x)\}p(x)$$

$$= \sum_x g(x)p(x) + \sum_x h(x)p(x)$$

$$= Eg(X) + Eh(X), \quad \text{by (5) again.}$$

Proof for continuous case.   The proof uses (6), and proceeds along similar lines, with integrals replacing sums.   □

Corollary: linear transformation. Let $Y = aX + b$. Then by (7),
$$EY = aEX + b.$$
□

Example 5.7.1: dominance. Suppose that $g(x) \leqslant c$, for some constant c, and all x. Then for the discrete random variable X

$$Eg(X) = \sum_x g(x)p(x)$$

$$\leqslant \sum_x cp(x), \quad \text{since } g(x) \leqslant c,$$

$$= c, \quad \text{since } \sum_x p(x) = 1.$$

The same argument works for a continuous random variable, so in either case we have shown that

(8) $$Eg(X) \leqslant c.$$
○

With all these new results, we can look with fresh eyes at the variance, briefly mentioned in section 5.6 and defined by (2) in that section.

Example 5.7.2: variance. Recall that in chapter 4 we defined the variance of a distribution $p(x)$ as

$$\sigma^2 = \sum_x (x - \mu)^2 p(x).$$

Comparison of this expression with (2) of section 5.6 and (5) in this section shows that

(9) $$\sigma^2 = E\{(X - \mu)^2\} = E\{X^2 - 2\mu X + \mu^2\}$$

$$= EX^2 - \mu^2$$

$$= EX^2 - (EX)^2.$$

In this new context we usually denote the variance by var X, and may write var $X = \sigma_X^2$.
○

The important point to notice here, as we did for expectation, is that if we write

(10) $$\sigma_X^2 = EX^2 - (EX)^2$$

then this definition holds for any random variable, whether discrete or continuous. The advantages of using random variables become ever more obvious.

Example 5.7.3: die. Let X be the number shown by rolling a fair die, numbered from 1 to 6. Then

$$EX^2 = \tfrac{1}{6}(1^2 + 2^2 + 3^2 + 4^2 + 5^2 + 6^2) = \tfrac{91}{6}$$

and so

$$\text{var } X = \tfrac{91}{6} - \left(\tfrac{7}{2}\right)^2 = \tfrac{35}{12}.$$
○

Example 5.7.4: change of location and scale.   Let X have mean μ and variance σ^2. Find the mean and variance of Y, where $Y = aX + b$.

Solution.   We have

$$E(aX + b) = aE(X) + b = a\mu + b$$

and

$$E(aX + b)^2 = a^2 EX^2 + 2ab EX + b^2.$$

Hence

$$\text{var } Y = E(aX + b)^2 - \{E(aX + b)\}^2 = a^2 \text{ var } X = a^2\sigma^2. \qquad\bigcirc$$

Here are some more examples and applications.

Example 5.7.5: normal density.   Let X have the standard normal density $\phi(x)$. Then $EX = 0$, and

$$\text{var } X = \int_{-\infty}^{\infty} x^2 \phi(x)\, dx = \left[\frac{xe^{-x^2/2}}{(2\pi)^{1/2}}\right]_{-\infty}^{\infty} + \int_{-\infty}^{\infty} \phi(x)\, dx$$

$$= 1.$$

Now let

$$Y = \mu + \sigma X.$$

We have shown in example 5.6.9 that Y has the $N(\mu, \sigma^2)$ density, namely

$$f_Y(y) = \frac{1}{(2\pi)^{1/2}\sigma} \exp\left\{-\frac{1}{2}\frac{(x-\mu)^2}{\sigma^2}\right\}.$$

By the law of the unconscious statistician (6) we can now calculate

(11) $$EY = E(\mu + \sigma X) = \mu$$

and

(12) $$\text{var } Y = E\{(Y - \mu)^2\} = E(\sigma^2 X^2) = \sigma^2.$$

You could verify this by explicit calculation using the density of Y, if you wished. The fact that the $N(\mu, \sigma^2)$ density has mean μ and variance σ^2 makes the notation even more transparent and reasonable. $\bigcirc$

It is by now obvious that we are very often interested simply in probabilities such as $P(X > x)$, or $P(|X| > x)$. These are simple to ask for, but frequently hard to find, or too complicated to be useful. One important use of expectation is to provide bounds for these probabilities (and many others).

Example 5.7.6: Markov's inequality.   Let X be any random variable; show that

(13) $$P(|X| \geq a) \leq \frac{E|X|}{a}, \quad a > 0.$$

Solution.   Let Y be the indicator of the event that $|X| \geq a$. Then it is always true that

$$aY \leq |X|.$$

Now using (8) yields the required result. $\bigcirc$

Example 5.7.7: Chebyshov's inequality.   Let X be any random variable; show that

$$P(|X| \geq a) \leq \frac{EX^2}{a^2}, \quad a > 0.$$

Solution.   We have

(14) $$P(|X| \geq a) = P(X^2 \geq a^2)$$

$$\leq \frac{E(X^2)}{a^2}$$

on using (13) applied to the random variable X^2. $\bigcirc$

Here is one final example of how we apply (5) and (6).

Example 5.7.8: fashion retailer.   A shop stocks fashionable items; the number that will be demanded by the public before the fashion changes is X, where $X > 0$ has distribution function $F(x)$. Each item sold yields a profit of £a; all those unsold when the fashion changes must be dumped at a loss of £b each. Since X is large, we shall assume that its distribution is well approximated by some density $f(x)$. If the shop stocks c of these items, what should the manager choose c to be, in order that the expected net profit is greatest?

Solution.   The net profit $g(c, X)$ garnered, when c are stocked and the demand is X, is given by

$$g(c, X) = \begin{cases} aX - b(c - X), & X \leq c, \\ ac, & X > c. \end{cases}$$

Hence by (6), we have approximately

$$Eg(c, X) = \int_{-\infty}^{c} \{ax - b(c - x)\} f(x)\, dx + \int_{c}^{\infty} acf(x)\, dx$$

$$= ac + (a + b) \int_{0}^{c} (x - c) f(x)\, dx.$$

The maximum of this function of c is found by equating its first derivative to zero; thus

$$0 = a + (a + b) \frac{d}{dc} \int_{0}^{c} (x - c) f(x)\, dx$$

$$= a - (a + b) \int_{0}^{c} f(x)\, dx$$

$$= a - (a + b) F(c).$$

Hence $F(c) = a/(a + b)$, and the manager should order around c items, where c is the point at which the distribution function F first reaches the level $a/(a + b)$, as x increases. $\bigcirc$

Exercises for section 5.7

1. Let X be binomial $B(n, \frac{1}{2})$, Y be binomial $B(2n, \frac{1}{4})$, and $Z = 2X$. Show that $EX = EY = \frac{1}{2}EZ$, and that $\operatorname{var} X = \frac{2}{3} \operatorname{var} Y = \frac{1}{8} \operatorname{var} Z$.

2. Let $h(x)$ be a non-negative function, and $a > 0$. Show that
$$P(h(X) \geq a) \leq \frac{Eh(X)}{a}.$$

3. Flip a fair coin repeatedly, and let X be the number of flips up to and including the first head.
    (a) Use Chebyshov's inequality to show that
    $$P(|X - 2| \geq 2) \leq \frac{1}{2}.$$
    Now show that $P(|X - 2| \geq 2) = \frac{1}{16}$.
    (b) Use Markov's inequality to show that
    $$P(X \geq 4) \leq \frac{1}{2}.$$
    What is $P(X \geq 4)$ exactly?

4. ***St Petersburg problem.***   You have to determine the fair entry fee for the following game. A fair coin is flipped until it first shows a head; if there have been X tails up to this point then the prize is $Y = \$2^X$. Find EY. Would anyone pay this entry fee to play?

5.8 CONDITIONAL DISTRIBUTIONS

Suppose we are considering some random variable X. Very often we may be told that X obeys some condition, or it may be convenient to impose conditions.

Example 5.8.1.   Let X be the lifetime of the lightbulb illuminating your desk. Suppose you know that it has survived for a time t up to now. What is the distribution of X, given this condition?

Except in very special cases, we must expect that the distribution of X given this condition is different from the unconditional distribution prevailing before you screw it in.   ○

This, and other obvious examples, lead us to define *conditional distributions*, just as similar observations led us to define conditional probability in chapter 2.

Once again it is convenient to deal separately with discrete and continuous random variables; conditional densities appear in section 5.9. Let X be discrete with probability distribution $p(x)$. Recall that there is an event $A_x = \{\omega : X(\omega) = x\}$, so that
$$p(x) = P(X = x) = P(A_x).$$
Now, given that some event B has occurred, we have the usual conditional probability
$$P(A_x|B) = P(A_x \cap B)/P(B).$$
It is therefore an obvious step to make the following

Definition.   The *conditional distribution* of X given B is denoted by $p_{X|B}(x|B)$, where
$$p_{X|B}(x|B) = P(A_x|B)$$
$$= P(X = x|B).$$
Sometimes we simply denote this by $p(x|B)$.   △

Of course, just as with events, it may happen that the conditional distribution of X given B is the same as the unconditional distribution of X. That is, we may have that for all x

(1) $$p(x|B) = p(x).$$

In this case we say that X is independent of B. This is of course consistent with our previous definition of independence, because (1) is equivalent to

$$P(A_x|B) = P(A_x \cap B)/P(B) = P(A_x),$$

which says that A_x and B are independent.

Here are some examples.

Example 5.8.2: key rule. By routine operations we see that

$$P(X \in C|B) = P\left(\bigcup_{x \in C} A_x \middle| B\right) = \sum_{x \in C} P(A_x \cap B)/P(B)$$

$$= \sum_{x \in C} p(x|B).$$

Thus conditional distributions obey the key rule, just like unconditional distributions; compare this with (4) of section 5.3. $\bigcirc$

Example 5.8.3. Let X be geometric with parameter p, and let B be the event that $X > a$. Then for $x > a$

$$P(X = x|B) = P(\{X = x\} \cap B)/P(B)$$

$$= pq^{x-1}/P(B) = pq^{x-1}/q^a$$

$$= pq^{x-a-1}.$$

This is still a geometric distribution! $\bigcirc$

Example 5.8.4. Let U be uniform on $\{1, \ldots, n\}$ and let B be the event that $a < U \leq b$, where $1 < a < b < n$. Then for $a < r \leq b$

$$P(U = r|B) = P(\{U = r\} \cap B)/P(B)$$

$$= \frac{1}{n} \bigg/ \frac{b-a}{n} = \frac{1}{b-a}.$$

This is still a uniform distribution! $\bigcirc$

Example 5.8.5. Let X be Poisson with parameter λ, and let B be the event $X \neq 0$. Then for $x > 0$

$$P(X = x|B) = \frac{e^{-\lambda}\lambda^x}{(1 - e^{-\lambda})x!}.$$

This is *not* a Poisson distribution, but it is still a distribution, for obviously

$$\sum_{x=1}^{\infty} \frac{e^{-\lambda}\lambda^x}{(1-e^{\lambda})x!} = 1.$$

This last result is generally true, and we single it out for special notice.

Lemma.   Conditional distributions are still probability distributions. This is easily seen as follows.

$$\sum_x P(X = x|B) = \sum_x P(\{X = x\} \cap B)/P(B)$$

$$= 1. \qquad \square$$

Since $p(x|B)$ is a distribution, it may have an expected value. Because of the condition that B occurs, it is naturally called *conditional expectation*.

Definition.   For a discrete random variable X and any event B, the *conditional expectation* of X given B is

(2) $$E(X|B) = \sum_x x p_{X|B}(x|B). \qquad \triangle$$

Note.   As usual we require that $E(|X\|B) < \infty$.

This definition turns out to be one of the most powerful and important concepts in probability, but at this stage we can only give a few simple illustrations of its use. Let us calculate a couple of examples.

Example 5.8.6.   You flip a fair coin three times; let X be the number of heads. Find the conditional expectation of X given that at least two heads are shown.

Solution.   Let B be the event that $X \geqslant 2$. Then
$$P(B) = \tfrac{3}{8} + \tfrac{1}{8} = \tfrac{1}{2},$$
and

$$p_{X|B}(2|B) = \frac{P(X = 2)}{P(B)} = \tfrac{3}{4},$$

$$p_{X|B}(3|B) = \frac{P(X = 3)}{P(B)} = \tfrac{1}{4},$$

$$p_{X|B}(x|B) = 0, \quad \text{for } x = 0 \text{ or } x = 1.$$

Hence
$$E(X|B) = 2 p_{X|B}(2|B) + 3 p_{X|B}(3|B)$$
$$= \tfrac{9}{4}.$$

Compare this with the unconditional $EX = \tfrac{3}{2}$.

Example 5.8.7: runs.   A biased coin shows a head with probability p, or a tail with probability $q = 1 - p$; it is flipped repeatedly. A run of heads is any unbroken sequence of heads, either until the first tail or after the last tail, or between any two tails. A run of tails is defined similarly. Find the distribution of the lengths of (i) the first run, and (ii) the second run.

Solution. The first flip is either H or T.
For (i): Let the first run have length X. Then

$$P(X = x|\text{first flip is } H) = p(x|H) = p^{x-1}q,$$

and

$$p(x|T) = q^{x-1}p.$$

Hence

$$p(x) = p(x|H)p + p(x|T)q = p^{x}q + q^{x}p, \quad x \geqslant 1.$$

For (ii): Let the second run have length Y. Then

$$p(y|H) = q^{y-1}p,$$

and

$$p(y|T) = p^{y-1}q.$$

Note that in deriving these, we have used the trivial observation that if the first flip shows heads, then the second run is of tails, and vice versa. Hence

$$p(y) = q^{y-1}p^2 + p^{y-1}q^2, \quad y \geqslant 1.$$

It is interesting that X and Y have different distributions. From the results above we may deduce immediately that

$$E(X|H) = \sum_{x=1}^{\infty} xp^{x-1}q = q^{-1}$$

and

$$E(X|T) = \sum_{x=1}^{\infty} xq^{x-1}p = p^{-1}.$$

Likewise

$$E(Y|H) = p^{-1} \quad \text{and} \quad E(Y|T) = q^{-1}. \qquad \bigcirc$$

A natural question is to ask how conditional expectation and unconditional expectation are related to each other. Let us consider a simple case first. For any discrete random variable X and event B we have

(3) $$p(x) = P(X = x) = P(X = x|B)P(B) + P(X = x|B^c)P(B^c).$$

Now multiply (3) by x and sum over all x, to give

(4) $$EX = \sum_{x} xp(x) = \sum_{x} xp(x|B)P(B) + \sum_{x} xp(x|B^c)P(B^c)$$

$$= E(X|B)P(B) + E(X|B^c)P(B^c).$$

This is a special case of the *partition rule* for conditional expectation. The following general rule is easily proved in the same way as (3).

Partition rule for expectation. Let X be a discrete random variable, and let $(B_r; r \geqslant 1)$ be a partition of Ω, which is to say that $B_j \cap B_k = \varnothing$ for $j \neq k$ and $\bigcup_r B_r = \Omega$. Then

(5) $$EX = \sum_r E(X|B_r)P(B_r).$$

Note that when X is an indicator,

$$X = I(A) = \begin{cases} 1 & \text{if } A \text{ occurs} \\ 0 & \text{otherwise;} \end{cases}$$

then setting $X = I$ in (5) yields

(6) $$P(A) = \sum_r P(A|B_r)P(B_r),$$

the partition rule for events, which we met in section 2.8.

Example 5.8.8: runs revisited. Recall our terminology in example 5.8.7: when you repeatedly flip a biased coin, the length of the first run is X and the length of the second run is Y.

From the results of that example we now see that

$$EX = pE(X|H) + qE(X|T) = \frac{p}{q} + \frac{q}{p},$$

whereas

$$EY = pE(Y|H) + qE(Y|T) = \frac{p}{p} + \frac{q}{q} = 2. \qquad \bigcirc$$

Conditional expectation offers a very neat way of analysing random variables that arise as a result of a sequence of independent actions, the archetype of which is, of course, flipping a coin or coins.

Example 5.8.9: a new way of finding the mean and variance of the geometric distribution. Of course we already know one way of doing this: you sum the appropriate series. The following is a typical application of conditional expectation.

We know that if a biased coin (showing a head with probability p) is flipped repeatedly, then the number of flips X up to and including the first head is geometric with parameter p. Let H and T denote the possible outcomes of the first flip. By the partition rule (4),

(7) $$EX = E(X|H)P(H) + E(X|T)P(T)$$

$$= pE(X|H) + qE(X|T).$$

Let us consider these terms. On the one hand, given H, we have immediately that $X = 1$. Hence

(8) $$E(X|H) = 1.$$

On the other hand, given T, we know that the number of further flips necessary to obtain a head has the same distribution as X. Hence

(9) $$E(X|T) = 1 + EX.$$

Therefore, using (7), (8), and (9),

$$EX = p + q(1 + EX)$$

and this gives $EX = p^{-1}$, which we have previously obtained by evaluating the sum in

$$EX = \sum_{r=1}^{\infty} rq^{r-1}p.$$

Now for the variance. By the partition rule,

(10) $$EX^2 = pE(X^2|H) + qE(X^2|T).$$

Once again, if a head occurs first flip then

$$E(X^2|H) = 1$$

and if a tail occurs first flip then

$$E(X^2|T) = E\{(1+X)^2\}$$

$$= 1 + 2EX + EX^2$$

$$= 1 + 2p^{-1} + EX^2.$$

Hence substituting into (10) gives

$$EX^2 = p + q(1 + 2p^{-1} + EX^2)$$

and

$$EX^2 = p^{-2}(2 - p),$$

which we have otherwise calculated as the sum of a series:

$$EX^2 = \sum_{r=1}^{\infty} r^2 q^{r-1}p.$$

Hence

$$\text{var } X = EX^2 - (EX)^2$$

$$= qp^{-2}. \qquad \bigcirc$$

There are many problems of this kind, which can be tricky if tackled head on but which are extremely simple if conditional expectation is used correctly. (It is perhaps for this reason that they are so often found in examinations.) We conclude this section with a few classic examples.

Example 5.8.10: quiz.   You are a contestant in a quiz show, answering a series of questions. You answer correctly with probability p, or incorrectly with probability q; you get \$1 for every correct answer, and you are eliminated when you first give two consecutive wrong answers. The questions are independent. Find the expected number of questions you attempt, and your expected total prize money.

Solution.   Let X be the number of questions answered, and Y your total prize money. To use conditional expectation we need a partition of the sample space; let C be the event that you answer a question correctly. Then an appropriate partition is supplied by the three events $\{C, C^cC, C^cC^c\}$, where

$$P(C) = p, \quad P(C^c C) = pq, \quad P(C^c C^c) = q^2.$$

Then, given C, the number of further questions you attempt has the same distribution as X. Hence

$$E(X|C) = 1 + EX.$$

Similarly

$$E(X|C^c C) = 2 + EX \quad \text{and} \quad E(X|C^c C^c) = 2.$$

Hence

$$E(X) = p(1 + EX) + pq(2 + EX) + q^2 2,$$

yielding

$$EX = q^{-2}(1 + q).$$

Likewise

$$EY = (1 + EY)p + (1 + EY)pq + 0,$$

yielding

$$EY = q^{-2}(1 - q^2). \qquad\qquad \bigcirc$$

Example 5.8.11.   Two archers (Actaeon and Baskerville) take it in turns to aim at a target; they hit the bull, independently at each attempt, with respective probabilities α and β. Let X be the number of shots until the first bull. What is EX?

Solution.   Let B denote a bull. Then using our by now familiar new method, we write

$$EX = E(X|B)\alpha + E(X|BB^c)(1 - \alpha)\beta + E(X|B^c B^c)(1 - \alpha)(1 - \beta)$$

$$= \alpha + 2(1 - \alpha)\beta + (2 + EX)(1 - \alpha)(1 - \beta).$$

Hence

$$EX = \frac{2 - \alpha}{\alpha + \beta - \alpha\beta}. \qquad\qquad \bigcirc$$

Example 5.8.12: Waldegrave's problem revisited.   Recall that a group of $n + 1$ players contest a sequence of rounds until one of them has beaten all the others. What is the expected duration of the game? We showed in example 2.11.10 that the number of rounds played is $1 + X$, where X is the number of flips of a coin until it first shows a sequence of $n - 1$ consecutive heads. Now by conditional probability, and independence of flips (with an obvious notation),

$$E(X) = \tfrac{1}{2}E(X|T) + \left(\tfrac{1}{2}\right)^2 E(X|HT) + \cdots$$

$$+ \left(\tfrac{1}{2}\right)^{n-1} E(X|H^{n-2}T) + \left(\tfrac{1}{2}\right)^{n-1} E(X|H^{n-1})$$

$$= \tfrac{1}{2}(1 + EX) + \left(\tfrac{1}{2}\right)^2 (2 + EX) + \cdots$$

$$+ \left(\tfrac{1}{2}\right)^{n-1}(n - 1 + EX) + \left(\tfrac{1}{2}\right)^{(n-1)}(n - 1)$$

Hence

$$\mathrm{E}X = \frac{\frac{1}{2} + 2\left(\frac{1}{2}\right)^2 + \cdots + (n-1)\left(\frac{1}{2}\right)^{n-1} + (n-1)\left(\frac{1}{2}\right)^{n-1}}{1 - \left\{\frac{1}{2} + \cdots + \left(\frac{1}{2}\right)^{n-1}\right\}}$$

$$= 2^n - 2.$$

So the required expectation, $\mathrm{E}(1 + X)$, is $2^n - 1$. We shall see an even neater way of doing this in exercise 4 of section 6.11. ○

Example 5.8.13: duration of gambler's ruin. Recall the gambler's ruin problem in which you gain or lose one point with equal probabilities $\frac{1}{2}$ at each bet. You begin with k points; if you ever reach zero points, or n points, then the game is over. All bets are independent. Let T_k be the number of bets until the game is over; show that $\mathrm{E}T_k = k(n - k)$.

Solution. Let W be the event that you win the first bet, and $\tau_k = \mathrm{E}T_k$. Then

$$\tau_k = \mathrm{E}(T_k|W)\mathrm{P}(W) + \mathrm{E}(T_k|W^c)\mathrm{P}(W^c)$$

$$= \tfrac{1}{2}(1 + \tau_{k+1}) + \tfrac{1}{2}(1 + \tau_{k-1})$$

$$= \tfrac{1}{2}\tau_{k+1} + \tfrac{1}{2}\tau_{k-1} + 1, \quad 0 < k < n.$$

Naturally $\tau_0 = \tau_n = 0$. Hence it is easy to verify that indeed

$$\tau_k = k(n - k). \qquad ○$$

Exercises for section 5.8

1. ***Tennis.*** Rod and Fred have reached 6–6 in their tie-break. Rod wins any point with probability ρ. Fred wins with probability ϕ, where $\rho + \phi = 1$.
   (a) Let X be the number of points until the first occasion when one or other wins two consecutive points. Find $\mathrm{E}X$.
   (b) Let Y be the number of points until the tie-break is won. Find $\mathrm{E}Y$.
   (c) Let L be the event that Rod wins the tie break. Find $\mathrm{E}(X|L)$ and $\mathrm{E}(Y|L)$.

2. ***Gamblers ruined unfairly.*** Suppose you win each point with probability p, or lose it with probability $q = 1 - p$; as always, bets are independent. Let τ_k be the expected number of bets until the game is over, given that you start with k points. (As usual the game stops when you first have either no points or n points.) Show that

$$\tau_k = p\tau_{k+1} + q\tau_{k-1} + 1$$

and deduce that for $p \neq q$

$$\tau_k = \left\{ k - n\left[\frac{1 - (q/p)^k}{1 - (q/p)^n}\right] \right\}(q - p)^{-1}.$$

5.9 CONDITIONAL DENSITY

We have found conditional probability mass functions to be very useful on many occasions. Naturally we expect conditional density functions to be equally useful. They are, but they require a slightly indirect approach. We start with the conditional distribution function.

Definition.   Let X have distribution function $F(x)$, and let B be an event. Then the *conditional distribution function* of X given B is

(1) $F_{X|B}(x|B) = P(X \leq x|B)$

$$= \frac{P(\{X \leq x\} \cap B)}{P(B)}.$$ $\triangle$

Sometimes we denote this simply by $F(x|B)$.

Example 5.9.1.   Let X be uniform on $(0, a)$, and let B be the event $0 \leq X \leq b$, where $b < a$. Find $F_{X|B}(x|B)$.

Solution.   By the definition, for $x \geq 0$,

(2) $F_{X|B}(x|B) = \dfrac{P(\{X \leq x\} \cap \{X \leq b\})}{P(X \leq b)}$

$$= \begin{cases} P(X \leq b)/P(X \leq b) & \text{if } b \leq x \\ P(X \leq x)/P(X \leq b) & \text{if } x \leq b \end{cases}$$

$$= \begin{cases} 1 & \text{if } b \leq x \\ \dfrac{x/a}{b/a} & \text{if } x \leq b \end{cases}$$

$$= x/b, \quad 0 \leq x \leq b.$$

This is just the uniform distribution on $(0, b)$. It is an important and intuitively natural result: a uniform random variable conditioned to lie in some subset B of its range is *still* uniform, but it is now uniform on B. $\bigcirc$

As usual, we can often get densities from distributions.

Definition.   Let X have a conditional distribution $F_{X|B}$ that is differentiable. Then X has *conditional density* $f_{X|B}(x|B)$ given by

(3) $f_{X|B}(x|B) = \dfrac{d}{dx} F_{X|B}(x|B) = F'_{X|B}(x|B).$ $\triangle$

Sometimes we denote this simply by $f(x|B)$.

Example 5.9.1 revisited.   Let X be uniform on $(0, a)$ and let B be the event $\{0 \leq X \leq b\}$. Then differentiating (2) gives the conditional density

$$f_{X|B}(x|B) = b^{-1}, \quad 0 \leq x \leq b.$$ $\bigcirc$

Just as in the discrete case, $f(x|B)$ satisfies the same key rule as any ordinary density.

Conditional key rule

$$P(X \in C|B) = \int_{x \in C} f(x|B)\, dx.$$

In particular, if $B \subseteq C$,

$$P(X \in C|B) = \int_{x \in B} f(x|B)\, dx = 1,$$

so $f(x|B)$ is indeed a probability density. An especially important example of this kind of thing arises when we consider the exponential density.

Example 5.9.2: lack of memory. Let X be exponential with parameter λ, and let B_t be the event that $X > t$. Show that the conditional density of $X - t$, given B_t, is also exponential with parameter λ.

Remark. The importance of this result is clear when we recall that the exponential density is a popular model for waiting times. Let X be the waiting time until your light bulb fails. Suppose X is exponential, and your light bulb has survived for a time t. Then the above result says that the further survival time is still exponential, as it was to begin with.

Roughly speaking, a component or device with this property cannot remember how old it is. Its future life has the same distribution at any time t, if it has survived until t.

Solution. Let $Y = X - t$. We introduce the conditional survival function

$$
\begin{aligned}
\overline{F}_{Y|B_t} &= 1 - F_{Y|B_t}(y|B_t) \\
&= P(Y > y|B_t) \\
&= P(\{Y > y\} \cap \{X > t\})/P(X > t) \\
&= P(X > y + t)/P(X > t) \\
&= e^{-\lambda(y+t)}/e^{-\lambda t} \\
&= e^{-\lambda y}.
\end{aligned}
$$

It follows that

(4) $$F_{Y|B_t}(y|B_t) = 1 - e^{-\lambda y}$$

and

(5) $$f_{Y|B_t}(y|B_t) = \lambda e^{-\lambda y}, \quad y \geq 0.$$

Hence $X - t$ is exponential, as claimed. $\quad\bigcirc$

Recall that, among discrete random variables, the geometric distribution also has this property, as you would expect.

Example 5.9.3: conditional survival. Let X be the lifetime of your light bulb, and let B be the event that it survives for a time a. Find the conditional distribution of X given B for each of the following distributions of X:

(i) $P(X \geqslant x) = x^{-1}, \quad x = 1, 2, 3, \ldots$

(ii) $P(X \geqslant x) = e^{-x(x-1)}, \quad x \geqslant 1.$

Solution.   For (i): Here $P(B) = a^{-1}$, so for $x \geqslant a$

$$P(X \geqslant x|B) = \frac{x^{-1}}{a^{-1}} = \frac{a}{x}, \quad x \geqslant a.$$

Thus

$$p(x|B) = P(X \geqslant x|B) - P(X \geqslant x+1|B) = \frac{a}{x(x+1)}.$$

For (ii): Likewise in this case

$$P(X \geqslant x|B) = e^{-x(x-1)}/e^{-a(a-1)} = \exp\{a(a-1) - x(x-1)\}. \qquad \bigcirc$$

At this point we note that a conditional density may also have a mean, called the *conditional expectation*. It is given by

$$(6) \qquad\qquad E(X|B) = \int_{-\infty}^{\infty} x f(x|B)\, dx.$$

This is just as important as the conditional expectation defined for discrete random variables in (2) of section 5.8, and has much the same properties. We explore some of these later on. Finally we record one natural and important special case. It may be that the continuous random variable X is independent of B; formally we write

Definition.   If, for all x, we have

$$(7) \qquad\qquad F(x|B) = P(X \leqslant x|B) = P(X \leqslant x) = F(x)$$

then we say X and B are independent. $\triangle$

This is essentially our usual definition, for it just says that the events B and $\{X \leqslant x\}$ are independent. Differentiating gives just what we would anticipate:

$$(8) \qquad\qquad f_{X|B}(x|B) = f_X(x),$$

whenever X and B are independent.

In any case we find from the key rule that

$$(9) \qquad\qquad P(\{X \in C\} \cap B) = P(X \in C)P(B),$$

whenever X and B are independent.

Exercises for Section 5.9

1. If X is an exponential random variable with parameter λ, show that $E(X|X > t) = t + \lambda^{-1}$. What is $\mathrm{var}(X|X > t)$?

2. Show that $E(X) = E(X|B)P(B) + E(X|B^c)P(B^c)$.

3. Show that if X and B are independent, then $E(X|B) = EX$.

5.10 REVIEW

This chapter has concerned itself with random variables and their properties, most importantly their distributions and moments.

Random variable. A random variable is a real-valued function defined on the sample space Ω. Formally we write $X(\omega) \in \mathbb{R}$, for $\omega \in \Omega$. The set of all possible values of a random variable X is called its *range*. If the range is countable then X is said to be *discrete*; otherwise, X could be *continuous*.

Distribution function. Every random variable X has a distribution function $F(x)$, where $F(x) = P(X \leqslant x)$. Discrete random variables have a probability distribution $p(x)$, and continuous random variables have a density $f(x)$, such that

Distributions

When X is discrete,

$$p(x) = P(X = x),$$

and $P(X \in A) = \sum_{x \in A} p(x),$

and when X is integer valued

$$p(x) = F(x) - F(x - 1).$$

When X is continuous,

$$f(x)h \simeq P(x < X \leqslant x + h)$$

for small h;

$$P(X \in A) = \int_{x \in A} f(x)\, dx, \quad \text{and}$$

$$f(x) = dF(x)/dx.$$

Expectation. Any random variable may have an expectation or mean EX; if $E(X) < \infty$ then

when X is discrete,

$$EX = \sum_x xp(x)$$

When $X \geqslant 0$ is integer valued

$$EX = \sum_{x=0}^{\infty} P(X > x)$$

$$= \sum_{x=0}^{\infty} \{1 - F(x)\}.$$

when X is continuous

$$EX = \int_{\infty}^{\infty} xf(x)\, dx$$

When $X > 0$ is continuous

$$EX = \int_0^{\infty} P(X > x)\, dx$$

$$= \int_0^{\infty} \{1 - F(x)\}\, dx.$$

Functions. Suppose that random variables X and Y are such that $Y = g(X)$ for some function g. Then

if both are discrete,

$$p_Y(y) = \sum_{x:\, g(x)=y} p_X(x).$$

if both are continuous,

$$f_Y(y) = \frac{d}{dy} \int_{x:\, g(x) \leqslant y} f_X(x)\, dx.$$

Expectation of functions.   If $Y = g(X)$ then the law of the unconcious statistician says that

if X is discrete   if both are continuous

$$EY = \sum_x g(x)p_X(x).$$ $$EY = \int_{-\infty}^{\infty} g(x)f_X(x)\,dx.$$

It follows that for *any* random variable X

$$E(ag(X) + bh(X)) = aEg(X) + bEh(X).$$

Variance.   For any random variable X, the variance σ^2 is defined to be

$$\operatorname{var} X = E\{(X - EX)^2\} = EX^2 - (EX)^2 = \sigma^2 \geq 0.$$

The number $\sigma > 0$ is called the standard deviation; we have in particular that

$$\operatorname{var}(aX + b) = a^2 \operatorname{var} X$$

Moments.   The kth moment of X is

$$\mu_k = EX^k, \quad k \geq 1;$$

usually we write $\mu_1 = \mu$. The kth central moment of X is

$$\sigma_k = E\{(X - EX)^k\}, \quad k \geq 2;$$

usually we write $\sigma_2 = \sigma^2$.

Inequalities.   For any random variable X, and $a > 0$,

(i)   Chebyshov: $P(|X| \geq a) \leq EX^2/a^2$,
(ii) Markov: $P(|X| \geq a) \leq E|X|/a$,
(iii) Dominance: If $g(x) \leq h(x)$ always, then $Eg(X) \leq Eh(X)$.

Conditioning.   Any event B in Ω may condition a random variable X, leading to a conditional distribution function

$$F_{X|B}(x|B) = P(X \leq x|B).$$

Since this is a distribution, it may have an expectation, called the conditional expectation;

in the discrete case,   in the continuous case,

$$E(X|B) = \sum_x xp(x|B)$$ $$E(X|B) = \int_x xf(x|B)$$

where   where

$$p(x|B) = P(X = x|B).$$ $$f(x|B) = \frac{d}{dx} F(x|B).$$

In all cases

$$E(X) = E(X|B)P(B) + E(X|B^c)P(B^c).$$

We give two tables of some common random variables with their associated characteristics, table 5.1 for the discrete case and table 5.2 for the continuous case.

Table 5.1. *Discrete random variables and their associated characteristics*

X	$p(x)$	μ	σ^2
indicator	$p(1) = p$ $p(0) = 1 - p = q$	p	pq
binomial $\mathrm{B}(n, p)$	$\binom{n}{x} p^x (1-p)^{n-x},$ $0 \leqslant x \leqslant n$	np	$np(1-p)$
geometric $p = 1 - q$	$pq^{x-1}, \quad x \geqslant 1$	p^{-1}	qp^{-2}
negative binomial $n, p = 1 - q$	$\binom{x-1}{n-1} p^n q^{x-n},$ $x \geqslant n$	np^{-1}	nqp^{-2}
hypergeometric a, b, n	$\dfrac{\binom{a}{x}\binom{b}{n-x}}{\binom{a+b}{n}}$	$n\dfrac{a}{a+b}$	$n\left(\dfrac{a+b-n}{a+b-1}\right)\dfrac{ab}{(a+b)^2}$
Poisson λ	$\dfrac{e^{-\lambda}\lambda^x}{x!}, \quad x \geqslant 0$	λ	λ
uniform $\{1, 2, \ldots, n\}$	$n^{-1}, \quad 1 \leqslant x \leqslant n$	$\tfrac{1}{2}(n+1)$	$\tfrac{1}{12}(n^2 - 1)$

Table 5.2. *Continuous random variables and their associated characteristics*

X	$f(x)$	$\mathrm{E}X$	$\mathrm{var}\,X$
uniform	$(b-a)^{-1}, \quad a \leqslant x \leqslant b$	$\tfrac{1}{2}(b+a)$	$\tfrac{1}{12}(b-a)^2$
exponential	$\lambda e^{-\lambda x}, \quad x \geqslant 0$	λ^{-1}	λ^{-2}
normal $\mathrm{N}(\mu, \sigma^2)$	$(2\pi)^{-1/2}\sigma^{-1}\exp\left\{-\dfrac{1}{2}\left(\dfrac{x-\mu}{\sigma}\right)^2\right\},$ $-\infty < x < \infty$	μ	σ^2
gamma	$x^{r-1}e^{-\lambda x}\dfrac{\lambda^r}{(r-1)!}, \quad x \geqslant 0$	$r\lambda^{-1}$	$r\lambda^{-2}$
beta $\beta(a, b)$	$\dfrac{(a+b-1)!}{(a-1)!(b-1)!}x^{a-1}(1-x)^{b-1},$ $0 \leqslant x \leqslant 1$	$\dfrac{a}{a+b}$	$\dfrac{ab(a+b)^2}{(a+b+1)}$
Rayleigh	$xe^{-x^2/2}, \quad x \geqslant 0$	$\left(\dfrac{\pi}{2}\right)^{1/2}$	$2 - \dfrac{\pi}{2}$

5.11 APPENDIX. DOUBLE INTEGRALS

For one continuous random variable, its probability distribution is supplied by integrating the density $f(x)$. Soon we shall consider two continuous random variables; in this case probabilities are given by integrating a joint density $f(x, y)$. Just as the integral of $f(x)$ may be interpreted as an area, the integral of $f(x, y)$ may be interpreted as a volume.

There are various ways of measuring volume; one way runs as follows. Let $\delta A_1, \ldots, \delta A_n$ be a collection of small areas which are disjoint, and whose union is C. Let (x_k, y_k) be a point in δA_k. Then the volume defined by the surface $f(x, y)$ above C is approximately

$$V = \sum_k f(x_k, y_k)\delta A_k.$$

As the δA_k become arbitrarily small, we obtain the double integral of f over C in the limit. (Many details have been omitted here.) It only remains to choose our coordinates, and the shapes of the areas δA_k. We consider two important cases.

Cartesian coordinates.   In this case it is very natural to let each δA_k be a small rectangle with sides having lengths denoted by δx_k and δy_k, as shown in figure 5.1. In the limit we obtain the required volume, denoted by

(1) $$V = \iint_C f(x, y)\, dx\, dy = \iint_C f(x, y)\, dy\, dx.$$

This notation is familiar from the one-dimensional case.

Polar coordinates.   In this case it is more natural to let each δA_k be a small curvilinear quadrilateral, as shown in figure 5.2. In this case $\delta A_k = r_k \delta r_k \delta \theta_k$, and we obtain the volume in the form

(2) $$V = \iint_C f(r, \theta)r\, dr\, d\theta.$$

Now we can again use the routines of the one-dimensional integral.

We give two examples, each with two solutions.

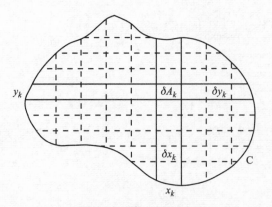

Figure 5.1. $\delta A_k = \delta x_k \delta y_k$.

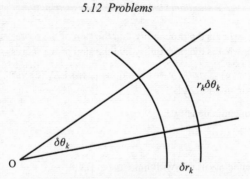

Figure 5.2. Polar coordinates with O as origin.

Example 5.11.1. Find the volume under

$$f(x, y) = 1, \quad 0 \leqslant x, y \leqslant 1.$$

(i) In Cartesian coordinates, by (1)

$$V = \int_0^1 \int_0^1 dx\, dy = \int_0^1 dx \int_0^1 dy = 1.$$

(ii) In polar coordinates f takes the form

$$f(r, \theta) = 1 \begin{cases} 0 \leqslant r \leqslant \sec \theta, & 0 \leqslant \theta \leqslant \pi/4 \\ 0 \leqslant r \leqslant \operatorname{cosec} \theta, & \pi/4 \leqslant \theta \leqslant \pi/2. \end{cases}$$

Then by (2)

$$V = 2 \int_0^{\pi/4} \int_0^{\sec \theta} r\, dr\, d\theta = \int_0^{\pi/4} \sec^2 \theta\, d\theta = [\tan \theta]_0^{\pi/4} = 1.$$

Example 5.11.2. Find the volume under

$$f(r, \theta) = \frac{1}{\pi}, \quad 0 \leqslant r \leqslant 1, 0 \leqslant \theta \leqslant 2\pi.$$

(i) In polar coordinates, by (2),

$$V = \int_0^1 \int_0^{2\pi} \frac{1}{\pi} r\, dr\, d\theta = \int_0^1 2r\, dr \int_0^{2\pi} \frac{1}{2\pi} d\theta = 1.$$

(ii) In Cartesian coordinates f takes the form

$$f(x, y) = \frac{1}{\pi}, \ |y| \leqslant \sqrt{1 - x^2}; \ |x| \leqslant 1.$$

Then by (1)

$$V = \int_{-1}^1 \int_{-\sqrt{1-x^2}}^{\sqrt{1-x^2}} \frac{1}{\pi} dy\, dx = \int_{-1}^1 \frac{2}{\pi} \sqrt{1 - x^2}\, dx = \frac{1}{\pi} \left[x\sqrt{1 - x^2} + \sin^{-1} x \right]_{-1}^1 = 1.$$

The point of these examples is that you can often save yourself a great deal of effort by choosing the most appropriate coordinates.

5.12 PROBLEMS

1. You roll 5 dice. Let X be the smallest number shown and Y the largest.
   (a) Find the distribution of X, and EX.
   (b) Find the distribution of Y, and EY.

2. (a) You roll 5 dice; what is the probability that the sum of the scores is 18 or more?
    (c) You roll 4 dice; what is the probability that the sum of the scores is 14 or more?

3. Let X have density $f(x) = cx^{-d}$, $x \geq 1$. Find (a) c, (b) EX, (c) $\text{var } X$. In each case state for what values of d your answer holds.

4. Let X have the exponential density. Find
    (a) $P(\sin X > \frac{1}{2})$,
    (b) EX^n, $\quad n \geq 1$.

5. Show that for any random variable with finite mean EX,
    $$(EX)^2 \leq (EX^2).$$

6. Which of the following can be density functions? For those that are, find the value of c, and the distribution function $F(x)$.

    (a) $f(x) = \begin{cases} cx(1-x) & 0 \leq x \leq 1 \\ 0 & \text{otherwise.} \end{cases}$

    (b) $f(x) = \begin{cases} cx^{-1} & x \geq 1 \\ 0 & \text{otherwise.} \end{cases}$

    (c) $f(x) = c \exp(-x^2 + 4x)$, $\quad -\infty < x < \infty$.
    (d) $f(x) = ce^x(1 + e^x)^{-2}$, $\quad -\infty < x < \infty$.

7. Which of the following can be distribution functions? For those that are, find the density.

    (a) $F(x) = \begin{cases} 1 - \exp(-x^2) & x \geq 0 \\ 0 & \text{otherwise.} \end{cases}$

    (b) $F(x) = \begin{cases} \exp(-x^{-1}), & x \geq 0 \\ 0 & \text{otherwise.} \end{cases}$

    (c) $F(x) = e^x(e^x + e^{-x})^{-1}$, $\quad -\infty < x < \infty$.
    (d) $F(x) = e^{-x^2} + e^x(e^x + e^{-x})^{-1}$, $\quad -\infty < x < \infty$.

8. Let X be Poisson with parameter λ. Find the distribution and expectation of X, given that X is odd.

9. Let X have the density
    $$f(x) = \frac{\exp x}{2 \sinh a}, \quad -a < x < a.$$
    Show that
    $$EX = a \coth a - 1,$$
    and
    $$\text{var } X = 1 - \left(\frac{a}{\sinh a}\right)^2.$$

10. Your dart is equally likely to hit any point of a circular dart board. Its height above the bull is Y (negative if below the bull), and its distance from the bull is R. Find the density and distribution of Y, and of R. What is ER?

11. An urn contains one carmine ball and one magenta ball. A ball is drawn at random; if it is carmine the game is over. If it is magenta then the ball is returned to the urn together with one extra magenta ball. This procedure is repeated until 10 draws have been made or a carmine ball is drawn, whichever is sooner. Let X be the number of draws. Find $p_X(x)$ and EX.

    Now suppose the game can only be terminated by the appearance of a carmine ball; let Y be the number of draws. Find the distribution $p_Y(y)$ and EY.

12. You are racing around a circular track C; the pits are at a point P on the circumference of C. Your car is equally likely to break down at any point B of the track. Let X be the distance from B to P in a straight line. Find the density, mean, and variance of X.

13. You fire a musket on the planet Zarg with muzzle velocity V, making an angle Θ with the horizontal ground. The musket ball strikes the ground at a distance $X = (V^2/g)\sin 2\Theta$ away, where g is the acceleration due to gravity; in Zarg units $g = 1$.
    (a) If V is constant and Θ is uniform on $[0, \pi/2]$, find the density of X.
    (b) If Θ is constant and V has density $f_V(x) = cx^2 e^{-x^2}$, $x > 0$, find the density of X. What is c in this case?

14. The diameter D of a randomly selected lead shot has density $f(x)$, $x > 0$. Find the density of the weight of a randomly selected lead shot. (Assume any shot is spherical.)

15. I try to open my door with one of the three similar keys in my pocket; one of them is the correct key, the other two will not turn. Let X be the number of attempts necessary if I choose keys at random from my pocket and drop those that fail to the ground. Let Y be the number of attempts necessary if I choose keys at random from my pocket and replace those that fail in my pocket. Find $\mathbb{E}X$ and $\mathbb{E}Y$.

16. You are at the origin between two walls lying at $x = \pm 1$. You shine your torch so that the beam makes an angle Θ with the line $x = 0$; Θ is uniformly distributed on $[0, 2\pi]$. Let Y be the y-coordinate of the point where the beam strikes a wall. Show that Y has a Cauchy density.

17. Let S be the speed of a randomly selected molecule in a gas. According to the kinetic theory of gases, S has probability density
$$f(s) = \alpha s^2 e^{-s^2/(2\sigma^2)}.$$
Find α. The kinetic energy of a molecule of mass m is $X = \frac{1}{2}mS^2$. Find the density of X.

18. Let X be a non-negative random variable with finite expectation, having distribution function $F(x)$ and density $f(x)$. Show that for $x > 0$,
$$x\{1 - F(x)\} \le \int_x^\infty xf(x)\,dx,$$
and deduce that as $x \to \infty$, $x\{1 - F(x)\} \to 0$. Hence show that $\mathbb{E}X = \int_0^\infty \{1 - F(x)\}\,dx$.

19. A fair three-sided die has its faces labelled 1, 2, 3. It is rolled repeatedly. Let X_n be the number of rolls until the sum of the numbers shown is at least n. Show that for $n \ge 4$
$$3\mathbb{E}X_n - \mathbb{E}X_{n-1} - \mathbb{E}X_{n-2} - \mathbb{E}X_{n-3} = 3.$$
Suppose now that the three faces are shown with respective probabilities p, q, and r. Write down the equivalent equation for the expectations $\mathbb{E}X_n$.

20. Let X be a random variable with $\mathbb{E}X^3 < \infty$. The *skewness* of X is given by
$$\mathrm{skw}\, X = \frac{\mathbb{E}(X - \mu)^3}{\sigma^3},$$
where $\mu = \mathbb{E}X$ and $\sigma^2 = \mathrm{var}\, X$.
    (a) If X is Bernoulli with parameter p, show that
    $$\mathrm{skw}\, X = (q - p)/(qp)^{1/2}.$$
    (b) For any random variable X show that
    $$\mathrm{skw}\, X = (\mathbb{E}X^3 - 3\mu\mathbb{E}X^2 + 2\mu^3)/\sigma^3.$$
    (c) If X is Poisson with parameter λ, show that $\mathrm{skw}\, X = \lambda^{-1/2}$.
    (d) If X is geometric with parameter p, $p(x) = q^{x-1}p$, $x \ge 1$, show that $\mathrm{skw}\, X = (1 + q)/q^{1/2}$.

21. Let X be a random variable with density $f(x) = cx^{a-1}e^{-x^a}$, $x \geq 0$.
    (a) Find c, and evaluate EX and var X.
    (a) Show that if $a > 1$, and s, $t > 0$, then

$$P(X > s + t \mid X > t) < P(X > s).$$

    What if $a < 1$?

22. **Beta density.**   Let X have density

$$f(x) = cx^{a-1}(1-x)^{b-1}, \quad 0 \leq x \leq 1.$$

    Find c, EX, and var X.

23. Let $F(x)$ be a distribution function, and r a positive integer. Show that the following are distribution functions:
    (a) $\{F(x)\}^r$,
    (b) $1 - \{1 - F(x)\}^r$,
    (c) $F(x) + \{1 - F(x)\}\log\{1 - F(x)\}$,
    (d) $\{F(x) - 1\}e + e^{1-F(x)}$.

24. My book has f pages, with n characters on each page. Each character is wrong with probability p, independently of the others. If I proofread the book once, I detect any error with probability δ independently of any other detections and of other proofreadings.
    Show that the number of errors remaining has a binomial distribution.
    I wish to proofread enough times that the chance of no errors remaining exceeds $\frac{1}{2}$. If $f = 2^8$, $n = 2^9$, $p = 2^{-8}$, $\delta = \frac{1}{2}$, show that 10 readings will do.

25. Let $X > 0$ have distribution $F(x)$. Show that

$$EX^2 = 2\int_0^\infty x\{1 - F(x)\}\, dx$$

    and that

$$EX^r = \int_0^\infty rx^{r-1}P(X > x)\, dx.$$

26. (a) Show that if X is a random variable with var $X = 0$, then, for some constant a,
    $P(X = a) = 1$.

27. Use Markov's inequality to show that, for any $t > 0$ and any random variable such that $E(e^{tX})$ exists,

$$P(X \geq a) \leq e^{-at}E(e^{tX}), \quad \text{for } a > 0.$$

    Deduce that

$$P(X \geq a) \leq \inf_{t > 0}\{e^{-at}\, Ee^{tX}\}.$$

    Hence show that if X is Poisson with parameter λ, and $x > \lambda$,

$$P(X \geq x) \leq e^{-\lambda}\left(\frac{\lambda e}{x}\right)^x.$$

28. A stretch of motorway is m miles long; a garage G is d miles from one end. A car is equally likely to break down at any point of the road. Let X be the distance from the garage to the breakdown. Find the distribution, density, mean, and variance of X. If you had a free choice, explain where you would choose to site G on this road, and why.

29. **Bertrand's paradox.**
    (a) Choose a point P at random inside a circle of radius a. Let X be the length of the chord of which P is the mid-point. Show that
    $$P(X > \sqrt{3}a) = \tfrac{1}{4}.$$
    (b) Choose two points independently at random on the perimeter of a circle of radius a. Let X be the length of the chord joining them. Show that
    $$P(X > \sqrt{3}a) = \tfrac{1}{3}.$$

30. Let $0 < a < m$, and let X be a random variable such that
    $$P(|X - EX| \leqslant m) = 1.$$
    Show that
    $$P(|X - EX| \geqslant a) \geqslant \frac{\operatorname{var} X - a^2}{m^2 - a^2}.$$

31. Let X have the Cauchy distribution, and let $Y = (1 + X^2)^{-1}$. Show that Y has the arcsin distribution.

32. Let $F(x)$ and $G(x)$ be continuous distribution functions such that $F(x) \geqslant G(x)$ for all x. Let U be uniform on (0, 1) and define $X = F^{-1}(U)$, $Y = G^{-1}(U)$. Show that $X \leqslant Y$, and find the distributions of X and Y.

6

Jointly distributed random variables

6.1 PREVIEW

In chapter 5 we looked at probability distributions of single random variables. But of course we often wish to consider the behaviour of two or more random variables together. This chapter extends the ideas of chapters 4 and 5, so that we can make probability statements about collections and sequences of random variables.

The most important instrument in this venture is the joint probability distribution, which we meet in section 6.2. We also define the concept of independence for random variables, and explore some consequences. Jointly distributed random variables have joint moments, and we look at the important ideas of covariance and correlation. Finally, we consider conditional distributions and conditional expectation in this new setting.

Prerequisites. We shall use one new technique in this chapter; see appendix 5.11 on double integrals.

6.2 JOINT DISTRIBUTIONS

A random variable $X(\omega)$ is a real-valued function on Ω. Often there will be several random variables of interest defined on Ω, and it may be important and useful to examine their joint behaviour. For example:

(i)   A meteorological station may record the wind speed and direction, air pressure, and the air temperature.
(ii)   Your physician may record your height, weight, blood pressure, cholesterol level, and more.
(iii)   The point count in the four hands dealt at bridge yields four random variables X_n, X_e, X_ω, X_s, with $X_n + X_e + X_\omega + X_s = 40$. The outcome of the deal depends on their joint distribution.

Just as for one random variable, we want to know the probabilities of the joint outcomes, and once again it is convenient to consider discrete and continuous random variables separately.

In general we introduce ideas and definitions for two random variables; extending these to larger collections is very easy, but also lengthy and tedious, and is therefore left to the reader as an exercise.

Looking back to chapter 4, we can see that we have already considered such pairs of random variables, when we looked at distributions in the plane. In that case $\Omega = \mathbb{R}^2$, and for each outcome $\omega(x, y)$, we set

$$(X(\omega), Y(\omega)) = (x, y).$$

We extend these simple ideas in the same way as we did in chapter 5.

Thus, let X and Y be a pair of discrete random variables defined on Ω. As usual we assume that X and Y take integer values, unless otherwise stated. As in chapter 5, the natural function to tell us how probability is distributed over the values of the pairs (X, Y) is the following.

Definition. The function

(1) $$p(x, y) = P(X = x, Y = y)$$

is the *joint probability distribution* of X and Y (the word 'joint' is usually omitted), and we may sometimes denote it by $p_{X,Y}(x, y)$, to avoid possible ambiguity. $\triangle$

Remark. As before, note that strictly speaking

$$P(X = x, Y = y) = P(A_x \cap A_y)$$

where $A_x = \{\omega: X(\omega) = x\}$, $A_y = \{\omega: Y(\omega) = y\}$. In general we need to use these underlying events no more than we did in Chapter 5; which is to say, hardly ever.

Obviously the distribution $p(x, y)$ satisfies

(2) $$0 \leqslant p(x, y) \leqslant 1$$

and

(3) $$\sum_{x,y} p(x, y) = 1.$$

Any function satisfying (2) and (3) is a joint or bivariate probability distribution. Note that, as usual, we specify any $p(x, y)$ by giving its values where it is not zero. Here are some simple examples of joint distributions.

Example 6.2.1: pair of dice. Two fair dice are rolled, yielding the scores X and Y. We know already that

$$p(x, y) = \tfrac{1}{36}, \quad 1 \leqslant x, y \leqslant 6 \qquad\qquad \bigcirc$$

Example 6.2.2: pair of indicators. Let X and Y be indicators, so that $X \in \{0, 1\}$ and $Y \in \{0, 1\}$. Then

$$(X, Y) \in \{(0, 0), (0, 1), (1, 0), (1, 1)\}$$

and the joint distribution is just the array

$$\left\{ \begin{array}{ll} p(0, 0), & p(0, 1) \\ p(1, 0), & p(1, 1) \end{array} \right\}. \qquad\qquad \bigcirc$$

Example 6.2.3: flipping a coin. Suppose a coin shows a head with probability p, or a tail with probability q. You flip the coin repeatedly. Let X be the number of flips until the first head, and Y the number of flips until the first tail. Then, obviously, the joint probability distribution of X and Y is

$$p(1, y) = p^{y-1}q, \quad y \geq 2,$$
$$p(x, 1) = pq^{x-1}, \quad x \geq 2.$$
○

Example 6.2.4: Bernoulli trials.   In n Bernoulli trials let X be the number of successes and Y the number of failures. Then $x + y = n$, and

$$p(x, y) = \frac{n!}{x!y!} p^x(1-p)^y, \quad 0 \leq x, y \leq n.$$
○

Example 6.2.5: de Moivre trials.   In n de Moivre (three-way) trials, let X, Y, and Z denote the number in each of the three possible categories. Then $x + y + z = n$, and

$$p(x, y, z) = P(X = x, Y = y, Z = z)$$

$$= \frac{n!}{x!y!z!} p_1^x p_2^y p_3^z,$$

where $p_1 + p_2 + p_3 = 1$. Note that the distribution in this example is trivariate.   ○

As before, questions about the joint behaviour of X and Y are answered by a key rule.

Key rule for joint distributions.   Let X and Y have joint distribution $p(x, y)$, and let C be a collection of possible values of (X, Y). Then

(4)
$$P((X, Y) \in C) = \sum_{x,y \in C} p(x, y).$$

The proof is essentially the same as that of (4) in section 5.3 and is left as a routine exercise. This is of course the same rule that we used in chapter 4 to look at distributions in the plane (allowing for changes in notation and emphasis).

Our first application of the key rule is exceedingly important and useful.

Marginal distributions.   Let X and Y have probability distributions $p_X(x)$ and $p_Y(y)$ respectively. If X and Y have joint distribution $p(x, y)$, then from (4)

(5)
$$p_X(x) = \sum_y p(x, y)$$

and

(6)
$$p_Y(y) = \sum_x p(x, y).$$

In each case the sum is taken over all possible values of y and x respectively; when calculated in this way these are sometimes called the *marginal distributions* of x and y. It is of course most important and useful that we can obtain them from $p(x, y)$.

Remark.   The use of the term 'marginal' is explained if we write the joint probabilities $p(x, y)$ in the form of an array. Then the distribution $p_X(x)$ of X is given by the column sums and the distribution $p_Y(y)$ of Y by the row sums; these are conveniently placed at the margins.

$$
\begin{array}{c|ccc}
p_Y(n) & p(1, n) & \cdots & p(m, n) \\
\vdots & \vdots & & \vdots \\
p_Y(2) & p(1, 2) & & p(m, 2) \\
p_Y(1) & p(1, 1) & \cdots & p(m, 1) \\
\hline
p(x, y) & p_X(1) & \cdots & p_X(m)
\end{array}
$$

Example 6.2.1 revisited: dice.   For two dice we know that the joint distribution is uniform; $p(x, y) = \frac{1}{36}$. Obviously and trivially

(7)
$$
p_X(x) = \sum_{y=1}^{6} p(x, y) = \frac{1}{6}.
$$
$\bigcirc$

Remark.   We have shown that $p(x, y)$ always yields the marginals $p_X(x)$ and $p_Y(y)$. However, the converse is not true. To see this, consider two experiments, (A) and (B).

(A) Flip a fair coin; let X be the number of heads and Y the number of tails. Then

$$
p_X(0) = p_X(1) = p_Y(0) = p_Y(1) = \tfrac{1}{2},
$$

with

$$
p(0, 1) = \tfrac{1}{2}, \quad p(0, 0) = 0; \quad p(1, 0) = \tfrac{1}{2}, \quad p(1, 1) = 0.
$$

(B) Flip two fair coins; let X be the number of heads shown by the first and Y the number of heads shown by the second. Then

$$
p_X(0) = p_X(1) = p_Y(0) = p_Y(1) = \tfrac{1}{2},
$$

which is the same as in (A). But

$$
p(0, 1) = p(0, 0) = p(1, 0) = p(1, 1) = \tfrac{1}{4},
$$

which is different from (A). In general the marginals do *not* determine the joint distribution. There is an important exception to this, which we examine in section 6.4.

Simple and empirical distributions are inevitably presented in the form of an array. In theoretical applications we usually have an algebraic representation, which occupies less space, saving trees and avoiding writer's cramp.

Example 6.2.6.   A pair of dice bear the numbers 1, 2, 3 twice each, on pairs of opposite faces. Both dice are rolled, yielding the scores X and Y respectively. Obviously

$$
p(j, k) = \tfrac{1}{9}, \quad \text{for } 1 \leqslant j, k \leqslant 3.
$$

We could display the joint probabilities $p(j, k)$ as a very dull array, if we wished.

Now suppose we roll these dice again, and consider the difference between their scores, denoted by U, and the sum of their scores, denoted by V. Then

$$
-2 \leqslant U \leqslant 2 \quad \text{and} \quad 2 \leqslant V \leqslant 6.
$$

We can calculate the joint distribution of U and V by running over all possible outcomes.

For example

$$p(0, 4) = P(U = 0, V = 4)$$

$$= P(\{2, 2\}) = \tfrac{1}{9}.$$

Eventually we produce the following array of probabilities:

V					
6	0	0	$\frac{1}{9}$	0	0
5	0	$\frac{1}{9}$	0	$\frac{1}{9}$	0
4	$\frac{1}{9}$	0	$\frac{1}{9}$	0	$\frac{1}{9}$
3	0	$\frac{1}{9}$	0	$\frac{1}{9}$	0
2	0	0	$\frac{1}{9}$	0	0
	-2	1	0	1	2 $\quad U$

We could write this algebraically, but it is more informative and appealing as shown. Observe that U and V both have triangular distributions, but U is symmetrical about 0 and V is symmetrical about 4. ◯

In practice we rarely display probabilities as an array, as the functional form is usually available and is of course much more compact.

Example 6.2.7.   Let X and Y have the joint distribution
$$p(x, y) = c(x + y), \quad 1 \leqslant x, y \leqslant n.$$
What is c? Find the marginal distributions.

Solution.   By (3),

$$1 = c \sum_{x,y}(x + y) = cn^2(n + 1).$$

Next,

$$p_X(x) = c \sum_{y=1}^{n}(x + y) = \frac{1}{n^2(n + 1)}\{nx + \tfrac{1}{2}n(n + 1)\}$$

$$= \frac{1}{n(n + 1)}\{x + \tfrac{1}{2}(n + 1)\}.$$

Likewise,

$$p_Y(y) = \frac{1}{n(n + 1)}\{y + \tfrac{1}{2}(n + 1)\}. \qquad ◯$$

Joint distributions are discovered by perfectly natural methods.

Example 6.2.8.   You roll three dice. Let X be the smallest number shown and Y the largest. Find the joint distribution of X and Y.

Solution.   Simple enumeration is sufficient here. For $x < y - 1$ there are three possibilities: the three dice show different values, or two show the larger, or two show the smaller. Hence

(8)
$$p(x, y) = \frac{6(y - x - 1)}{216} + \frac{3 + 3}{216} = \frac{y - x}{36}, \quad x < y - 1.$$

For $x = y - 1$ there are two possibilities, and

$$p(x, y) = \frac{3 + 3}{216} = \frac{1}{36}, \quad x = y - 1.$$

For $x = y$, there is one possibility, so

$$p(x, y) = \frac{1}{216}, \quad x = y.$$

It is easy for you to check that $\sum_{x,y} p(x, y) = 1$, as it must be. $\quad\bigcirc$

Sometimes joint distributions are empirical.

Example 6.2.9: Benford's distribution for significant digits. Suppose you take a large volume of numerical data, such as can be found in an almanac or company accounts. Pick a number at random and record the first two significant digits, which we denote by (X, Y). It is found empirically that X has the distribution

(9)
$$p(x) = \log_{10}\left(1 + \frac{1}{x}\right), \quad 1 \leqslant x \leqslant 9.$$

As we noted in example 4.2.3 it has recently been proved that there are theoretical grounds for expecting this result. Likewise it is found empirically, and theoretically, that the pair (X, Y) has the joint distribution

$$p(x, y) = \log_{10}\left(1 + \frac{1}{10x + y}\right), \quad 1 \leqslant x \leqslant 9; 0 \leqslant y \leqslant 9.$$

Of course, we find the marginal distribution of X to be

$$p_X(x) = \sum_{y=0}^{9} p(x, y) = \log_{10}\left(\prod_{y=0}^{9} \frac{10x + y + 1}{10x + y}\right) = \log_{10}\left(1 + \frac{1}{x}\right),$$

which is (9). However, the marginal distribution of Y is rather repulsive. $\quad\bigcirc$

Obtaining the marginals in this way is attractive and useful, but the key rule can be applied to find more interesting probabilities than just the marginals. The point is that in most applications of interest, the region C (see equation (4)) is determined by the joint behaviour of X and Y. For example, to find $P(X = Y)$ we set

$$C = \{(x, y): x = y\};$$

to find $P(X > Y)$ we set

$$C = \{(x, y): x > y\};$$

and so on. Here is an example.

Example 6.2.10. Let X and Y have the joint distribution

$$p(x, y) = c\lambda^x \mu^y, \quad x, y \geqslant 1; 0 < \lambda, \mu < 1.$$

Find (i) the value of c, (ii) $P(X > Y)$, (iii) $P(X = Y)$.

Solution.   For (i): As usual c is determined by

$$1 = c \sum_{x=1}^{\infty} \sum_{y=1}^{\infty} \lambda^x \mu^y = \frac{c\lambda\mu}{(1-\lambda)(1-\mu)}.$$

For (ii): By the key rule,

(10) $$P(X > Y) = \sum_{y=1}^{\infty} \sum_{x=y+1}^{\infty} p(x, y) = \sum_{y=1}^{\infty} c\mu^y \frac{\lambda^{y+1}}{1-\lambda}$$

$$= \frac{c\lambda}{1-\lambda} \left(\frac{\mu\lambda}{1-\mu\lambda} \right)$$

$$= \frac{\lambda(1-\mu)}{1-\lambda\mu}.$$

For (iii): Also by the key rule,

$$P(X = Y) = \sum_{x=1}^{\infty} c(\lambda\mu)^x = \frac{(1-\lambda)(1-\mu)}{1-\lambda\mu}.$$ ◦

Finally we note that just as a single random variable X has a distribution function $F(x) = P(X \leq x)$, so too do jointly distributed random variables have joint distribution functions.

Definition.   Let X and Y have joint distribution $p(x, y)$. Then their *joint distribution function* is

(11) $$F(x, y) = \sum_{i \leq x} \sum_{j \leq y} p(i, j) = P(X \leq x, Y \leq y).$$ △

Once again we can find $p(x, y)$ if we know $F(x, y)$, though it is not quite so simple as it was for one random variable:

(12) $$p(x, y) = F(x, y) - F(x, y - 1)$$

$$- F(x - 1, y) + F(x - 1, y - 1).$$

The proof of (12) is easy on substituting (11) into the right-hand side.

Example 6.2.11.   You roll r dice. Let X be the smallest number shown and Y the largest. Find the joint distribution of X and Y.

Solution.   We could use distribution functions directly, but it is neater to use an identity similar to (12), proved in the same way. That is,

$$p(x, y) = P(X \geq x, Y \leq y) - P(X \geq x + 1, Y \leq y)$$

$$- P(X \geq x, Y \leq y - 1) + P(X \geq x + 1, Y \leq y - 1).$$

Now, by independence of the dice, for $1 \leq x, y \leq 6$

$$P(X \geq x, Y \leq y) = \left(\frac{y - x + 1}{6} \right)^r, \quad x \neq y.$$

Hence

$$p(x, y) = \begin{cases} 6^{-r}\{(y - x + 1)^r - 2(y - x)^r + (y - x - 1)^r\}, & y \neq x \\ 6^{-r}, & y = x. \end{cases}$$

When $r = 3$ we recover (8), of course. ○

Exercises for section 6.2

1. In example 6.2.11, what happens to $p(x, y)$ as $r \to \infty$? Explain.
2. Let X and Y have the joint distribution

    $$p(x, y) = n^{-2}, \quad 1 \leqslant x, y \leqslant n.$$

    Find $P(X > Y)$ and $P(X = Y)$.

3. **Voter paradox.** Let X, Y, Z have the joint distribution

    $$p(2, 1, 3) = p(3, 2, 1) = p(1, 3, 2) = \tfrac{1}{3}.$$

    Show that

    $$P(X > Y) = P(Y > Z) = P(Z > X) = \tfrac{2}{3}.$$

    Explain why this is called the *voter paradox*, by considering an election in which voters are required to place three parties in strict order of preference.

6.3 JOINT DENSITY

Just as discrete random variables may be jointly distributed, so may continuous random variables be jointly distributed. And, just as in the discrete case, we need a function to tell us how likely the various possibilities are. Our requirements are satisfied by the following

Definition. The random variables X and Y are said to be jointly continuous, with *joint density* $f(x, y)$, if for all $a < b$ and $c < d$

$$(1) \qquad P(a < X < b, c < Y < d) = \int_c^d \int_a^b f(x, y) \, dx \, dy. \qquad \triangle$$

This is the natural extension of the definition of density for one random variable, which is

$$(2) \qquad P(a < X < b) = \int_a^b f(x) \, dx.$$

In (2) the integral represents the area under the curve $f(x)$; in (1) the double integral represents the volume under the surface $f(x, y)$. It is clear that $f(x, y)$ has properties similar to those of $f(x)$, that is,

$$(3) \qquad f(x, y) \geqslant 0,$$

and

$$(4) \qquad \int_{-\infty}^{\infty} \int_{-\infty}^{\infty} f(x, y) \, dx \, dy = 1.$$

And, most importantly, we likewise have the

Key rule for joint densities. Let X and Y have joint density $f(x, y)$. The probability that the point (X, Y) lies in some set C is given by

(5) $$P((X, Y) \in C) = \iint_C f(x, y)\, dx\, dy.$$

The integral represents the volume under the surface $f(x, y)$ above C. (There is a technical necessity for C to be a set nice enough for this volume to be defined, but that will not bother us here.) It is helpful to observe that (5) is completely analogous to the rule for discrete random variables

$$P((X, Y) \in C) = \sum_{(x,y) \in C} p(x, y).$$

We shall often find that expressions for jointly continuous random variables just take the form of the discrete case with $\sum$ replaced by $\int$, as was the case for single random variables. You may find this helpful in developing insight and intuition in trickier calculations. Of course this is only an informal guide; it is still true that for continuous random variables

(6) $$P(X = x, Y = y) = 0.$$

Here are some simple examples.

Example 6.3.1.   Let X and Y have joint density
$$f(x, y) = cxy, \quad 0 \leqslant x, y \leqslant 1.$$

What is c?

Solution.   We know $f \geqslant 0$, so $c \geqslant 0$. Now by (4)
$$\int_0^1 \int_0^1 cxy\, dx\, dy = c \int_0^1 x\, dx \int_0^1 y\, dy = \frac{c}{4} = 1$$

Hence $c = 4$.   ○

Example 6.3.2.   Let X and Y have joint density
$$f(x, y) = cxy, \quad 0 \leqslant x < y \leqslant 1.$$

What is c?

Solution.   Again $c \geqslant 0$. By (4)
$$\int_0^1 \int_x^1 cxy\, dy\, dx = \int_0^1 cx \int_x^1 y\, dy\, dx$$
$$= \int_0^1 \frac{cx}{2}(1 - x^2)\, dx = \frac{c}{8} = 1.$$

Hence $c = 8$. This is also obvious from example 6.3.1, by symmetry.   ○

One important example is worth emphasizing as a definition.

Definition.   Let the region A have area $|A|$. Then X and Y are said to be *jointly uniform* on A if they have joint density

(7) $$f(x, y) = \begin{cases} |A|^{-1}, & (x, y) \in A \\ 0 & \text{elsewhere.} \end{cases}$$ △

In general, just as for discrete random variables, use of the key rule provides us with any required probability statement about X and Y. In particular we note the

Marginal densities. If X and Y have density $f(x, y)$ then X has density

(8)
$$f_X(x) = \int_{-\infty}^{\infty} f(x, y) \, dy$$

and Y has density

(9)
$$f_Y(y) = \int_{-\infty}^{\infty} f(x, y) \, dx.$$

These both follow from a trivial application of the key rule (5). As usual, the ideas and definitions that we give for pairs of random variables extend naturally but tediously to larger collections.

Here are several examples to illustrate what we have said.

Example 6.3.3. Let X and Y have joint density
$$f(x, y) = c(x + y), \quad 0 \leqslant x, \, y \leqslant 1.$$
Find (i) the value of c, (ii) $P(X > Y)$, (iii) the marginal density $f_X(x)$, (iv) $P(Y < \frac{1}{2})$, (v) $P(X < Y^2)$.

Solution. For (i): By condition (4) we have
$$1 = \int_0^1 \int_0^1 c(x + y) \, dx \, dy = c\left(\int_0^1 x \, dx + \int_0^1 y \, dy \right) = c.$$

For (ii): By symmetry, $P(X > Y) = P(Y > X) = \frac{1}{2}$.
For (iii): By (8),
$$f_X(x) = \int_0^1 (x + y) \, dy = x + \tfrac{1}{2}, \quad 0 \leqslant x \leqslant 1.$$
For (iv): Likewise $f_Y(y) = y + \frac{1}{2}$, and so
$$P(Y < \tfrac{1}{2}) = \int_0^{1/2} (y + \tfrac{1}{2}) \, dy = \tfrac{3}{8}.$$

For (v): By the key rule,
$$P(X < Y^2) = \iint_{0 \leqslant x < y^2 \leqslant 1} (x + y) \, dx \, dy$$
$$= \int_0^1 \int_0^{y^2} (x + y) \, dx \, dy = \int_0^1 \tfrac{1}{2} y^4 + y^3 \, dy$$
$$= \tfrac{7}{20}. \qquad \qquad \bigcirc$$

Example 6.3.4. Suppose X and Y have the joint density
$$f(x, y) = c(x + y), \quad 0 \leqslant x + y \leqslant 1; \, x, \, y > 0.$$
What is c?

Solution.   As usual we must have

$$1 = \iint f(x, y)\, dx\, dy = \int_0^1 \int_0^{1-y} c(x + y)\, dx\, dy$$

$$= \int_0^1 c\left\{\frac{(1 - y)^2}{2} + y(1 - y)\right\} dy = \frac{c}{3}.$$

So $c = 3$.   ○

Example 6.3.5.   Let (X, Y) be uniformly distributed over the circle centred at the origin, radius 1. Find (i) the joint density of X and Y, (ii) the marginal density of X.

Solution.   For (i): Since the circle has area π,

$$f(x, y) = \frac{1}{\pi}, \quad 0 \leqslant x^2 + y^2 \leqslant 1.$$

For (ii): As usual,

$$f_X(x) = \int_{-\infty}^{\infty} f(x, y)\, dy = \int_{-(1-x^2)^{1/2}}^{+(1-x^2)^{1/2}} \frac{1}{\pi}\, dy = \frac{2}{\pi}(1 - x^2)^{1/2}, \quad |x| \leqslant 1. \qquad ○$$

Now let us consider an unbounded pair of jointly distributed random variables.

Example 6.3.6.   Let (X, Y) have joint density given by

$$f(x, y) = \lambda\mu e^{-\lambda x - \mu y}, \quad 0 \leqslant x, \, y < \infty.$$

Then $f_X(x) = \lambda e^{-\lambda x}$ and $f_Y(y) = \mu e^{-\mu y}$, so X and Y are random variables with exponential densities having parameters λ and μ respectively.

Hence, for example, we may calculate

$$(10) \qquad P(Y > X) = \iint_{y > x} \lambda\mu e^{-\lambda x - \mu y}\, dx\, dy$$

$$= \int_0^{\infty} \int_x^{\infty} \lambda\mu e^{-\mu y}\, dy\, e^{-\lambda x}\, dx$$

$$= \int_0^{\infty} \lambda e^{-\mu x - \lambda x}\, dx$$

$$= \frac{\lambda}{\lambda + \mu}. \qquad ○$$

As in the case of a single continuous random variable, the distribution function is often useful.

Definition.   Let X and Y have joint density $f(x, y)$. The *joint distribution function* of X and Y is denoted by $F(x, y)$, where

$$(11) \qquad F(x, y) = P(X \leqslant x, Y \leqslant y) = \int_{-\infty}^{x} \int_{-\infty}^{y} f(u, v)\, dv\, du. \qquad \triangle$$

It is related to the density by differentiation also,

(12)
$$\frac{\partial^2}{\partial x \partial y} F(x, y) = f(x, y),$$

and it yields the marginal distributions of X and Y:
$$F_X(x) = F(x, \infty), \quad F_Y(y) = F(\infty, y).$$
Obviously $0 \le F(x, y) \le 1$, and F is non-decreasing as x or y increases. Furthermore, as in the discrete case we have

(13) $P(a \le X \le b, c \le Y \le d) = F(b, d) - F(a, d) - F(b, c) + F(a, c).$

The distribution function turns out to be most useful when we come to look at functions of jointly distributed random variables. For the moment we just look at a couple of simple examples.

Example 6.3.7.   Let X and Y have joint density $f(x, y) = x + y, 0 \le x, y \le 1$. Then for $0 \le x, y \le 1$

$$F(x, y) = \int_0^y \int_0^x (u + v) \, du \, dv = \tfrac{1}{2} xy(x + y).$$

For $0 \le x \le 1, y > 1$,

$$F(x, y) = \int_0^1 \int_0^x (u + v) \, du \, dv = \tfrac{1}{2} x(x + 1).$$

For $0 \le y \le 1, x > 1$,

$$F(x, y) = \tfrac{1}{2} y(y + 1).$$

Obviously for $x, y > 1$ we have $F(x, y) = 1$. $\circ$

Example 6.3.8.   Let X and Y have the joint distribution function
$$F(x, y) = 1 - \tfrac{1}{2} x^{-2} y^{-2} \{ x^2 + y^2 + (xy - 1)(x + y) \}, \quad x, y \ge 1.$$
What is the joint density of X and Y? Find also the marginal densities of X and Y.

Solution.   Routine differentiation shows that
$$f(x, y) = \frac{\partial^2 F}{\partial x \partial y} = \frac{x + y}{x^3 y^3}, \quad x, y \ge 1.$$
Remembering that $F_X(x) = F(x, \infty)$, we find that
$$F_X(x) = 1 - \frac{x + 1}{2x^2}, \quad x > 1,$$

and so
$$f_X(x) = \frac{dF_X}{dx} = \frac{2 + x}{2x^3}, \quad x > 1.$$

Likewise

$$f_Y(y) = \frac{2 + y}{2y^3}, \quad y > 1.$$ $\circ$

Example 6.3.4 revisited.   If X and Y have joint density
$$f(x, y) = 3(x + y), \ 0 \leqslant x + y \leqslant 1; \quad 0 < x, \, y,$$
then their joint distribution is
$$F(x, y) = \begin{cases} \frac{3}{2}(x^2 y + y^2 x), & 0 \leqslant x + y \leqslant 1 \\ \frac{1}{2}(x + y)(3 - x^2 - y^2 + xy) - 1, & 1 \leqslant x + y \leqslant 2, \, x < 1, \, y < 1. \end{cases}$$

Exercises for section 6.3

1. Let X and Y have the joint density
$$f(x, y) = e^{-y}, \quad 0 \leqslant x < y < \infty.$$
    Find the joint distribution function of X and Y, and the marginal density and distribution functions.

2. If U and V are jointly continuous, show that $P(U = V) = 0$. Let X be uniform on $[0, 1]$ and let $Y = X$. Obviously $P(X = Y) = 1$, and X and Y are continuous. Is there any contradiction here? Explain.

3. Let $F(x, y) = 1 - e^{-xy}, \ 0 \leqslant x, \, y < \infty$. This is zero on the axes and increases to 1 as x and y increase to infinity. Is $F(x, y)$ a joint distribution function?

6.4 INDEPENDENCE

The concept of independence has been useful and important on many previous occasions. Recall that events A and B are independent if
$$(1) \qquad\qquad\qquad P(A \cap B) = P(A)P(B).$$
In section 5.9 we noted that the event A and the random variable X are independent if, for any C,
$$(2) \qquad\qquad\qquad P(\{X \in C\} \cap A) = P(X \in C)P(A).$$
It therefore comes as no surprise that the following definition is equally useful and important.

Definition.   Let X and Y have joint distribution function $F(x, y)$. Then X and Y are *independent* if and only if for all x and y
$$(3) \qquad\qquad\qquad F(x, y) = F_X(x)F_Y(y). \qquad\qquad\qquad \triangle$$

Remark.   We can relate this to our basic concept of independence in (1) by noting that (3) says
$$P(B_x \cap B_y) = P(B_x)P(B_y),$$
where $B_x = \{\omega: X(\omega) \leqslant x\}$ and $B_y = \{\omega: Y(\omega) \leqslant y\}$.

As usual, the general statement (3) implies different special forms for discrete and continuous random variables.

Discrete case.   If X and Y have the joint discrete distribution $p(x, y)$, then X and Y are independent if, for all x, y,
$$(4) \qquad\qquad\qquad p(x, y) = p_X(x)p_Y(y).$$

Continuous case. If X and Y have joint density $f(x, y)$, then they are independent if, for all x, y,

(5) $$f(x, y) = f_X(x)f_Y(y).$$

In any case the importance of independence lies in the special form of the key rule.

Key rule for independent random variables. If X and Y are independent then, for any events $\{X \in A\}$ and $\{Y \in B\}$,

(6) $$P(X \in A, Y \in B) = P(X \in A)P(Y \in B).$$

The practical implications of this rule explain why independence is mostly employed in two converse ways:

(i) We assume that X and Y are independent, and use the rule to find their joint behaviour.
(ii) We find that the joint distribution of X and Y satisfies (1), and deduce that they are independent; this simplifies all future calculations.

Of these, (i) is the more usual. Note that all these ideas and definitions are extended in obvious and trivial ways to any sequence of random variables. Here are some simple examples.

Example 6.4.1. Pick a card at random from a conventional pack of 52 cards. Let X denote the suit (in bridge order, so $X(C) = 1$, $X(D) = 2$, $X(H) = 3$, $X(S) = 4$), and Y the rank with aces low (so $1 \leqslant Y \leqslant 13$). Then for any x, y

$$P(X = x, Y = y) = \tfrac{1}{52} = \tfrac{1}{4} \times \tfrac{1}{13}$$

$$= P(X = x)P(Y = y).$$

So the rank and suit are independent, which is of course already known to you. ○

Example 6.4.2. Pick a point (X, Y) uniformly at random in the rectangle $0 \leqslant x \leqslant a$, $0 \leqslant y \leqslant b$. Then

$$f(x, y) = \frac{1}{ab}$$

The marginal densities are

$$f_X(x) = \int_0^b f(x, y)dy = \frac{1}{a},$$

and

$$f_Y(y) = \frac{1}{b}.$$

Hence

$$f(x, y) = \frac{1}{ab} = f_X(x)f_Y(y),$$

and X and Y are independent. ○

In fact it follows from our definition of independence that if the joint distribution

$F(x, y)$ factorizes as the product of a function of x and a function of y, for all x, y, then X and Y are independent.

However, a little thought is needed in applying this result.

Example 6.4.3: Bernoulli trials.   In n trials the joint distribution of the number of successes X and the number of failures Y is

$$p(x, y) = n!\, \frac{p^x\,(1-p)^y}{x!\quad y!}.$$

This *looks* like a product of functions of x and y, but of course it is not, because it is only valid for $x + y = n$. Here X and Y are not independent.   ○

Example 6.4.4.   Let X and Y have joint density

$$f(x, y) = \pi^{-1}$$

for $(x, y) \in C$, where C is the unit circle. This is a uniform density and resembles example 6.4.2, but here of course X and Y are not independent. To see this, calculate the marginal densities

$$f_X(x) = \int_{-(1-x^2)^{1/2}}^{+(1-x^2)^{1/2}} \pi^{-1}\, dy = 2\pi^{-1}(1 - x^2)^{1/2}$$

and

$$f_Y(y) = 2\pi^{-1}(1 - y^2)^{1/2}.$$

Obviously $f(x, y) \neq f_X(x) f_Y(y)$.   ○

Example 6.4.5.   Let X and Y be discrete, with

$$p(x, y) = \theta^{x+y+2}, \quad 0 \leqslant x, y < \infty.$$

Here $p(x, y)$ does factorize for all x and y, so X and Y may be independent, provided that p is a distribution. If we calculate

$$p_X(x) = \sum_{y=0}^{\infty} \theta^{x+y+2} = \theta^{x+2}(1 - \theta)^{-1}$$

and likewise

$$p_Y(y) = \theta^{y+2}(1 - \theta)^{-1},$$

then we find

$$p(x, y) = \theta^{x+y+2} = p_X(x) p_Y(y) = \theta^{x+y+4}(1 - \theta)^{-2}$$

provided that

$$\theta^2 = (1 - \theta)^2$$

which entails $\theta = \frac{1}{2}$. If $\theta \neq \frac{1}{2}$, then $p(x, y)$ is not a distribution.   ○

Example 6.4.6: independent normal random variables.   Let X and Y be independent standard normal random variables. Then, by definition, their joint density is

(7) $$f(x, y) = f_X(x) f_Y(y) = \frac{1}{2\pi}\, \exp\{-\tfrac{1}{2}(x^2 + y^2)\}.$$

The most striking property of this density is that it has rotational symmetry about the origin. That is to say, in polar coordinates with $r^2 = x^2 + y^2$ we have

$$f(x, y) = \frac{1}{2\pi} \exp(-\tfrac{1}{2}r^2).$$

This does not depend on the angle $\theta = \tan^{-1}(y/x)$. Roughly speaking, the point (X, Y) is equally likely to lie in any direction from the origin. Hence, for example,

$$P(0 < Y < X) = P((X, Y) \text{ lies in the first octant})$$

$$= \tfrac{1}{8}.$$

As a bonus, we can use (7) to prove what we skipped over in chapter 5, namely that

$$\int_{-\infty}^{\infty} \exp(-\tfrac{1}{2}x^2)\, dx = \sqrt{2\pi}.$$

To see this, let X and Y be independent with the same density

$$f(x) = c \exp(-\tfrac{1}{2}x^2).$$

Then (X, Y) has joint density

$$f(x, y) = c^2 \exp\{-\tfrac{1}{2}(x^2 + y^2)\}.$$

Since $f(x, y)$ is a density, we have by appendix 5.11 that

$$1 = \iint f(x, y)\, dx\, dy = \iint c^2 \exp(-\tfrac{1}{2}r^2)\, r\, dr\, d\theta$$

$$= c^2 \int_0^{2\pi} d\theta \int_0^{\infty} r e^{-r^2/2}\, dr$$

$$= c^2 2\pi$$

as required. ◯

Example 6.4.7: independent uniform random variables. Let $X, Y,$ and Z be independent, and each be uniformly distributed on $(0, 1)$. Let U be the smallest of the three, W the largest, and V the intermediate, so that $U < V < W$. Find the joint density of $U, V,$ and W.

Solution. Let $F(u, v, w)$ be the joint distribution of $U, V,$ and W. By a slight extension of (13) in section 6.3, or by use of inclusion–exclusion, or simply by inspection, we see that

$$F(u, v, w) = P(U < u < V < v < W < w)$$

$$+ F(u, v, v) + F(u, u, v) - F(u, u, u).$$

Now (U, V, W) must be one of the six permutations of X, Y, Z, depending on whether $X < Y < Z$, or $Y < X < Z$, or $Z < Y < X$, etc. For any of these permutations, say the first,

$$P(U < u < V < v < W < w) = P(X < u < Y < v < Z < w)$$

$$= u(v - u)(w - v)$$

by the independence of X, Y, and Z. Since there are six such disjoint possibilities, we have

(8) $F(u, v, w) = 6u(v - u)(w - v) + F(u, v, v) + F(u, u, v) - F(u, u, u).$

Now we obtain the required density by differentiating (8) with respect to u, v, and w. Hence

$$f(u, v, w) = 6, \quad 0 < u < v < w < 1. \qquad \bigcirc$$

As an application of this, suppose that three points are placed at random on $[0, 1]$, independently. What is the probability that no two are within a distance $\frac{1}{4}$ of each other? By the key rule, this is

$$\iiint_{\substack{v-u>1/4, \\ w-v>1/4}} f(u, v, w)\, du\, dv\, dw = \int_0^{1-1/2} \int_{u+1/4}^{1-1/4} \int_{v+1/4}^1 6\, dw\, dv\, du = \tfrac{1}{8},$$

after a little calculation.

We conclude by noting that any collection of random variables $X_1, \ldots, X_n$ is said to be (mutually) independent if for all $x_1, \ldots, x_n$, we have

(9) $F(x_1, \ldots, x_n) = F_{X_1}(x_1) \cdots F_{X_n}(x_n).$

That is to say, the joint distribution is the same as the product of all the marginal distributions. It follows that disjoint subsets of a collection of independent random variables are also independent.

Exercises for section 6.4

1. Let X, Y, and Z be independently and uniformly distributed over $[0, 1]$. Show that $P(Z \leqslant XY) = \frac{1}{4}$.

2. Let X and Y have joint distribution
    $$p(x, y) = \theta^{x+y+2}, \quad 0 \leqslant x < y < \infty.$$
    For what values (if any) of θ is this possible? Find the marginal distributions of X and Y. Are X and Y independent?

3. Let X and Y have joint density
    $$f(x, y) = 1, \quad 0 \leqslant x, y \leqslant 1.$$
    Let $U = \min\{X, Y\}$, $V = \max\{X, Y\}$. Show that (U, V) has the joint density
    $$f(u, v) = 2, \quad 0 \leqslant u \leqslant v \leqslant 1.$$

4. Let X, Y, and Z be independently distributed on $[0, 1]$ with common density $f(x)$. Let U be the smallest, V the next, and W the largest. Show that U, V, and W have joint density
    $$f(u, v, w) = 6f(u)f(v)f(w), \quad 0 < u < v < w < 1.$$

6.5 FUNCTIONS

We have seen in section 5.7 that it is very easy to deal with functions of a single random variable. Most problems in real life involve functions of more than one random variable, however, and these are more interesting.

Example 6.5.1.   Your steel mill rolls a billet of steel from the furnace. It is 10 metres long but, owing to the variations to be expected in handling several tons of white-hot metal, the height and width are random variables X and Y. The volume is the random variable $Z = 10XY$.

What can we say of Z? ○

The general problem for functions of two random variables amounts to this: let the random variable

$$Z = g(X, Y)$$

be a function of X and Y; what is the distribution of Z?

As we did for single random variables it is convenient to deal separately with the discrete and continuous cases; also, as it was for single random variables, the answer to our question is supplied by the key rules for joint distributions.

Discrete case.   Let the discrete random variables X, Y, and Z satisfy

$$Z = g(X, Y).$$

Then by (4) of section 6.2 we have

(1) $$p_Z(z) = P(Z = z) = \sum_{\substack{x, y: \\ z = g(x, y)}} p(x, y).$$

Continuous case.   Let the random variables X, Y, and Z satisfy $Z = g(X, Y)$, where X and Y have density $f(x, y)$. Then by (5) of section 6.3,

(2) $$F_Z(z) = P(Z \leq z) = \iint_{\substack{x, y: \\ g(x, y) \leq z}} f(x, y) \, dx \, dy.$$

When Z is also continuous, its density $f_Z(z)$ is easily obtained by differentiating (2) above.

These ideas are best grasped by inspection of examples.

Example 6.5.2.   (i) Suppose two numbers X and Y are picked at random from $\{1, 2, \ldots, 49\}$, without replacement. What is the distribution of $Z = \max\{X, Y\} = X \vee Y$?

(ii) Lottery revisited. Let Z be the largest of six numbers picked from $\{1, 2, \ldots, 49\}$ in a draw for the lottery. What is the distribution of Z?

Solution.   For (i): We know that the joint distribution of (X, Y) is

$$p(x, y) = \begin{cases} \frac{1}{49} \times \frac{1}{48}, & x \neq y \\ 0, & x = y. \end{cases}$$

Hence, by (1), Z has distribution

$$p(z) = \sum_{x \lor y = z} p(x, y)$$

$$= \sum_{x=1}^{z-1} p(x, z) + \sum_{y=1}^{z-1} p(z, y)$$

$$= (z-1)\binom{49}{2}^{-1}, \quad 2 \leqslant z \leqslant 49$$

For (ii): With an obvious notation,

$$p(x_1, \ldots, x_6) = \begin{cases} \frac{1}{49} \times \frac{1}{48} \times \frac{1}{47} \times \frac{1}{46} \times \frac{1}{45} \times \frac{1}{44} & x_i \neq x_j, \, 1 \leqslant i, j \leqslant 6 \\ 0 & \text{otherwise.} \end{cases}$$

Using (1) again gives, after a little calculation,

$$p(z) = \binom{z-1}{5} \bigg/ \binom{49}{6}, \quad 6 \leqslant z \leqslant 49. \qquad \circ$$

Equation (2) can be used for similar purposes with continuous random variables.

Example 6.5.3.   Let X and Y have joint density
$$f(x, y) = e^{-y}, \quad 0 < x < y < \infty.$$
Find the density of $Z = \max\{X, Y\}$.

Solution.   By (5),

$$P(Z \leqslant z) = \iint_{x \lor y \leqslant z} f(x, y) \, dx \, dy.$$

Since $X < Y$ always, this calculation is rather an easy one. In fact $Z = Y$, so Z has the density

$$f(z) = f_Y(z) = z e^{-z}, \quad z > 0. \qquad \circ$$

The function of X and Y that is most often of interest is their sum.

Example 6.5.4: sum of uniform random variables.   Let X and Y be independent and uniform on $[0, 1]$. What is the density of $Z = X + Y$?

Solution.   As with functions of one continuous random variable, it is convenient to use the distribution function. Thus, by the key rule,

$$F_Z(z) = P(X + Y \leqslant z) = \iint_{x+y \leqslant z} f(x, y) \, dx \, dy,$$

which is just the volume of the wedge cut off by the plane $x + y = z$. Either by elementary geometry, or by plodding through the integrals, you find

$$F_Z(z) = \begin{cases} \frac{1}{2} z^2, & 0 \leqslant z \leqslant 1 \\ 1 - \frac{1}{2}(2 - z)^2, & 1 \leqslant z \leqslant 2. \end{cases}$$

Hence, differentiating,

$$f_Z(z) = \begin{cases} z, & 0 \leqslant z \leqslant 1 \\ 2 - z, & 1 \leqslant z \leqslant 2. \end{cases}$$

This is a triangular density. ○

Example 6.5.5. Let X and Y have the joint density derived in example 6.4.6, so that

$$f(x, y) = \frac{1}{2\pi} \exp\{-\tfrac{1}{2}(x^2 + y^2)\}.$$

What is the distribution of $Z = (X^2 + Y^2)^{1/2}$?

Solution. From (2) we have that

$$F_Z(z) = P(Z \leqslant z) = \iint_{\substack{x, y: \\ x^2 + y^2 \leqslant z^2}} \frac{1}{2\pi} \exp\{-\tfrac{1}{2}(x^2 + y^2)\} \, dx \, dy$$

$$= \int_0^z \exp(-\tfrac{1}{2}r^2) \, r \, dr$$

$$= 1 - e^{-z^2/2}, \quad z > 0.$$

Since we can differentiate this, we find that Z has density

$$f_Z(z) = z \exp(-\tfrac{1}{2}z^2). \qquad ○$$

Often the direct use of independence allows calculations to be carried out simply, without using (1) or (2).

Example 6.5.6. Let X and Y be independent geometric random variables with parameters α and β respectively. Find the distribution of $Z = X \wedge Y$, their minimum.

Solution. Arguing directly,

$$P(Z \leqslant z) = 1 - P(Z > z)$$

$$= 1 - P(X > z, Y > z)$$

$$= 1 - P(X > z)P(Y > z), \quad \text{by independence}$$

$$= 1 - (1 - \alpha)^z(1 - \beta)^z.$$

Hence

$$P(Z = z) = \{(1 - \alpha)(1 - \beta)\}^{z-1}\{1 - (1 - \alpha)(1 - \beta)\}, \quad z \geqslant 1.$$

Thus Z is also geometric. ○

Joint distributions of two or more functions of several random variables are obtained in much the same way, only with a good deal more toil and trouble. We look at a few simple examples here; a general approach to transformations of continuous random variables is deferred to section 6.13.

Example 6.5.7. Let X and Y be independent and geometric, with parameters α and β respectively. Define

$$U = \min\{X, Y\} = X \wedge Y,$$

$$V = \max\{X, Y\} = X \vee Y,$$

$$W = V - U.$$

Find the joint probability distribution of U and V, and of U and W. Show that U and W are independent of each other.

Solution. X and Y are independent, so

$$p_{X,Y}(x, y) = (1 - \alpha)^{x-1}(1 - \beta)^{y-1}\alpha\beta, \quad 1 \leqslant x, y < \infty.$$

Hence, since either $X \leqslant Y$ or $Y \leqslant X$,

$$P(U = j, V = k) = \{(1 - \alpha)^{j-1}(1 - \beta)^{k-1}$$

$$+ (1 - \alpha)^{k-1}(1 - \beta)^{j-1}\}\alpha\beta, \quad j \leqslant k.$$

Likewise

$$P(U = j, W = k) = \{(1 - \alpha)^{j-1}(1 - \beta)^{j+k-1}$$

$$+ (1 - \alpha)^{j+k-1}(1 - \beta)^{j-1}\}\alpha\beta$$

$$= \{(1 - \alpha)(1 - \beta)\}^{j-1}\{(1 - \beta)^{k} + (1 - \alpha)^{k}\}\alpha\beta.$$

Hence U and W are independent, by (4) of section 6.4.   ○

A similar result is true for exponential random variables; this is important in more advanced probability.

Example 6.5.8. Let X and Y be independent exponential random variables, with parameters λ and μ respectively. Define

$$U = X \wedge Y,$$

$$W = X \vee Y - U.$$

Find the joint density of U and W, and show that they are independent.

Solution. The joint density of X and Y is

$$f(x, y) = \lambda\mu e^{-\lambda x - \mu y}, \quad 0 \leqslant x, y < \infty.$$

The probability $P(U > u, W \leqslant w)$ is given by the key rule:

$$P(U > u, W \leqslant w) = \iint_{\substack{x, y: x \wedge y > u \\ x \vee y - u \leqslant w}} f(x, y)\, dx\, dy.$$

The region of integration is in two parts; either

$$u < y < x < y + w \quad \text{or} \quad u < x < y < x + w.$$

We can calculate this as follows:

$$\int_u^\infty \int_y^{y+w} f \, dx \, dy + \int_u^\infty \int_x^{x+w} f \, dy \, dx$$

$$= \int_u^\infty \mu e^{-\mu y} \left\{ e^{-\lambda y} - e^{-\lambda(y+w)} \right\} dy + \int_u^\infty \lambda e^{-\lambda x} \left\{ e^{-\mu x} - e^{-\mu(x+w)} \right\} dx$$

$$= e^{-(\mu+\lambda)u} - \frac{\mu}{\lambda+\mu} e^{-(\lambda+\mu)u-\lambda w} - \frac{\lambda}{\lambda+\mu} e^{-(\lambda+\mu)u-\mu w}.$$

Hence we can find the joint density by differentiation, yielding

$$f(u, w) = \frac{\lambda\mu}{\lambda+\mu} (e^{-\lambda w} + e^{-\mu w})(\lambda + \mu) e^{-(\lambda+\mu)u}.$$

Thus U and W, where U is exponential with parameter $\lambda + \mu$ and W has density

$$f_W(w) = \frac{\lambda\mu}{\lambda+\mu} (e^{-\lambda w} + e^{-\mu w}), \quad 0 \leqslant w < \infty,$$

are independent by (5) of section 6.4. ○

If we are sufficiently careful and persistent, we can establish even more surprising results in this way. Here is one final illustration.

Example 6.5.9. Let X and Y be independent and exponential, both with parameter 1. Define

$$U = X + Y,$$

$$V = \frac{X}{X+Y}.$$

Find the joint density of U and V. Deduce that U and V are independent, and find their marginal density functions.

Solution. Here the joint density of X and Y is given by

$$f(x, y) = e^{-(x+y)}, \quad 0 \leqslant x, y < \infty.$$

Let us seek the probability $P(U \geqslant u, V \geqslant v)$. In the XY-plane this corresponds to

(3) $$P\left(X + Y \geqslant u, \frac{X}{X+Y} \geqslant v\right) = P\left(X + Y \geqslant u, \ Y \leqslant \frac{1-v}{v} X\right).$$

This probability is given by the integral of f over the infinite shaded area in Figure 6.1. Hence (3) becomes

(4) $$\int_0^{u-uv} \int_{u-y}^\infty e^{-x-y} \, dx \, dy + \int_{u-uv}^\infty \int_{vy/(1-v)}^\infty e^{-x-y} \, dx \, dy$$

$$= \int_0^{u-uv} e^{-u} \, dy + \int_{u-uv}^\infty e^{-y} e^{-vy/(1-v)} \, dy$$

$$= e^{-u}(u - uv + 1 - v).$$

Differentiating (4) with respect to u and v yields the joint density of U and V as

$$f(u, v) = u e^{-u}, \quad 0 \leqslant v \leqslant 1, \quad 0 \leqslant u < \infty.$$

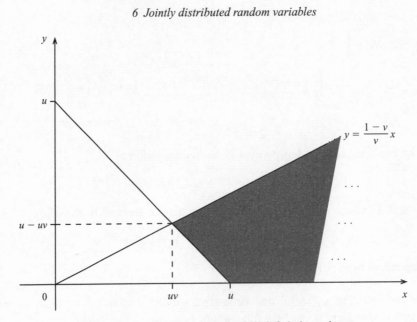

Figure 6.1. The shaded area extends indefinitely as shown.

Hence, surprisingly, we find that U and V are independent, where U has a gamma density and V is uniform on $(0, 1)$.   ◯

Exercises for section 6.5

1. Let X, Y, and Z be independent and uniform on $[0, 1]$, so that their joint density is $f(x, y, z) = 1$. Show that the density of their sum $W = X + Y + Z$ is
$$f_W(w) = \begin{cases} \frac{1}{2}w^2, & 0 \leqslant w \leqslant 1 \\ \frac{3}{4} - \left(w - \frac{3}{2}\right)^2, & 1 \leqslant w \leqslant 2 \\ \frac{1}{2}(3 - w)^2, & 2 \leqslant w \leqslant 3. \end{cases}$$

2. Let Y be the minimum of six numbers drawn from $\{1, 2, \ldots, 49\}$ for the lottery. Find the distribution of Y.

3. Let X and Y be geometric and independent, with parameters α and β respectively, as in example 6.5.6. Show that the distribution of $Z = |X - Y|$ is
$$p(z) = \begin{cases} \dfrac{\alpha\beta}{\alpha + \beta - \alpha\beta}\{(1 - \alpha)^z + (1 - \beta)^z\}, & z > 0 \\ \dfrac{\alpha\beta}{\alpha + \beta - \alpha\beta}, & z = 0. \end{cases}$$

6.6 SUMS OF RANDOM VARIABLES

In the previous section we looked in a general way at how to find the probability distributions of various functions of random variables. In practical applications it most often turns out that we are interested in the sum of random variables. For example:

- The success or failure of an insurance company or bank depends on the cumulative sum of payments in and out.

- We have noted above that practical estimates or measurements very often use the sample mean of observations X_r, that is,

$$\overline{X} = \frac{1}{n} \sum_{r=1}^{n} X_r.$$

- Quality control often concerns itself with the total sum of errors or defective items in some process.

And so on; think of some more yourself. In this section we therefore look at various ways of finding the distributions of sums of random variables. We begin with some easy examples; in particular, we first note that in a few cases we already know the answer.

Example 6.6.1: binomial sum. Let X be the number of successes in m independent Bernoulli trials and Y the number of successes in n further Bernoulli trials, all with parameter p. Then if $Z = X + Y$, it follows that Z is the number of successes in $m + n$ Bernoulli trials. However, by this construction we know that X is binomial B(m, p), Y is binomial B(n, p), and their sum Z is binomial B($m + n$, p). ○

Example 6.6.2: craps. Let X and Y be independent and uniform on $\{1, 2, 3, 4, 5, 6\}$; define $Z = X + Y$. Then Z is just the sum of the scores of two dice, and we know that

$$f(z) = \begin{cases} \frac{1}{36}(z - 1), & 2 \leqslant z \leqslant 7 \\ \frac{1}{36}(13 - z), & 7 < z \leqslant 12. \end{cases}$$ ○

This is useful but limited; the time has come to give a general approach to this problem. As with much else, the answer is supplied by the key rule for joint distributions.

Sum of discrete random variables. Let X and Y have joint distribution $p(x, y)$, and let $Z = X + Y$. Then, by (1) of section 6.5,

$$(1) \qquad p_Z(z) = P(X + Y = z) = \sum_{\substack{x, y: \\ x+y=z}} p(x, y)$$

$$= \sum_{x} p(x, z - x).$$

Convolution rules. When X and Y are independent we know that $p(x, y) = p_X(x)p_Y(y)$, and (1) takes the most important form

$$(2) \qquad p_Z(z) = \sum_{x} p_X(x)p_Y(z - x).$$

When X and Y are also non-negative, the sum reduces to

$$(3) \qquad p_Z(z) = \sum_{x=0}^{z} p_X(x)p_Y(z - x).$$

Here are the equivalent results for the sum of continuous random variables.

Sum of continuous random variables.   When X and Y are continuous, and $Z = X + Y$, the rules above take the following forms.

Probability density of a sum.   If the pair (X, Y) has density $f(x, y)$, the density of $Z = X + Y$ is given by

$$(4) \qquad\qquad f_Z(z) = \int_{-\infty}^{\infty} f(x, z - x)\, dx.$$

Convolution rules.   When X and Y are independent then

$$(5) \qquad\qquad f_Z(z) = \int_{\infty}^{\infty} f_X(x) f_Y(z - x)\, dx.$$

And when X and Y are non-negative this becomes

$$(6) \qquad\qquad f_Z(z) = \int_{0}^{z} f_X(x) f_Y(z - x)\, dx.$$

These rules are extremely plausible, since they are just the integral versions of the summation rules that apply to discrete random variables. Their plausibility can be reinforced by an informal consideration of the kind we have used before, as follows.

Roughly speaking, since $Z = X + Y$, when dx and dz are small

$$P(x < X \leqslant x + dx,\ z < Z \leqslant z + dz) = P(x < X \leqslant x + dx,\ z - x \leqslant Y \leqslant z - x + dz)$$

$$\simeq f(x, z - x)\, dx\, dz.$$

Now integrating with respect to x gives the marginal distribution of Z:

$$f_Z(z)\, dz \simeq P(z < Z \leqslant z + dz) = \left\{ \int f(x, z - x) dx \right\} dz,$$

which agrees with (4).

This supplies an intuitive feeling for the truth of the convolution rules; if you require a proof, it runs as follows.

Proof of (4).   Let X and Y have joint density $f(x, y)$, with $Z = X + Y$. Then by the key rule for joint densities, (5) of section 6.3,

$$(7) \qquad\qquad F_Z(z) = P(Z \leqslant z) = \iint_{x + y \leqslant z} f(x, y)\, dx\, dy$$

$$= \int_{x=-\infty}^{\infty} \int_{y=-\infty}^{z-x} f(x, y)\, dy\, dx.$$

Now differentiating this with respect to z, and recalling the fundamental theorem of calculus in the more general form (see appendix 4.14), gives

$$(8) \qquad\qquad f_Z(z) = \int_{x=-\infty}^{\infty} f(x, z - x)\, dx,$$

as required. $\square$

Here are some examples. In each of the following X and Y are independent, with $Z = X + Y$.

Example 6.6.3: binomial sum. If X is binomial B(n, p) and Y is binomial B(m, p), then by (2)

$$p_Z(z) = \sum_{x=0}^{z} \binom{n}{x} p^x q^{n-x} \binom{m}{z-x} p^{z-x} q^{m-z+x}$$

$$= p^z q^{m+n-z} \sum_{x=0}^{z} \binom{n}{x} \binom{m}{z-x}$$

$$= \binom{m+n}{z} p^z q^{m+n-z}.$$

Thus Z is binomial B$(m+n, p)$, as we knew already. $\quad\bigcirc$

Example 6.6.4: Poisson sum. If X and Y are Poisson, with parameters λ and μ respectively, then by (2)

$$p_Z(z) = \sum_{x=0}^{z} \frac{\lambda^x}{x!} e^{-\lambda} \frac{\mu^{z-x}}{(z-x)!} e^{-\mu} = \frac{e^{-(\lambda+\mu)}}{z!} \sum_{x=0}^{z} \binom{z}{x} \lambda^x \mu^{z-x}$$

$$= \frac{e^{-(\lambda+\mu)}}{z!} (\lambda + \mu)^z.$$

Thus Z is Poisson with parameter $\lambda + \mu$. $\quad\bigcirc$

Example 6.6.5: geometric sum. Let X and Y be independent and geometric with parameter p. Find the mass functions of $U = X + Y$ and $V = X - Y$.

Solution. By (2),

$$p_U(u) = \sum_{1}^{u-1} p_X(x) p_Y(u-x)$$

$$= \sum_{1}^{u-1} q^{x-1} p q^{u-x-1} p$$

$$= (u-1) q^{u-2} p, \quad u \geqslant 2,$$

which is a negative binomial distribution. Now likewise, for $v > 0$,

$$p_V(v) = \sum_{x-y=v} p_X(x) p_Y(y) = \sum_{y=1}^{\infty} p_X(v+y) p_Y(y),$$

$$= p^2 \sum_{y=1}^{\infty} q^{v+y-1} q^{y-1} = p^2 q^{v-2} \frac{q^2}{1-q^2}$$

$$= \frac{pq^v}{1+q}.$$

For $v < 0$ we have

$$p_V(v) = \sum_{x-y=v} p_X(x)p_Y(y) = \sum_{x=1}^{\infty} p_X(x)p_Y(x-v)$$

$$= \sum_{x=1}^{\infty} p^2 q^{x-1} q^{x-v-1} = p^2 \frac{q^{-v-2}q^2}{1-q^2}$$

$$= \frac{pq^{-v}}{1+q}.$$

Finally, for $v = 0$,

$$p_V(0) = \sum_{x=1}^{\infty} p^2 q^{2x-2} = \frac{p^2}{1-q^2}$$

$$= \frac{p}{1+q}.$$ ○

Example 6.6.6: uniform sum.   Let X and Y be uniform on $(0, a)$. Then by (6)

$$f_Z(z) = \int_0^z f_X(x) f_Y(z-x) dx$$

$$= \begin{cases} \int_0^z a^{-2} \, dx = za^{-2} & 0 \leqslant z \leqslant a \\ \int_{z-a}^a a^{-2} \, dx = (2a-z)a^{-2}, & a \leqslant z \leqslant 2a, \end{cases}$$

which is the triangular density on $(0, 2a)$, as we already know.   ○

Example 6.6.7: exponential sum.   Let X and Y be exponential with parameter λ. Then by (6)

$$f_Z(z) = \int_0^z \lambda^2 e^{-\lambda x} e^{-\lambda(z-x)} \, dx = \lambda^2 z e^{-\lambda z},$$

which is a gamma distribution.   ○

Next we consider the sum of two independent normally distributed random variables. It is worth recalling that the normal density was first encountered in chapter 4 as an approximation to the binomial distribution, and that the sum of two independent binomial random variables (with the same parameter p) was found to be binomial in example 6.6.3. We should therefore expect that the sum of two independent normal random variables is itself normal. This is so, as we now see.

Example 6.6.8: normal sum.   Let X and Y be independent and normal with zero mean, having variances σ^2 and τ^2 respectively. We let $\sigma^2 + \tau^2 = \omega^2$. Then we have from

(5), as usual,

$$f_Z(z) = \int_{-\infty}^{\infty} \frac{1}{2\pi\sigma\tau} \exp\left\{ -\frac{x^2}{2\sigma^2} - \frac{(z-x)^2}{2\tau^2} \right\} dx$$

$$= \frac{1}{(2\pi\omega^2)^{1/2}} \int_{-\infty}^{\infty} \frac{\omega}{(2\pi)^{1/2}\sigma\tau} \exp\left\{ -\frac{\omega^2}{2\sigma^2\tau^2}\left(x - \frac{\sigma^2}{\omega^2}z\right)^2 \right\} dx$$

$$\times \exp\left(-\frac{z^2}{2\omega^2}\right)$$

$$= \frac{1}{(2\pi\omega^2)^{1/2}} \exp\left(-\frac{z^2}{2\omega^2}\right).$$

This last step follows because the integrand is just the

$$N\left(\frac{\sigma^2 z}{\omega^2}, \frac{\sigma^2\tau^2}{\omega^2}\right)$$

density, which when integrated over $\mathbb{R}$ gives 1, as always.

Thus we have shown that if X and Y are independent normal random variables, with zero mean and variances σ^2 and τ^2, then $X + Y$ is also normal, with variance $\sigma^2 + \tau^2$. ○

The same argument will give the distribution function or the survival function of a sum of random variables. For example, let X and Y be discrete and independent, with $Z = X + Y$. Then

$$(9) \qquad F_Z(z) = P(Z \leqslant z) = P(X + Y \leqslant z)$$

$$= \sum_y P(X \leqslant z - y)P(Y = y)$$

$$= \sum_y F_X(z - y)p_Y(y).$$

Likewise

$$(10) \qquad \overline{F}_Z(z) = P(Z > z) = \sum_y \overline{F}_X(z - y)p_Y(y).$$

As an application, let us reconsider Pepys' problem, which appeared in exercise (5) at the end of section 2.6. We can now solve a more general case.

Example 6.6.9: extended Pepys' problem. Show that if A_n has $6n$ dice and needs at least n sixes, then A_n has an easier task than A_{n+1}. That is to say, the chance of at least n sixes in $6n$ rolls is greater than the chance of at least $n + 1$ sixes in $6(n + 1)$ rolls.

Solution. We can divide A_{n+1}'s rolls into two groups, one of size $6n$, yielding X sixes, and the other of size 6, yielding Y sixes. Write $Z = X + Y$. Then we need to show that

$$P(Z \geqslant n + 1) \leqslant P(X \geqslant n).$$

Write $p(n) = P(X = n)$, $p_Y(n) = P(Y = n)$, and $\overline{F}(n) = P(X \geqslant n)$. We know that X is binomial $B\left(6n, \frac{1}{6}\right)$ and Y is binomial $B\left(6, \frac{1}{6}\right)$. From what we have proved about the binomial distribution, we have

(11) $\qquad p(n-5) \leqslant p(n-4) \leqslant p(n-3) \leqslant p(n-2) \leqslant p(n-1) \leqslant p(n),$

and

(12) $\qquad\qquad\qquad\qquad\qquad EY = 1.$

Now, by conditioning on Y, we obtain

$$P(Z \geqslant n+1) = P(X+Y \geqslant n+1)$$

$$= \sum_{r=0}^{6} P(X+Y \geqslant n+1 | Y = r) p_Y(r)$$

$$= \sum_{r=0}^{6} P(X \geqslant n+1-r) p_Y(r)$$

$$= \{\overline{F}(n) - p(n)\} p_Y(0) + \overline{F}(n) p_Y(1) + \cdots$$

$$+ \{\overline{F}(n) + p(n-1) + p(n-2)$$

$$+ p(n-3) + p(n-4) + p(n-5)\} p_Y(6).$$

By (11) we then have

$$P(Z \geqslant n+1) \leqslant \{\overline{F}(n) - p(n)\} p_Y(0) + \cdots$$

$$+ \{\overline{F}(n) + 4p(n)\} p_Y(5) + \{\overline{F}(n) + 5p(n)\} p_Y(6)$$

$$= \overline{F}(n) + p(n) \sum_{r=0}^{6} (r-1) p_Y(r)$$

$$= \overline{F}(n), \quad \text{by (12)}$$

$$= P(X \geqslant n)$$

as required. $\qquad\qquad\qquad\qquad\qquad\qquad\qquad\qquad\qquad\qquad\qquad\bigcirc$

Exercises for section 6.6

1. Deduce from example 6.6.9 that

$$\left(\tfrac{5}{6}\right)^6 \sum_{r=0}^{n} \binom{6(n+1)}{r} 5^{-r} \geqslant \sum_{r=0}^{n-1} \binom{6n}{r} 5^{-r}.$$

   Can you think of any other way of proving this inequality?

2. Let X and Y have joint density $f(x, y) = e^{-y}$, $0 < x < y < \infty$. Find the density of $Z = X + Y$.

3. Let $X_1, \ldots, X_n$ be independent geometric random variables with parameter p. Show that $Z = \sum_{r=1}^{n} X_r$ has the negative binomial distribution

$$p(z) = \binom{z-1}{n-1} p^n (1-p)^{z-n}, \quad z \geqslant n.$$

   (Hint: Either use induction, or think a bit.)

4. Let $X_1, \ldots, X_n$ be independent exponential random variables with parameter λ. Show that

$$Z = \sum_{r=1}^{n} X_r$$

has the gamma density

$$f(z) = \lambda^n z^{n-1} e^{-\lambda z}/(n-1)!$$

(Same hint as for exercise 3.)

6.7 EXPECTATION; THE METHOD OF INDICATORS

In the previous sections we have looked at the distribution of functions of two or more random variables. It is often useful and interesting to know the expected value of such functions, so the following results are very important.

Extended laws of the unconscious statistician. Let X, Y, and Z be random variables such that

$$Z = g(X, Y).$$

If X and Y are discrete with joint distribution $p(x, y)$, then

(1) $$EZ = \sum_x \sum_y g(x, y) p(x, y).$$

If X and Y are continuous with joint density $f(x, y)$, then

(2) $$EZ = \int_x \int_y g(x, y) f(x, y) \, dx \, dy.$$

(Note that expected values are always assumed to exist unless we explicitly remark otherwise.) The proofs of these two results are very similar to those used in the case of functions of a single random variable, and we omit them. One of the most commonly used application of these rules is in the following result.

Addition rule for expectation. For any two random variables X and Y with a joint distribution, we have

(3) $$E(X + Y) = EX + EY.$$

The proof is easy. Suppose X and Y are discrete; then by (1)

$$E(X + Y) = \sum_x \sum_y (x + y) p(x, y)$$

$$= \sum_x \sum_y x p(x, y) + \sum_y \sum_x y p(x, y)$$

$$= \sum_x x p_X(x) + \sum_y y p_Y(y), \quad \text{by (6) of section 6.2}$$

$$= EX + EY, \quad \text{by (3) of section 5.6.}$$

If X and Y have a density, then just replace $\sum$ by $\int$ in this proof, and p by f. Obviously it follows that if $X_1, X_2, \ldots, X_n$ are jointly distributed, then

(4) $$\mathrm{E}\sum_{r=1}^{n} X_r = \sum_{r=1}^{n} \mathrm{E}X_r. \qquad \square$$

Example 6.7.1: dice.   You add the scores X_r from n rolls of a die. By the above,

$$\mathrm{E}\sum_{r=1}^{n} X_r = \sum_{r=1}^{n} \mathrm{E}X_1$$

$$= \sum_{r=1}^{n} \frac{7}{2}, \quad \text{by (4)}$$

$$= \frac{7n}{2}.$$

The important thing about this trivial example is that the calculation is extremely easy, though the actual probability distribution of $\sum_{r=1}^{n} X_r$ is extremely complicated.   ○

Example 6.7.2: waiting for r successes.   Suppose you undergo a sequence of Bernoulli trials with $\mathrm{P}(S) = p$; let T be the number of trials until the rth success. What is $\mathrm{E}T$?

Solution.   We know from exercise 3 at the end of section 6.6 that T has a negative binomial distribution, so

$$\mathrm{E}T = \sum_{n=r}^{\infty} n\binom{n-1}{r-1}(1-p)^{n-r}p^r.$$

Summing this series is feasible but dull. Here is a better way. Let X_1 be the number of trials up to and including the first success, X_2 the *further* number of trials to the second success, and so on for $X_3, X_4, \ldots, X_r$. Then each of $X_1, X_2, \ldots$ has a geometric distribution with parameter p and mean p^{-1}. Hence

$$\mathrm{E}T = \mathrm{E}\sum_{k=1}^{r} X_k = rp^{-1}. \qquad ○$$

Example 6.7.3: coupon collecting.   Each packet of some ineffably dull and noxious product contains one of d different types of flashy coupon. Each packet is independently equally likely to contain any of the d types. How many packets do you expect to need to buy until the moment when you first possess all d types?

Solution.   Let T_1 be the number of packets bought until you have one type of coupon, T_2 the further number required until you have two types of coupon, and so on up to T_d. Obviously $T_1 = 1$. Next, at each purchase you obtain a new type with probability $(d-1)/d$, or not with probability $1/d$. Hence T_2 is geometric, with mean $d/(d-1)$.

Likewise, T_r is geometric with mean $d/(d-r+1)$, for $1 \leqslant r \leqslant d$. Hence the expected number of packets purchased is

(5) $$\mathrm{E}(T_1 + T_2 + \cdots + T_d) = \sum_{r=1}^{d} \mathrm{E}T_r = \sum_{r=1}^{d} d/(d-r+1). \qquad ○$$

Here is an example involving a density.

Example 6.7.4. A dart hits a plane target at the point (X, Y) where X and Y have density

$$f(x, y) = \frac{1}{2\pi} \exp\{-\tfrac{1}{2}(x^2 + y^2)\}.$$

Let R be the distance of the dart from the bullseye at the origin. Find ER^2.

Solution. Of course we could find the density of R, and then the required expectation. It is much easier to note that X and Y are each $N(0, 1)$ random variables, and then

$$ER^2 = E(X^2 + Y^2) = EX^2 + EY^2$$

$$= \operatorname{var} X + \operatorname{var} Y$$

$$= 2. \qquad\qquad \bigcirc$$

We have looked first at the addition law for expectations because of its paramount importance. But, of course, the law of the unconscious statistician works for many other functions of interest. Here are some examples.

Example 6.7.5. Let X and Y have density

$$f(x, y) = \pi^{-1}, \ x^2 + y^2 \leqslant 1.$$

Find (i) $E\{(X^2 + Y^2)^{1/2}\}$, (ii) $E|X \wedge Y|$, (iii) $E(X^2 + Y^2)$, (iv) $E\{X^2/(X^2 + Y^2)\}$.

Solution. It is clear that polar coordinates are going to be useful here. In each case by application of (2) we have
For (i):

$$E\{(X^2 + Y^2)^{1/2}\} = \int_0^{2\pi}\int_0^1 \pi^{-1} r^2 \, dr \, d\theta$$

$$= \frac{2}{3}.$$

For (ii): By symmetry the answer is the same in each octant, so

$$E|X \wedge Y| = 8\iint_{0 < y < x} yf(x, y) \, dx \, dy = \frac{8}{\pi}\int_0^{\pi/4}\int_0^1 r^2 \sin\theta \, dr \, d\theta$$

$$= \frac{8}{3\pi}\left(1 - \frac{1}{\sqrt 2}\right).$$

For (iii):

$$E(X^2 + Y^2) = \int_0^{2\pi}\int_0^1 \pi^{-1} r^3 \, dr \, d\theta = \frac{1}{2}.$$

For (iv): By symmetry $E\{X^2/(X^2 + Y^2)\} = E\{Y^2/(X^2 + Y^2)\}$, and their sum is 1. Hence

$$E\{X^2/(X^2 + Y^2)\} = \frac{1}{2}. \qquad\qquad \bigcirc$$

In concluding this section we first recall a simple but important property of indicator random variables.

Indicator property.   If X is an indicator then
(6) $$EX = P(X = 1).$$
This elementary fact is the basis of a surprisingly versatile and useful technique:

The method of indicators.   The essential idea is best illustrated by examples.

Example 6.7.6: non-homogeneous Bernoulli trials.   Suppose a complicated structure has n different elements that either fail, or work as desired. What is the expected number η of working elements?

Solution.   Let X_i be the indicator of the event that the ith element works. Since the elements are *different* we have $EX_i = p_i$, where p_i is not necessarily equal to p_j, for $i \neq j$. Nevertheless

(7) $$\eta = E \sum_{i=1}^{n} X_i = \sum_{i=1}^{n} p_j.$$

The point here is that we do not need to know anything about the joint distribution of the failures of elements; we need only know their individual failure rates in this structure in order to find η.   ○

Corollary: binomial mean.   In the special case when $p_i = p$ for all i, we know that the number of elements working is a binomial random variable X with parameters n and p. Hence $EX = np$, as we showed more tediously in chapter 4.

This idea will also supply the mean of other sampling distributions discussed in chapter 4.

Example 6.7.7: hypergeometric mean.   Suppose n balls are drawn at random without replacement from an urn containing f fawn balls and m mauve balls. What is the expected number of fawn balls removed? As usual, $n \leqslant f \wedge m$.

Solution.   Let Y be the number of fawn balls and let X_r be the indicator of the event that the rth ball drawn is fawn. Then

$$EX_r = f/(m+f)$$

and the answer is

$$EY = E \sum_{r=1}^{n} X_r = nf/(m+f).$$

Since we know that Y has a hypergeometric distribution, this shows that

(8) $$EY = \sum_{y} yP(Y = y) = \sum_{y=0}^{n} y \binom{f}{y} \binom{m}{n-y} \Big/ \binom{m+f}{n}$$

$$= nf/(m+f).$$

You may care to while away an otherwise idle moment in proving this by brute force. ○

In a number of important and useful examples we have a function of a set of indicators that is *itself* an indicator. This can provide a neat way of calculating probabilities.

Example 6.7.8: inclusion–exclusion.   For $1 \leqslant r \leqslant n$, let I_r be the indicator of the event A_r. Then if we define X by

$$X = 1 - \prod_{r=1}^{n}(1 - I_r)$$

we see that X is an indicator. In fact

$$X = \begin{cases} 1 & \text{if at least one of the } A_r \text{ occurs} \\ 0 & \text{otherwise.} \end{cases}$$

Hence, expanding the product and taking expectations,

$$\begin{aligned} \mathrm{E}X &= \mathrm{P}\left(\bigcup_{r=1}^{n}\right) \\ &= \mathrm{E}\left\{1 - \prod_{r=1}^{n}(1 - I_r)\right\} \\ &= \sum_{r=1}^{n}\mathrm{P}(A_r) - \sum_{r<s}\mathrm{P}(A_r \cap A_s) + \cdots + (-1)^{n+1}\mathrm{P}\left(\bigcap_{r=1}^{n}A_r\right). \end{aligned}$$ ○

Indicators can be used to prove the following useful and familiar result.

Tail-sum lemma.   Let X be a non-negative integer-valued random variable. Then

(9)
$$\mathrm{E}X = \sum_{r=0}^{\infty}\mathrm{P}(X > r).$$

Proof.   Let I_r be the indicator of the event that $X > r$. Then

(10)
$$X = \sum_{r=0}^{\infty}I_r.$$

(To see this just note that if $X = k$ then

$$I_0 = I_1 = \cdots = I_{k-1} = 1, \text{ and } 0 = I_k = I_{k+1} = \cdots)$$

Taking expectations of each side of (10) yields (9). Note that (9) may also be written

$$\mathrm{E}X = \sum_{r=1}^{\infty}\mathrm{P}(X \geqslant r).$$ □

Here is an illustration.

Example 6.7.9.   An urn contains c cobalt balls and d dun balls. They are removed without replacement; let X be the number of dun balls removed before the first cobalt ball. Then $X \geqslant x$ if and only if the first x balls are dun. The probability of this is

$$\binom{d}{x}\bigg/\binom{c+d}{x}.$$

Hence

$$EX = \sum_{x=1}^{d}\binom{d}{x}\bigg/\binom{c+d}{x}.$$

This is continued in exercise 3.

We conclude this section with yet another look at the method of indicators.

Example 6.7.10: matching.   The first n integers are drawn at random out of a hat (or urn, if you prefer). Let X be the number of occasions when the integer drawn in the rth place is in fact r. Find var X.

Solution.   This is tricky if you try to use the distribution of X, but simple using indicators. Let I_r indicate that r is drawn in the rth place. Then

$$X = \sum_{r=1}^{n} I_r.$$

Furthermore

$$EI_r = \frac{1}{n} \quad \text{and} \quad E(I_r I_s) = \frac{1}{n(n-1)}, \quad r \neq s.$$

Hence

$$\text{var } X = E(X^2) - (EX)^2$$

$$= E\{(\textstyle\sum_{r=1}^{n} I_r)^2\} - 1$$

$$= nE(I_r^2) + n(n-1)E(I_r I_s) - 1$$

$$= 1 + 1 - 1 = 1.$$

Exercises for section 6.7

1. ***Coupon collecting; example 6.7.3 continued.***   Suppose there are d different types of coupon, and you collect n coupons. Show that the expected number of different types of coupon you have in your collection is $d\{1 - (1 - d^{-1})^n\}$.

2. ***Boole's inequality.***   Use indicators to show that for events $A_1, \ldots, A_n$

$$P\left(\bigcup_{r=1}^{n} A_r\right) \leq \sum_{r=1}^{n} P(A_r).$$

3. Let $X_1, \ldots, X_n$ have the same mean μ, and set $S_m = \sum_{r=1}^{m} X_r$. What is $E(S_r/S_k)$? Hence show using example 6.7.9 that

$$\sum_{x=1}^{d}\binom{d}{x}\bigg/\binom{c+d}{x} = d/(c+1).$$

4. We have $E(X + Y) = EX + EY$, for any X and Y. Show that it is not necessarily true that
$$\text{median } (X + Y) = \text{median } X + \text{median } Y,$$
nor is it necessarily true that
$$\text{mode } (X + Y) = \text{mode } X + \text{mode } Y.$$

6.8 INDEPENDENCE AND COVARIANCE

We have seen on many occasions that independence has useful and important conse-quences for random variables and their distributions. Not surprisingly, this is also true for expectation. The reason for this is the following

Product rule for expectation. When X and Y are independent, we have
(1)
$$E(XY) = EX \, EY.$$
To see this when X and Y are discrete, recall that
(2)
$$p(x, y) = p_X(x)p_Y(y)$$
for independent discrete random variables. Then

$$E(XY) = \sum_x \sum_y xyp(x, y)$$

$$= \sum_x xp_X(x) \sum_y yp_Y(y), \quad \text{by (2)}$$

$$= EX \, EY.$$

When X and Y are jointly continuous the proof just replaces summations by integrals. $\square$

A very common application of this result is the following

Corollary: variance of a sum of independent random variables. Let X and Y be independent. Then by definition
(3)
$$\text{var}(X + Y) = E\{X + Y - E(X + Y)\}^2$$

$$= E(X - EX)^2 + E(Y - EY)^2$$

$$+ 2E\{(X - EX)(Y - EY)\}.$$

Now by (1), because X and Y are independent,
$$E\{(X - EX)(Y - EY)\} = \{E(X - EX)\}\{E(Y - EY)\} = 0.$$
Hence
(4)
$$\text{var}(X + Y) = \text{var } X + \text{var } Y.$$
It is easy to see that this is true for the sum of any number of independent random variables; that is, if $X_1, X_2, \ldots$ are independent then
(5)
$$\text{var}\left(\sum_{i=1}^n X_i\right) = \sum_{i=1}^n \text{var } X_i.$$
Here is a very simple illustration.

Example 6.8.1: binomial variance.   If $\{X_1, \ldots, X_n\}$ is a collection of independent Bernoulli trials then we know that their sum Y has a binomial distribution. By (5),

$$\operatorname{var} Y = \sum_{r=1}^{n} \operatorname{var} X_r = \sum_{r=1}^{n} pq = npq. \qquad \bigcirc$$

Compare this with your previous derivation. Here is a slightly more complicated application.

Example 6.8.2: the coupon collector's problem.   Suppose a population is known to contain equal numbers of n different types of individual. You repeatedly pick a member of the population at random, with replacement, until you have seen at least one of all n types. Let X be the number of individuals you had to inspect; what is the mean and variance of X?

For reasons of tradition and convention, this is known as the coupon collector's problem. If each box of some product contains a coupon, and there are n different types of coupon, and each box is independently equally likely to contain any type, then X is the number of boxes you need to buy and open to get the complete set of all n different types.

Solution.   Obviously your first box supplies you with the first coupon of your set. Let N_1 be the number of boxes you need to open to get a coupon different from that in the first box. The probability that any coupon is the same as your first is $1/n$, the probability that it is different is $(n-1)/n$. Boxes are independent. Hence

$$P(N_1 = x) = \left(\frac{1}{n}\right)^{x-1} \left(\frac{n-1}{n}\right), \quad x \geqslant 1,$$

which is geometric with parameter $(n-1)/n$. Now let N_2 be the further number of boxes you need to open to obtain the third coupon of your set. The same line of argument shows that

$$P(N_2 = x) = \left(\frac{2}{n}\right)^{x-1} \left(\frac{n-2}{n}\right), \quad x \geqslant 1,$$

which is geometric with parameter $(n-2)/n$.

Continuing in this way yields a series of geometric random variables $(N_k; \ 1 \leqslant k \leqslant n-1)$ with respective parameters $(n-k)/n$. Obviously the process stops at N_{n-1}, because this yields the nth and final member of your complete set. Also

$$X = 1 + N_1 + \cdots + N_{n-1},$$

and furthermore the random variables $N_1, \ldots, N_{n-1}$ are independent, because the boxes are. Hence

$$EX = 1 + EN_1 + \cdots + EN_{n-1}$$

$$= 1 + \frac{n}{n-1} + \frac{n}{n-2} + \cdots + \frac{n}{1}$$

$$= n \sum_{k=1}^{n} \frac{1}{k}.$$

The variance is given by

$$(6) \qquad \text{var } X = \text{var } 1 + \text{var } N_1 + \cdots + \text{var } N_{n-1}$$

$$= 0 + \frac{1}{n} \bigg/ \left(\frac{n-1}{n}\right)^2 + \frac{2}{n} \bigg/ \left(\frac{n-2}{n}\right)^2 + \cdots + \frac{n-1}{n} \bigg/ \left(\frac{1}{n}\right)^2$$

$$= n \left\{ \frac{1}{(n-1)^2} + \cdots + \frac{n-2}{2^2} + \frac{n-1}{1^2} \right\}$$

$$= n^2 \sum_{k=1}^{n-1} \frac{1}{k^2} - n \sum_{1}^{n-1} \frac{1}{k}.$$

It is shown in calculus books that as $n \to \infty$,

$$\sum_{k=1}^{n} \frac{1}{k} - \log n \to \gamma,$$

where γ is called Euler's constant, and

$$\sum_{k=1}^{n} \frac{1}{k^2} \to \frac{\pi^2}{6}.$$

Hence if we are collecting a large number of 'coupons', we find as $n \to \infty$

$$\frac{\text{E}X}{n \log n} \to 1$$

and

$$\frac{\text{var } X}{n^2} \to \frac{\pi^2}{6}. \qquad\qquad ○$$

When X and Y are not independent, the product rule does not necessarily hold, so it is convenient to make the following definition.

Definition. The *covariance* of two random variables X and Y is $\text{cov}(X, Y)$ where

$$(7) \qquad\qquad \text{cov}(X, Y) = \text{E}\{(X - \text{E}X)(Y - \text{E}Y)\}$$

$$= \text{E}(XY) - \text{E}X \, \text{E}Y = \text{cov}(Y, X). \qquad △$$

Thus for any pair of random variables, from (3),

$$\text{var}(X + Y) = \text{var } X + \text{var } Y + 2 \, \text{cov}(X, Y)$$

More generally, for any sum,

$$\text{var}\left(\sum_{r=1}^{n} X_r\right) = \sum_{r=1}^{n} \text{var } X_r + 2 \sum_{j < k} \text{cov}(X_j, X_k).$$

Things are obviously simpler when $\text{cov}(X_j, X_k) = 0$, so we make the following

Definition. If $\text{cov}(X, Y) = 0$, then X and Y are said to be *uncorrelated*. $\qquad △$

Thus independent random variables are uncorrelated, by (1).

At this point we note that while independent X and Y have zero covariance, the converse is *not* true.

Example 6.8.3.   Let X be any bounded non-zero random variable with a distribution symmetric about zero; that is to say, $p(x) = p(-x)$, or $f(x) = f(-x)$. Let $Y = X^2$. Then Y is not independent of X, but nevertheless

$$\text{cov}(X,\, Y) = \text{E}X^3 - \text{E}X\,\text{E}X^2 = 0. \qquad \bigcirc$$

Despite this, their covariance is clearly a rough and ready guide to the mutual dependence of X and Y. An even more useful guide is the correlation function.

Definition.   The *correlation function* $\rho(X,\, Y)$ of X and Y is

$$\rho(X,\, Y) = \frac{\text{cov}(X,\, Y)}{(\text{var}\, X\,\text{var}\, Y)^{1/2}}. \qquad \triangle$$

This may appear unnecessarily complicated, but the point is that if we change the location and scale of X and Y then ρ is essentially unchanged, since

(8) $$\rho(aX + b,\, cY + d) = \text{sign}(ac)\,\rho(X,\, Y),$$

where

$$\text{sign}(x) = \begin{cases} 1, & x > 0 \\ 0, & x = 0 \\ -1, & x < 0. \end{cases}$$

Thus ρ is scale-free, but covariance is not, because

(9) $$\text{cov}(aX + b,\, cX + d) = ac\,\text{cov}(X,\, Y).$$

Here are two simple routine examples.

Example 6.8.4.   Let X and Y have joint density

$$f(x,\, y) = \frac{1}{\pi}, \quad x^2 + y^2 \leqslant 1.$$

Then, easily, $\text{E}X = \text{E}Y = \text{E}(XY) = 0$, by symmetry, and so X and Y are uncorrelated. They are not independent, of course.

Now let $U = |X|$, and $V = |Y|$. Then

$$\text{E}U = \text{E}|X| = \frac{4}{\pi}\int_0^1\int_0^{\pi/2} r^2 \cos\theta \, dr \, d\theta = \frac{4}{3\pi} = \text{E}V$$

and

$$\text{E}(UV) = \text{E}(|XY|) = \frac{4}{\pi}\int_0^1\int_0^{\pi/2} r^3 \cos\theta \sin\theta \, dr \, d\theta = \frac{1}{2\pi}.$$

Hence

$$\text{cov}(U,\, V) = \frac{1}{2\pi} - \frac{16}{9\pi^2} < 0. \qquad \bigcirc$$

Example 6.8.5. Roll two dice, yielding the scores X and Y respectively. By independence $\mathrm{cov}(X, Y) = 0$. Now let $U = \min\{X, Y\}$ and $V = \max\{X, Y\}$, with joint distribution

$$p(u, v) = \begin{cases} \frac{1}{18}, & u < v \\ \frac{1}{36}, & u = v. \end{cases}$$

Then, using (9) of section 6.7 and some arithmetic, we find that $\mathrm{E}V = \frac{91}{36}$. Also $\mathrm{E}V = 7 - \mathrm{E}U = \frac{161}{36}$, with $\mathrm{E}(UV) = \mathrm{E}(XY) = \mathrm{E}X\,\mathrm{E}Y$, and so

$$\mathrm{cov}(U, V) = \left(\tfrac{35}{36}\right)^2. \qquad \bigcirc$$

The use of covariance extends our range of examples of the use of indicators.

Example 6.8.6. Of b birds, a number r are ringed. A sample of size n is to be taken from the b birds. Let X be the number in the sample with rings if sampling is with replacement, and Y the number with rings if sampling is without replacement. Show that

$$\mathrm{var}\,Y = \frac{b - n}{b - 1}\,\mathrm{var}\,X.$$

Solution. If sampling is with replacement, we know that X is binomial $\mathrm{B}(n, r/b)$, so

$$\mathrm{var}\,X = n\frac{r}{b}\left(1 - \frac{r}{b}\right).$$

If sampling is without replacement, let I_k be the indicator of the event that the kth bird is ringed. Now for any such Bernoulli trial we know that

$$\mathrm{E}I_k = \frac{r}{b} \quad \text{and} \quad \mathrm{var}\,I_k = \frac{r}{b}\left(1 - \frac{r}{b}\right).$$

Also, we calculate

$$\mathrm{E}(I_j I_k) = \mathrm{P}(I_j = 1 | I_k = 1)\mathrm{P}(I_k = 1)$$

$$= \frac{r}{b}\left(\frac{r - 1}{b - 1}\right), \quad j \neq k,$$

so that

(10) $$\mathrm{cov}(I_j, I_k) = -\frac{r}{b^2}\left(\frac{b - r}{b - 1}\right), \quad j \neq k.$$

Hence finally

$$\mathrm{var}\,Y = \mathrm{var}\left(\sum_{k=1}^{n} I_k\right) = \sum_{k=1}^{n} \mathrm{var}\,I_k + 2\sum_{j < k} \mathrm{cov}(I_j, I_k)$$

$$= n\frac{r}{b}\left(1 - \frac{r}{b}\right) - n(n - 1)\frac{r}{b^2}\left(\frac{b - r}{b - 1}\right)$$

$$= n\frac{r}{b}\left(1 - \frac{r}{b}\right)\left(\frac{b - n}{b - 1}\right)$$

as required. $\qquad \bigcirc$

We end this section with a look at one of the most important results in probability, the so-called law of large numbers.

At various points throughout the book we have remarked that given a set $X_1, \ldots, X_n$ of independent observations or readings, which we take to be random variables, their average

$$(11) \qquad \overline{X}_n = n^{-1} \sum_{r=1}^{n} X_r$$

is a quantity of natural interest. We used this idea to motivate our interest in, and definitions of, expectation. In the case when the X's are indicators, we also used this expression to motivate our ideas of probability. In that case $\overline{X}_n$ is just the relative frequency of whatever the X's are indicating.

In both cases we claimed that empirically, as $n \to \infty$, the averages $\overline{X}_n$ tend to settle down around some mean value. It is now time to justify that assertion. Obviously, we need to do so, for if $\overline{X}_n$ in (11) did not ever display this type of behaviour, it would undermine our theory, to say the least.

We remind ourselves that the X's are independent and identically distributed, with mean μ and variance $\sigma^2 < \infty$. Then we have the following so-called

Law of large numbers.   For any $\varepsilon > 0$, as $n \to \infty$

$$(12) \qquad P(|\overline{X}_n - E\overline{X}_n| > \varepsilon) \to 0.$$

That is to say as n increases, the probability that $\overline{X}_n$ differs from $E\overline{X}_n$ by more than any given amount (however small) tends to zero. This is of course extremely gratifying, as it is consistent with what we assumed in the first place.

Proof.   The proof of (12) is very simple. Recall Chebyshov's inequality

$$P(|X - EX| > \varepsilon) \leq \frac{\operatorname{var} X}{\varepsilon^2}.$$

Using this gives

$$(13) \qquad P(|\overline{X}_n - E\overline{X}_n| > \varepsilon) \leq \{\operatorname{var}(n^{-1} \sum_{r=1}^{n} X_r)\} \varepsilon^{-2}$$

$$= \sigma^2 n^{-1} \varepsilon^{-2}, \quad \text{by independence}$$

$$\to 0$$

as $n \to \infty$, for any given $\varepsilon > 0$.   □

Here is an important application. Let X_r be the indicator of an event A. Thus $\overline{X}_n$ is the relative frequency of occurrence of A in a sequence of Bernoulli trials, and $E\overline{X}_n = P(A)$. By (13) we have

$$(14) \qquad P(|\overline{X}_n - P(A)| > \varepsilon) \leq \frac{P(A)\{1 - P(A)\}}{n\varepsilon^2}.$$

Thus as $n \to \infty$ the relative frequency of occurrence of the event A does indeed settle down around $P(A)$, in the sense that we stated above.

And, in general, the fact that

$$\operatorname{var}\overline{X}_n = \frac{\operatorname{var}X_1}{n}$$

supports our intuitive idea that when we take averages, their dispersion decreases as the size of the set being averaged increases. However, it is important to note two things about these conclusions. The first is that if the X_i do not have finite variance then good behaviour of averages is not always to be expected, as the theorem does not necessarily hold.

The second point to note is another gambler's fallacy. If (for example) some number x fails to come up for a prolonged spell at roulette, or in a lottery, then one hears it said that 'x must be more likely next time, by the law of averages'. Of course this statement is not true, and the law of large numbers offers no support for it. If A is the event that x turns up, then (14) says only that the relative frequency of occurrences of x settles down. The actual absolute fluctuations in the frequency of x about its expected value will tend to get larger.

For example, it is easy for you to show that given any number m, no matter how large, with probability unity x will eventually fail to appear on m successive occasions (*exercise*).

Note finally that our result (12) is often called the weak law of large numbers, to distinguish it from another called the strong law of large numbers; we do not discuss the strong law here.

Exercises for section 6.8

1. Let $EX = EY$ and $\operatorname{var}X = \operatorname{var}Y$. Show that
    (a) $\operatorname{cov}(X + Y, X - Y) = 0$ and
    (b) $\operatorname{cov}(X + Y, X/(X + Y)) = 0$, if X and Y have the same distribution.

2. Let X and Y have the joint distribution
    $$p(x, y) = \frac{1}{1 + 2n(n + 1)}, \quad |x - y| \leqslant n; \; |x + y| \leqslant n.$$
    Show that X and Y are dependent and that $\operatorname{cov}(X, Y) = 0$.

3. Recall that $\operatorname{skw}X = E(X - \mu)^3/\sigma^3$.
    Show that if $X_1, \ldots, X_n$ are independent and identically distributed then
    $$\operatorname{skw}\sum_{i=1}^{n}X_i = n^{-1/2}\operatorname{skw}X_1.$$
    Deduce that if X is binomial B(n, p), then
    $$\operatorname{skw}X = \frac{q - p}{(npq)^{1/2}}.$$

4. Show that for any pair of random variables X and Y
    $$E\{(XY)^2\} \leqslant EX^2\,EY^2.$$
    This is the Cauchy–Schwarz inequality. (Hint: Consider Z^2 where $Z = sX - tY$.) Deduce that $|\rho(X, Y)| \leqslant 1$.

5. Let $X_1, \ldots, X_n$ be independent and identically distributed with mean μ and variance σ^2. Define $\overline{X} = n^{-1}\sum_{r=1}^{n}X_r$. Show that
    $$\operatorname{cov}(\overline{X}, X_r - \overline{X}) = 0, \quad 1 \leqslant r \leqslant n.$$

6. 'You can never foretell what any one man will do, but you can say with precision what an average number of men will be up to'. Attributed to Sherlock Holmes by A. Conan Doyle.
   (a) Has Conan Doyle said what he presumably meant to say?
   (b) If not, rephrase the point correctly.

7. Let X and Y be independent. Is it ever true that

$$\text{var}(XY) = \text{var}\, X \, \text{var}\, Y?$$

6.9 CONDITIONING AND DEPENDENCE, DISCRETE CASE

Having dealt with independent random variables, it is now natural to ask what happens when random variables are *not* independent. Here is a simple example.

Example 6.9.1.   You roll a fair die, which shows X. Then you flip X fair coins, which show Y heads. Clearly, as always,

$$P(X = 1) = \tfrac{1}{6}.$$

However, suppose we observe that $Y = 2$. Now it is obviously impossible that $X = 1$. Knowledge of Y has imposed conditions on X, and to make this clear we use an obvious notation and write

$$P(X = 1|Y = 2) = 0. \qquad \bigcirc$$

What can we say in general about such conditional probabilities? Recall from chapter 2 that for events A and B

(1) $$P(A|B) = P(A \cap B)/P(B).$$

Also recall from section 5.8 that for a discrete random variable X and an event B,

(2) $$P(X = x|B) = P(\{X = x\} \cap B)/P(B).$$

The following definition is now almost self-evident.

Definition.   Let X and Y be discrete random variables, with distribution $p(x, y)$. Then the *conditional distribution of X given Y* is denoted by $p(x|y)$, where

(3) $$p(x|y) = P(X = x, Y = y)/P(Y = y)$$
$$= p_{X,Y}(x, y)/p_Y(y). \qquad \triangle$$

Remark.   Of course this is just (1) written in terms of random variables, since

$$P(X = x|Y = y) = P(A_x|A_y)$$
$$= P(A_x \cap A_y)/P(A_y)$$

where $A_x = \{\omega: X(\omega) = x\}$, $A_y = \{\omega: Y(\omega) = y\}$. All the above definition really comprises is the name and notation.

In view of this connection, it is not surprising that the partition rule also applies.

Partition rule for discrete distributions.   Let X and Y have joint distribution $p(x, y)$. Then we have

(4) $$p_X(x) = \sum_y p(x|y)p_Y(y).$$

To prove this, just multiply (3) by $p_Y(y)$ and sum over y. Of course, it also follows directly from the partition rule in chapter 2, provided always that X and Y are defined on the same sample space. $\qquad\square$

Here are some examples to show the ways in which (3) and (4) are commonly applied.

Example 6.9.1 continued. You roll a die, which shows X, and flip X coins, which show Y heads. Find $p(x, y)$ and $p_Y(y)$.

Solution. Given $X = x$, the number of heads is binomial $B\left(x, \frac{1}{2}\right)$. That is to say,

$$p(y|x) = 2^{-x}\binom{x}{y}, \quad 0 \leq y \leq x.$$

Hence

$$p(x, y) = p(y|x)p_X(x)$$

$$= \tfrac{1}{6} \times 2^{-x}\binom{x}{y}, \quad 0 \leq y \leq x;\ 1 \leq x \leq 6.$$

Finally

$$p_Y(y) = \sum_{x=y}^{6} \tfrac{1}{6} \times 2^{-x}\binom{x}{y}. \qquad\bigcirc$$

Example 6.9.2. Let X and Y be independent geometric random variables, each with parameter p. Suppose $Z = X + Y$. Find the distribution of X given Z.

Solution. First we recall that

$$p_Z(z) = \sum_{x=1}^{z-1} p_X(x)p_Y(z - x) = \sum_{x=1}^{z-1} q^{x-1}pq^{z-x-1}p$$

$$= (z - 1)p^2q^{z-2}, \quad z \geq 2.$$

Hence

(5) $$p(x|z) = \frac{p_{X,Z}(x, z)}{p_Z(z)} = \frac{p_{X,Y}(x, z - x)}{p_Z(z)}$$

$$= \frac{p^2q^{x-1}q^{z-x-1}}{(z - 1)p^2q^{z-2}} = \frac{1}{z - 1}, \quad 1 \leq x \leq z - 1.$$

Thus, given $Z = z$, X is uniformly distributed over the possible range $1 \leq x \leq z - 1$. $\quad\bigcirc$

As an amusing example of the use of a conditional mass function, consider the following.

Example 6.9.3: Poisson number of Bernoulli trials. Suppose we perform n Bernoulli trials, yielding U successes and V failures. We know U and V are dependent. Suppose now we perform N Bernoulli trials where N is Poisson with parameter λ, yielding X successes and Y failures. Show that the number of successes X and the number of failures Y are independent.

Solution.   We have

(6) $$P(X = x, Y = y) = P(X = x, Y = y, N = x + y)$$

$$= P(X = x, Y = y | N = x + y)P(N = x + y)$$

$$= \frac{(x + y)!}{x! y!} p^x q^y \frac{e^{-\lambda} \lambda^{x+y}}{(x + y)!}$$

$$= \frac{(\lambda p)^x}{x!} e^{-\lambda p} \frac{(\lambda q)^y}{y!} e^{-\lambda q}, \quad x, y \geqslant 0.$$

This factorizes for all x and y, so X and Y are independent, being Poisson with parameters λp and λq respectively.   ○

Next we return to the probability $p(x|y)$ defined in (3), and stress the point that the conditional distribution of X given Y is indeed a distribution. That is to say,

(7) $$p(x|y) \geqslant 0$$

and

(8) $$\sum_x p(x|y) = 1.$$

The first of these is obvious; to see the second write

$$\sum_x p(x|y) = \sum_x \frac{p(x, y)}{p_Y(y)} = \frac{p_Y(y)}{p_Y(y)} = 1.$$

Even more importantly we have the

Conditional key rule

(9) $$P(X \in C | Y = y) = \sum_{x \in C} p(x|y).$$

Thus $p(x|y)$ has all the properties of the probability distributions that we have used in earlier chapters. It is therefore not at all surprising to find that, like them, this distribution may have an expectation. For obvious reasons, it is called the conditional expectation.

Definition.   The *conditional expectation* of X given $Y = y$ is

(10) $$E(X | Y = y) = \sum_x x p(x|y),$$

with the usual proviso that $\sum_x |x| p(x|y) < \infty$.   △

Naturally, expectation is related to conditional expectation in much the same way as probability is related to conditional probability.

Theorem.   Let X and Y be discrete random variables. Then

(11) $$E(X) = \sum_y E(X|Y = y)p_Y(y).$$

Proof.   Multiply (10) by $p_Y(y)$ and sum over y. Then

$$\sum_y E(X|Y = y)p_Y(y) = \sum_y \sum_x xp(x|y)p_Y(y)$$

$$= \sum_x x \sum_y p(x, y)$$

$$= \sum_x xp_X(x)$$

$$= EX. \qquad \square$$

We give two simple illustrations of this. Further interesting examples follow later. We note that there is a *conditional law of the unconscious statistician*, that is,

$$E(g(X)|Y = y) = \sum_x g(x)p(x|y).$$

Furthermore,

$$Eg(X, Y) = \sum_y E(g(X, Y)|Y = y)p_Y(y),$$

a result which is often very useful.

Example 6.9.5: dice.   Your roll two dice; let U be the minimum and V the maximum of the two numbers shown. Then, as we know,

$$p(u, v) = \begin{cases} \frac{2}{36}, & 1 \leqslant u < v \leqslant 6 \\ \frac{1}{36}, & u = v. \end{cases}$$

Hence as usual

$$p_U(u) = \sum_v p(u, v) = \tfrac{1}{36}(13 - 2u), \quad 1 \leqslant u \leqslant 6$$

and

$$p_V(v) = \sum_u p(u, v) = \tfrac{1}{36}(2v - 1), \quad 1 \leqslant v \leqslant 6.$$

Thus, for example

$$p(u|v) = \begin{cases} \dfrac{2}{2v - 1}, & u < v \\ \dfrac{1}{2v - 1}, & u = v. \end{cases}$$

Therefore

$$E(U|V = v) = \sum_{u=1}^{v} u p(u|v) = \frac{v^2}{2v - 1}$$

and of course

$$E(U) = \sum_{v=1}^{6} E(U|V = v) p(v)$$

$$= \sum_{v=1}^{6} \tfrac{1}{36} v^2 = \tfrac{91}{36}.$$

which we obtained in a more tedious way earlier.   ○

Example 6.9.6: random sum.   Suppose that we can regard insurance claims as independent random variables X_1, X_2, ..., having a common distribution $p(x)$. Suppose the number of claims next year were to be N, where N is independent of the X_i. Find the expected total of next year's claims.

Solution.   We can denote this total by S_N, where

$$S_N = \sum_{r=0}^{N} X_r.$$

Then, by (11),

$$E(S_N) = \sum_{n} E(S_N|N = n) \, p_N(n)$$

$$= \sum_{n} E(S_n) \, p_N(n)$$

$$= \sum_{n} n E X_1 \, p_N(n)$$

$$= E X_1 E N.$$

It is interesting to note that this result still holds under much weaker conditions. If we suppose only that the X_i have a common mean, and are independent of N, then the above proof is valid as it stands. Even more remarkably, we can allow N to depend on the X_i, as we shall see in the next example.   ○

We conclude this section with a remarkable extension of the idea of a 'random sum', which we looked at in the above example. In that case the number of terms N in the sum S_N was independent of the summands X_1, X_2, ... However, this is often not the case. For example, suppose you are a trader (or gambler), who makes a profit of X_1, X_2, ... on a sequence of deals until your retirement after the Nth deal. This index N is a random variable, and it can depend only on your previous deals. That is to say, you may retire because $X_1 + X_2 + \cdots + X_N > \$10^9$, and you decide to take up golf; or you may retire because $S_N < -\$10^9$, and you are ruined (or in gaol for fraud). You cannot choose to retire on the basis of future deals. Either way, the event $\{N > k\}$ that you continue trading after the kth deal depends only on X_1, ..., X_k. Then the following amazing result is true.

Example 6.9.7: Wald's equation. If the X_i are independent and identically distributed with finite mean, N is integer valued with $\mathrm{E}N < \infty$, and $\{N > k\}$ is independent of $X_{k+1}, X_{k+2}, \ldots$ then

(12)
$$\mathrm{E}\sum_{r=1}^{N} X_r = \mathrm{E}N\,\mathrm{E}X_1.$$

To prove this, let I_{k-1} be the indicator of the event $\{N > k - 1\}$. Then I_{k-1} is independent of X_k, and we can write

$$\mathrm{E}\sum_{r=0}^{N} X_r = \mathrm{E}\sum_{k=1}^{\infty} X_k I_{k-1}, \quad \text{because } I_{k-1} = 0 \text{ for } N < k$$

$$= \sum_{k=1}^{\infty} \mathrm{E}(X_k I_{k-1}), \quad \text{because } E|X_k| < \infty$$

$$= \sum_{k=1}^{\infty} \mathrm{E}X_k\,\mathrm{E}I_{k-1}, \quad \text{by independence}$$

$$= \sum_{k=1}^{\infty} \mathrm{E}X_1\,\mathrm{P}(N > k - 1)$$

$$= \mathrm{E}X_1\,\mathrm{E}N. \qquad\qquad \bigcirc$$

The principal application of this is to gambling; in a casino it is always the case that $\mathrm{E}X_k < 0$. It follows that no matter what system you play by, when you stop you have $\mathrm{E}S_N < 0$. You must expect to lose.

Exercises for section 6.9

1. Show that X and Y are independent if and only if the conditional distribution of X given $Y = y$ is the same as the marginal distribution of X for all y. When X and Y are independent show that $\mathrm{E}(X|Y = y) = \mathrm{E}X$.

2. In Example 6.9.7 find $p(v|u)$ and $\mathrm{E}(V|U = u)$. Also find $\mathrm{E}(UV)$ by evaluating
$$\mathrm{E}(UV) = \sum_v v\mathrm{E}(U|V = v)p_V(v).$$
What is $\mathrm{cov}(U, V)$?

3. Let $X_1, \ldots, X_n$ be independent and identically distributed, and let N be non-negative and independent of the X_i. Show that
$$\mathrm{var}\sum_{r=1}^{N} X_r = (\mathrm{E}X_1)^2\,\mathrm{var}\,N + \mathrm{E}N\,\mathrm{var}\,X_1.$$

4. ***Gambler's ruin.*** Suppose you are a gambler playing a sequence of fair games for unit stakes, with initial fortune k. Let N be the first time at which your fortune is either 0 or n, $0 \leqslant k \leqslant n$; at this time you stop playing. Use Wald's equation to show that the probability that you stop with a fortune of size n is k/n.

5. *Derangements and matching.*   There are n coats belonging to n people, who make an attempt to leave by choosing a coat at random. Those who have their own coat can leave; the rest hang the coats up at random, and then make another attempt to leave by choosing a coat at random. Let N be the number of attempts until everyone leaves. Use the method of proving Wald's equation to show that $EN = n$. (Hint: Recall that the expected number of matches in a derangement of n objects is 1.)

6.10 CONDITIONING AND DEPENDENCE, CONTINUOUS CASE

For discrete random variables X and Y with distribution $p(x, y)$, we have established that

(1) $$p_{X|Y}(x|y) = p(x, y)/p_Y(y),$$

(2) $$p_X(x) = \sum_y p_{X|Y}(x|y)p_Y(y),$$

(3) $$P(X \in B|Y = y) = \sum_{x \in B} p_{X|Y}(x|y),$$

(4) $$E(X|Y = y) = \sum_x xp(x|y),$$

(5) $$E(X) = \sum_y E(X|Y = y)p_Y(y).$$

It is natural next to consider the case when X and Y are jointly continuous random variables with density $f(x, y)$. Of course we cannot use the elementary arguments that give (1), because $P(Y = y) = 0$ for all y. Nevertheless, as we shall show, slightly more complicated reasoning will supply the following very appealing definition and results parallelling (1)–(5).

Definition.   Let X and Y have density $f(x, y)$. Then for $f_Y > 0$ the *conditional density* of X given $Y = y$ is defined by

(6) $$f_{X|Y}(x|y) = \frac{f(x, y)}{f_Y(y)}. \qquad\qquad \triangle$$

Just as in the discrete case, we can recover the unconditional density by integrating $f_{X|Y}(x|y) \equiv f(x|y)$:

(7) $$f_X(x) = \int_{-\infty}^{\infty} f(x|y)f_Y(y)\, dy.$$

Note crucially that $f(x|y)$ is indeed a density, as it is non-negative and

$$\int_{-\infty}^{\infty} f(x|y)\, dx = \frac{1}{f_Y(y)} \int_{-\infty}^{\infty} f(x, y)\, dx = 1.$$

Therefore it has the properties of a density, including the key rule

(8) $$P(X \in C|Y = y) = \int_{x \in C} f_{X|Y}(x|y)\, dx,$$

and it may have an expectation

(9)
$$E(X|Y = y) = \int_{-\infty}^{\infty} x f_{X|Y}(x|y)\, dx.$$

From (6) and (9) we find that in general

$$EX = \int_{-\infty}^{\infty} E(X|Y = y) f_Y(y)\, dy.$$

In particular, if we let X be the indicator of the event that $X \leqslant x$, this proves that

(10)
$$P(X \leqslant x) = \int_{-\infty}^{\infty} P(X \leqslant x|Y = y) f_Y(y)\, dy.$$

Alternatively you can prove (10) by simply integrating (7) over $(-\infty, x]$.

It is extremely pleasing, and convenient in remembering them, that (6)–(9) are essentially exactly the same as (1)–(4); in each case $\sum$ is replaced by $\int$, as usual. We give some theoretical background to (6)–(9) a little later on; first it is helpful to look at simple examples to show what is going on, and develop understanding.

Example 6.10.1.   Let X and Y be uniform on the triangle $0 \leqslant x \leqslant y \leqslant 1$, so that $f(x, y) = 2, 0 \leqslant x \leqslant y \leqslant 1$. First we can easily calculate the marginals:

$$f_X(x) = \int_x^1 2\, dy = 2(1 - x) \quad f_Y(y) = \int_0^y 2\, dx = 2y.$$

Hence, by (6) we have the conditional densities:

(11)
$$f_{X|Y}(x|y) = \frac{f}{f_Y} = \frac{1}{y}, \quad 0 \leqslant x \leqslant y,$$

(12)
$$f_{Y|X}(y|x) = \frac{f}{f_X} = \frac{1}{1 - x}, \quad x \leqslant y \leqslant 1.$$

Of course this just confirms what intuition and inspection of figure 6.2 would tell us: given $Y = y$, X is clearly uniform on $[0, y]$; and, given $X = x$, Y is clearly uniform on $[x, 1]$.

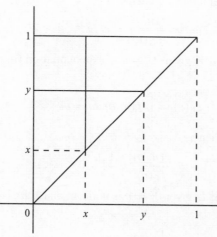

Figure 6.2. The point (X, Y) is picked uniformly at random in the triangle $\{(0, 0), (0, 1), (1, 1)\}$.

Hence, from (9) and (11),

$$E(X|Y = y) = \int_0^y \frac{x}{y}\, dx = \frac{y}{2},$$

and from (9) and (12)

(13) $$E(Y|X = x) = \int_x^1 \frac{y}{1-x}\, dy = \frac{1-x}{2}.$$

Again, this is intuitively obvious. Further, given any conditional density we can work out probabilities of interest in the usual way by (8):

$$P(X > \tfrac{1}{2}|Y = y) = \int_{1/2}^y f_{X|Y}(x|y)\, dx$$

$$= \int_{1/2}^y y^{-1}\, dx$$

$$= \begin{cases} y^{-1}(y - \tfrac{1}{2}), & y > \tfrac{1}{2} \\ 0 & \text{otherwise.} \end{cases} \qquad \bigcirc$$

Example 6.10.2.   Let the random variables U and V have joint density

$$f(u, v) = e^{-v}, \quad \text{for } 0 < u < v < \infty.$$

Find the marginals $f_U(u)$, $f_V(v)$, and the conditional densities $f_{U|V}(u|v)$ and $f_{V|U}(v|u)$. What is the density of $Y = V - U$?

Solution.   By (8) of section 6.3,

$$f_U(u) = \int_u^\infty f(u, v)\, dv = e^{-u}.$$

So U is exponential with parameter 1. Then

(14) $$f_{V|U} = \frac{f(u, v)}{f_U(u)} = e^{u-v}, \quad v > u$$

so the conditional density of V given $U = u$ is exponential on (u, ∞), with parameter 1. Again

(15) $$f_V(v) = \int_0^v f(u, v)\, du = v e^{-v}.$$

This is a gamma density. Then

(16) $$f_{U|V} = \frac{f(u, v)}{f_V(v)} = \frac{1}{v}, \quad 0 < u < v,$$

so the conditional density of U given $V = v$ is uniform on $(0, v)$.
    Finally, by (10),

$$P(V - U \leqslant y) = \int_0^\infty P(V - U \leqslant y|U = u) f_U(u)\, du.$$

Now

(17)
$$P(V - U \leqslant y | U = u) = \int_u^{y+u} f_{V|U}(v|u) \, dv$$

$$= e^{+u}(e^{-u} - e^{-y-u}) = 1 - e^{-y}.$$

Therefore

$$P(V - U \leqslant y) = \int_0^{\infty} (1 - e^{-y}) e^{-u} \, du$$

$$= 1 - e^{-y}.$$

Since this is the distribution function of an exponential random variable, it follows that $Y = V - U$ has density e^{-y}. Since (17) does not depend on u, it is also true that U and $V - U$ are independent.   ○

Remark: link with Poisson process. We have noted above that times of occurrence of earthquakes, meteorite strikes, and other rare random events are well described by a Poisson process. This has the property that intervals between events are independent exponential random variables. In view of what we have proved above, we have that U is the time of the first event, and V the time of the second event, in such a process.

Equation (16) then has the following interpretation. Given that the time of the second event in such a sequence is V, the time of the first event is uniform on $(0, V)$: *it was equally likely to have been any time in* $(0, V)$!

Sometimes the use of a conditional density offers a slightly different approach to calculations that we can already do.

Example 6.10.3. Let X and Y be independent exponential random variables with parameters λ and μ respectively. Then by (10)

$$P(X < Y) = \int_0^{\infty} P(X < Y | Y = y) f_Y(y) \, dy$$

$$= \int_0^{\infty} (1 - e^{-\lambda y}) \mu e^{-\mu y} \, dy$$

$$= 1 - \frac{\mu}{\lambda + \mu}$$

$$= \frac{\lambda}{\lambda + \mu},$$

which we could alternatively find using a routine double integral.   ○

Before we conclude this section, we return to supply additional reasons for making the definition (6). Let us consider the event

$$B = \{y < Y \leqslant y + h\}.$$

Now by ordinary conditional probability, it follows that the conditional distribution function of X given B is

(18) $F_{X|B}(x|B) = P(X \leqslant x|B)$

$$= \frac{P(X \leqslant x, \, y < Y \leqslant y + h)}{P(B)} = \frac{F(x, \, y + h) - F(x, \, y)}{F_Y(y + h) - F_Y(y)},$$

where $F(x, y)$ is the joint distribution function of X and Y, and $F_Y(y)$ is the distribution function of Y.

Now if we let $h \downarrow 0$ in (18) this supplies an attractive candidate for the conditional distribution function of X, given $Y = y$. We get

(19) $F(x|Y = y) = \lim_{h \downarrow 0} F_{X|B}(x|B)$

$$= \frac{1}{f_Y(y)} \int_{-\infty}^{x} f(u, \, y) \, du.$$

As usual, the derivative of the distribution function (when it exists) is the density. Thus differentiating (19) we obtain

$$f(x|Y = y) = f(x, \, y)/f_Y(y),$$

which is (6). Because this *is* a density, it may have an expectation given by

(20) $E(X|Y = y) = \int x f(x|Y = y) \, dx$

when the integral exists. Hence, multiplying by $f_Y(y)$ and integrating over y, we obtain

(21) $E(X) = \int E(X|Y = y) f_Y(y) \, dy.$

In particular we can often use this to find probabilities of interest, as follows.

Let X be the indicator of any event A; in practice A is determined by the joint behaviour of X and Y. Then $EX = P(A)$ and $E(X|Y = y) = P(A|Y = y)$. Hence, substituting in (21) we obtain what may be called the

Continuous partition rule

(22) $P(A) = \int_{\mathbb{R}} P(A|Y = y) f_Y(y) \, dy.$

The name is explained by a glance at the original partition rule in section 2.8. We can use this rule to supply a slightly different derivation of the convolution rule for the distribution of sums of continuous random variables.

Example 6.10.4: convolution rule.   Let X and Y have density $f(x, y)$, and set $Z = X + Y$. Then

$$F_Z(z) = P(Z \le z)$$

$$= P(X + Y \le z)$$

$$= \int P(X + Y \le z | Y = y) f_Y(y) \, dy, \quad \text{by (22)}$$

$$= \int P(X + y \le z | Y = y) f_Y(y) \, dy$$

$$= \int F_{X|Y=y}(z - y|y) f_Y(y) \, dy.$$

Differentiating with respect to z gives

$$f_Z(z) = \int f_{X|Y=y}(z - y|y) f_Y(y) \, dy$$

$$= \int f(z - y, y) \, dy. \qquad \bigcirc$$

This is the equivalent for continuous random variables of the expression which we derived for discrete random variables.

Exercises for section 6.10

1. (a) Let X and Y be independent and non-negative, and identically distributed with density $f(x)$ and distribution $F(x)$. Let $U = \min\{X, Y\}$ and $V = \max\{X, Y\}$. Find the conditional density of U given V, and find $E(U|V = v)$.
    (b) A satellite has two transmitters; their lifetimes are independent and exponential with parameter λ. The satellite operates if they both work. Transmissions cease at time t. Show that the expected failure time of the first to fail is

    $$\frac{1}{\lambda} + \frac{t}{1 - e^{\lambda t}}.$$

    (It was launched at time 0.)

2. Let X and Y be independent exponential random variables with parameter λ. Find the density of X given $X + Y = z$.

3. The random variable R is uniform in $[0, 1)$. Given that $R = \rho$, the random variable X is binomial with parameters n and ρ.
    (a) Show that the unconditional probability that $X = k$ is uniform,

    $$P(X = k) = (n + 1)^{-1}, \quad 0 \le k \le n.$$

    (b) What is the conditional density of R given that $X = k$?

6.11 APPLICATIONS OF CONDITIONAL EXPECTATION

As we have seen, conditional distributions and conditional expectations can be very useful. This usefulness is even more marked in more advanced work, so we now introduce a new and compact notation for conditional expectation.

The essential idea that underlies this is the fact that *conditional expectation is a random variable*. This sounds like a paradox, but just recall the definitions. First let X and Y be discrete; then given $Y = y$ we have

(1) $$E(X|Y = y) = \sum_x xp(x, y)/p_Y(y).$$

When X and Y are continuous we have

(2) $$E(X|Y = y) = \int_x xf(x, y)/f_Y(y)\, dx.$$

In both cases (1) and (2), we observe that $E(X|Y = y)$ is a function of y. Let us denote it by $\psi(y)$, that is

(3) $$E(X|Y = y) = \psi(y).$$

But y ranges over all the possible values of Y, and therefore $\psi(Y)$ is a function of Y. And as we noted in section 5.2, a function of a random variable is a random variable! For clarity and consistency we can now write

$$\psi(Y) = E(X|Y)$$

on the understanding that $\psi(Y)$ is the function of Y that takes the value $E(X|Y = y)$ when $Y = y$.

As a random variable, $\psi(Y)$ may have an expectation, of course; and we know from sections 6.9 and 6.10 that this expectation is EX. For convenience, we repeat the argument here: suppose X and Y are discrete, then

$$EX = \sum_y E(X|Y = y)p_Y(y)$$

$$= \sum_y \psi(y)p_Y(y)$$

$$= E\psi(Y), \quad \text{by the law of the unconscious statistician}$$

$$= E\{E(X|Y)\}.$$

Exactly the same argument works when X and Y are continuous; we just replace sums by integrals. Thus in our new notation we have found that in *any* case

(4) $$EX = E\{E(X|Y)\}.$$

The really attractive and useful thing about (4) is that it holds for *any* pair of random variables whatever their type; discrete, continuous, or both, and even for those that are neither. Previously we have had to deal separately with all these types. With equation (4) we can see that the idea of random variables has really rewarded us with economical and helpful ways of displaying and proving important results.

Furthermore, (4) enables us to write the expected value of products in a neat way. Obviously, for any function $g(Y)$ of Y we have

$$E\{g(Y)|Y\} = g(Y).$$

Hence, using (4), we have

(5) $$E(XY) = E\{Y E(X|Y)\},$$

and this can often simplify the calculation of covariances and correlations, and other expectations.

Here are some examples.

Example 6.11.1: potatoes. A sack contains n potatoes. The weight of the rth potato is a random variable X_r (in kilos), where the X_r are independent and identically distributed. The sack of potatoes weighs 100 kilos, and you remove m potatoes at random. What is the expected weight of your sample of size m?

Solution. We set $\sum_{r=0}^{n} X_r = S_n$, $\sum_{r=0}^{m} X_r = S_m$. Then the question asks for $E(S_m|S_n)$, when $S_n = 100$. Now it is clear (essentially by symmetry) that $E(X_j|S_n) = E(X_k|S_n)$, for all j and k. Hence

$$E(X_i|S_n) = \frac{S_n}{n},$$

$$E(S_m|S_n) = \sum_{i=1}^{m} E(X_i|S_n) = \frac{m}{n} S_n.$$

Thus the expected weight of your sample is $100 m/n$ kilos. ○

Example 6.11.2: thistles. A thistle plant releases X seeds, where X is Poisson with parameter λ. Each seed independently germinates with probability p; the total crop is Y thistle seedlings. Find $\text{cov}(X, Y)$ and $\rho(X, Y)$.

Solution. By (5),

$$E(XY) = E\{E(XY|X)\} = E\{X E(Y|X)\}.$$

The total crop number Y, now conditional on X, is binomial with parameters X and p. Hence

$$E(Y|X) = Xp$$

and

(6) $$E(Y^2|X) = Xp(1 - p) + X^2 p^2.$$

Thus

$$E(XY) = E(X^2 p) = (\lambda^2 + \lambda)p,$$

giving

$$\text{cov}(X, Y) = \lambda p^2 + \lambda p - EX EY = \lambda p.$$

Finally, using (6),

$$\text{var } Y = E\{E(Y^2|X)\} - (EY)^2$$

$$= \lambda p.$$

Hence

$$\rho(X, Y) = \frac{\text{cov}(X, Y)}{(\text{var } X \text{ var } Y)^{1/2}} = \frac{\lambda p}{(\lambda^2 p)^{1/2}} = \sqrt{p}. \qquad ○$$

Exercises for section 6.11

1. Show that
    $$\text{var } Y = E\{\text{var}(Y|X)\} + \text{var}\{E(Y|X)\}.$$

2. For any two jointly distributed random variables X and Y, show that for any function $g(\cdot)$
    $$E[\{X - g(Y)\}^2] \geqslant E[\{X - E(X|Y)\}^2] = E\{\text{var}(X|Y)\}.$$
    For this reason, $E(X|Y)$ is the minimum mean squared error predictor of X, given Y. Show also that $E\{\text{var}(X|Y)\} \leqslant (1 - \rho^2)\text{var } X$, where ρ is the correlation between X and Y. (Hint: Apply Cauchy–Schwarz to $XE(Y|X)$.)

3. Let X and Y be a pair of indicators with joint distribution
    $$p(0, 0) = a, \quad p(0, 1) = b, \quad p(1, 0) = c, \quad p(1, 1) = d.$$
    Show that
    $$E(X|Y) = \frac{c}{a + c} + \frac{ad - bc}{(a + c)(b + d)} Y.$$

4. ***Waldegrave's problem again.***   Suppose a coin shows heads with probability p, and you flip it repeatedly. Let X_n be the number of flips until it first shows a run of n consecutive heads. Show that
    $$E(X_n|X_{n-1}) = X_{n-1} + 1 + qEX_n.$$
    Deduce that
    $$EX_n = p^{-1} + p^{-1}EX_{n-1}$$
    $$= \sum_{k=1}^{n} p^{-k}.$$
    Hence derive the result of example 5.8.12.

5. Let X and Y have the joint density
    $$f(x, y) = cx(y - x)e^{-y}, \quad 0 \leqslant x \leqslant y < \infty.$$
    (a) Find c.
    (b) Show that
    $$f(x|y) = 6x(y - x)y^{-3}, \quad 0 \leqslant x \leqslant y,$$
    $$f(y|x) = (y - x)\exp\{-(y - x)\}, \quad x \leqslant y < \infty.$$
    (c) Deduce that $E(X|Y) = \frac{1}{2}Y$ and $E(Y|X) = X + 2$.

6.12 BIVARIATE NORMAL DENSITY

We know well by now that if X and Y are independent $N(0, 1)$ random variables, then they have joint density

(1) $$\phi(x, y) = \frac{1}{2\pi} \exp\{-\tfrac{1}{2}(x^2 + y^2)\}.$$

Very often it is necessary to consider random variables that are separately normal but *not* independent. With this in mind, define

(2) $$U = \sigma X,$$

(3) $$V = \tau\rho X + \tau(1 - \rho^2)^{1/2} Y,$$

where $\sigma, \tau > 0$ and $|\rho| \leq 1$. Then, from what we know already about sums of independent normal random variables, we find that U is $N(0, \sigma^2)$ and V is

$$N(0, \tau^2 \rho^2 + \{\tau(1 - \rho^2)^{1/2}\}^2) = N(0, \tau^2).$$

Thus U and V are also separately normal. But what about the joint distribution of U and V? They are clearly not independent! Proceeding in the usual way, we calculate

$$F(u, v) = P(U \leq u, V \leq v)$$

$$= P(\sigma X \leq u, \tau\rho X + \tau(1 - \rho^2)^{1/2} Y \leq v)$$

$$= \int_{-\infty}^{u/\sigma} \int_{-\infty}^{(u - \tau\rho x)/\{\tau(1-\rho^2)^{1/2}\}} \phi(x, y) \, dy \, dx$$

where $\phi(x, y)$ is given by (1). Now make the change of variable $\tau(1 - \rho^2)^{1/2} y = w - \tau\rho x$, to obtain

$$F(u, v) = \int_{-\infty}^{u/\sigma} \int_{-\infty}^{v} \phi\left(x, \frac{w - \tau\rho x}{\tau(1 - \rho^2)^{1/2}}\right) \frac{dw \, dx}{\tau(1 - \rho^2)^{1/2}}$$

Differentiating with respect to u and v gives the density $f(u, v)$ (we are using the fundamental theorem of calculus here),

(4) $$f(u, v) = \phi\left(\frac{u}{\sigma}, \frac{v - \tau\rho u}{\tau(1 - \rho^2)^{1/2}}\right) \frac{1}{\sigma\tau(1 - \rho^2)^{1/2}}$$

$$= \frac{1}{2\pi\sigma\tau(1 - \rho^2)^{1/2}} \exp\left\{-\frac{1}{2}\left(\frac{u^2}{\sigma^2} - \frac{2\rho u v}{\sigma\tau} + \frac{v^2}{\tau^2}\right) \frac{1}{(1 - \rho^2)}\right\}.$$

The density $f(u, v)$ in (4) is called the *bivariate normal density centred at the origin*. When $\sigma = \tau = 1$, it is called the *standard bivariate normal density* with parameter ρ. (We often call these *binormal* densities, for brevity.) By a translation

$$u^* = \mu + u, \quad v^* = v + v$$

we can centre it at any other point (μ, ν) in the plane, but we shall not do so unless it is necessary.

Given a bivariate distribution such as (4), it is natural to seek the value of things like

$$E(U|V), \quad E(V|U), \quad E(UV), \quad \text{cov}(U, V), \quad \rho(U, V)$$

and so on. This task can be accomplished using $f(u, v)$ by routine integrations, but these turn out to be fairly messy and tedious. It is often neater and more illuminating to use the initial representation (2) and (3) in terms of X and Y, whose behaviour we fully comprehend. For example, from (2) and (3),

$$E(UV) = \sigma\tau\rho \, EX^2 + \sigma\tau(1 - \rho^2)^{1/2} \, E(XY)$$

$$= \sigma\tau\rho$$

since X and Y are independent normal distributions, each $N(0, 1)$. Hence

(5) $$\rho(U, V) = \frac{E(UV)}{(\text{var } U \text{ var } V)^{1/2}} = \rho.$$

This explains why we introduced the parameter ρ in the way we did, in (2) and (3). Equally easily from (2) and (3) we see that, conditional on $U = u$,

$$V = \frac{\tau\rho u}{\sigma} + \tau(1 - \rho^2)^{1/2} Y.$$

Hence given $U = u$, V is normally distributed with mean $\tau\rho u/\sigma$ and variance $\tau^2(1 - \rho^2)$. That is to say,

(6) $$f_{V|U}(v|u) = \frac{1}{\{2\pi\tau^2(1 - \rho^2)\}^{1/2}} \exp\left\{-\frac{1}{2\tau^2(1 - \rho^2)}\left(v - \frac{\tau\rho u}{\sigma}\right)^2\right\}.$$

Now suppose that we require the distribution of $Z = aU + bV$, for constants a and b. From (2) and (3) we find

$$Z = (a\sigma + b\tau\rho)X + b\tau(1 - \rho^2)^{1/2} Y.$$

Hence Z is normal with variance

(7) $$\omega^2 = (a\sigma + b\tau\rho)^2 + b^2\tau^2(1 - \rho^2)$$

$$= a^2\sigma^2 + 2ab\sigma\tau\rho + b^2\tau^2.$$

You will appreciate the merits of this approach much more if you spend a while seeking to derive $f_Z(z)$ from (4) using a convolution integral.

The bivariate normal density gives rise to others, in the same way as the univariate case. Here is one example.

Definition.   If $\log X$ and $\log Y$ have jointly a bivariate normal density, then X and Y have a *bivariate log normal* density.   △

Example 6.12.1: body–mass index.   Suppose a randomly selected person's weight W and height H have a bivariate log normal density. Find the density of the body–mass index $Z = WH^{-2}$.

Solution.   First, remark that $\log Z = \log W - 2\log H$. By assumption, $\log W$ and $\log H$ have a bivariate normal density, centred at (μ, v) say, with parameters σ, τ, and ρ. Hence $\log Z$ has a normal density with mean $\mu - 2v$ and variance

$$\sigma^2 - 2\sigma\tau\rho + 4\tau^2.$$

Thus Z is log normal. (See problem 24.)   ○

Finally we note the important point that the joint density $f(u, v)$ in (4) factorizes as the product of separate functions of u and v, if and only if $\rho = 0$. This tells us that U and V are independent if and only if $\rho = 0$. That is, normal random variables are independent if and only if they are uncorrelated, unlike most other bivariate distributions.

Remark.   The bivariate normal density is of great importance in the development of probability and statistics. Its use by F. Galton around 1885, to explain previously baffling features of inheritance, marked a breakthrough of huge significance. The problem is as follows. The height of men (aged 21 say) is well known to be approximately normally distributed. And then the height of their sons, on reaching the age of 21, is found to have exactly the same approximately normal distribution. But there is a difficulty here, for it is a matter of simple observation that taller parents tend to have taller offspring, and shorter parents tend to have shorter offspring. So, before measuring anything, one might expect the spread (or variance) of the heights of successive generations to increase. But it does not.

Galton resolved this dilemma by finding the distribution of the sons' heights *conditional* on that of their parents. He found empirically that this was normal, with a mean intermediate between the population mean and their parent's height. He called this phenomenon *regression to the mean* and described it mathematically by assuming that the heights of fathers and sons were jointly binormal about the population mean. Then the conditional densities behaved exactly in accordance with the observations. It would be hard to overestimate the importance of this brilliant analysis.

One curious and interesting consequence is the following.

Example 6.12.2: doctor's paradox.   Suppose your doctor measures your blood pressure. If its value X is high (where $X = 0$ is average), you are recalled for a further measurement, giving a second value Y. On the reasonable assumption that X and Y have approximately a standard bivariate normal distribution, the conditional density of Y given $X \gg 0$ is seen from (6) to have mean ρX, which is less than X. It seems that merely to revisit your doctor makes you better, whether or not you are treated for your complaint. This result may partly explain the well-known placebo effect.   ○

Exercises for section 6.12

1. Let X and Y be independent N(0, 1) random variables, and let $U = \mu + \sigma X$ and $V = v + \tau \rho X + \tau (1 - \rho^2)^{1/2} Y$. What is the conditional density of V given $U = u$?

2. Let U and V be the heights of a randomly selected parent and child. Centred at the average height, in appropriate units, U and V have approximately a standard bivariate normal density with correlation ρ and $\sigma^2 = \tau^2 = 1$.
    (a) Show that the probability that both parent and child are above average height is
    $$P(U > 0,\ V > 0) = \frac{1}{4} + \frac{1}{2\pi} \sin^{-1} \rho$$
    (b) Show that
    $$P(0 < V < U) = \frac{1}{8} + \frac{1}{4\pi} \sin^{-1} \rho.$$
    (c) Show that the expected height above the average of the taller of the two is
    $$E(\max\{U,\ V\}) = \left(\frac{1 - \rho}{\pi} \right)^{1/2}.$$
    (d) Show that $E(\max\{U,\ V\})^2 = 1$. (Note: *No* integration is required.)

3. Let X and Y have the standard bivariate normal density, with parameters ρ, σ, and τ. Let
    $$U = X \cos \theta + Y \sin \theta, \qquad V = Y \cos \theta - X \sin \theta.$$
    Find the values of θ such that U and V are independent.

4. Let X and Y have joint density $f(x, y) = \pi^{-1} \exp\{-\frac{1}{2}(x^2 + y^2)\}$ for $xy > 0$; $f = 0$ otherwise. Show that X and Y are normally distributed. Are they jointly normally distributed? Find $\rho(X,\ Y)$.

5. Let X and Y have a bivariate normal distribution centred at the origin. Show that
    (a)
    $$E(X | Y = y) = \frac{\text{cov}(X,\ Y)}{\text{var } Y} y = \frac{\rho \sigma}{\tau} y,$$

(b) $$\mathrm{var}(X|Y=y) = \sigma^2(1-\rho^2),$$

(c) $$\mathrm{E}(X|X+Y=v) = \frac{\sigma^2+\rho\sigma\tau}{\sigma^2+2\rho\sigma\tau+\tau^2}v$$

(d) $$\mathrm{var}(X|X+Y=v) = \frac{\sigma^2\tau^2(1-\rho^2)}{\tau^2+2\rho\sigma\tau+\sigma^2}.$$

(Hint: use exercise 1; no integrations are required.)

6.13 CHANGE-OF-VARIABLES TECHNIQUE; ORDER STATISTICS

At several places in earlier sections it has been necessary or desirable to find the joint distribution of a pair of random variables U and V, where U and V are defined as functions of X and Y,

(1) $$U = u(X, Y), \quad V = v(X, Y).$$

In many cases of interest X and Y have joint density $f(x, y)$; the question is, what is the joint density $f_{U,V}(u, v)$?

We have always succeeded in answering this, because the transformations were mostly linear (or bilinear), which simplifies things. (The most recent example was in the previous section, when we considered U and V as linear combinations of the normal random variables X and Y.) Otherwise, the transformation was to polars. Not all transformations are linear or polar, and we therefore summarize a general technique here. We consider two dimensions for simplicity.

The proofs are not short, and we omit them, but it is worth remarking that in general it is intuitively clear what we are doing. The point about $f(x, y)$ is that $f(x, y)\delta x\,\delta y$ is roughly the probability that (X, Y) lies in the small rectangle

$$R = (x, x+\delta x) \times (y, y+\delta y).$$

The joint density of U and V has the same property, so we merely have to rewrite $f(x, y)\delta x\,\delta y$ in terms of $u(x, y)$ and $v(x, y)$. The first bit is easy, because

$$f(x, y) = f(x(u, v), y(u, v)).$$

The problem arises in finding out what the transformation does to the rectangle R. We have seen one special case: in polars, when $x = r\cos\theta$ and $y = r\sin\theta$, we replace $\delta x\,\delta y$ by $r\,\delta r\,\delta\theta$. In general, the answer is given by the following.

Change of variables.   Let S and T be sets in the plane. Suppose that $u = u(x, y)$ and $v = v(x, y)$ define a one–one function from S to T with unique inverses $x = x(u, v)$ and $y = y(u, v)$ from T to S. Define the determinant

(2) $$J(u, v) = \begin{vmatrix} \dfrac{\partial x}{\partial u} & \dfrac{\partial y}{\partial u} \\[2mm] \dfrac{\partial x}{\partial v} & \dfrac{\partial y}{\partial v} \end{vmatrix} = \frac{\partial x}{\partial u}\frac{\partial y}{\partial v} - \frac{\partial x}{\partial v}\frac{\partial y}{\partial u},$$

where all the derivatives are continuous in T. Then the joint density of $U = u(X, Y)$ and $V = v(X, Y)$ is given by

(3) $$f_{U,V}(u, v) = f_{X,Y}(x(u, v), y(u, v))|J(u, v)|.$$

Informally we see that the rectangle R with area $\delta x \, \delta y$, has become a different shape, with area $|J(u, v)| \delta u \, \delta v$.

Example 6.13.1: bivariate normal. From (2) and (3) in section 6.12 we see that

$$x(u, v) = \frac{u}{\sigma} \quad \text{and} \quad y(u, v) = \frac{v - \tau\rho\sigma^{-1}u}{\tau(1 - \rho^2)^{1/2}}$$

Hence

$$|J| = \begin{vmatrix} \sigma^{-1} & 0 \\ \dfrac{-\rho}{\sigma(1 - \rho^2)^{1/2}} & \dfrac{1}{\tau(1 - \rho^2)^{1/2}} \end{vmatrix} = \frac{1}{\sigma\tau(1 - \rho^2)^{1/2}}$$

and

$$f(u, v) = \frac{1}{2\pi\sigma\tau(1 - \rho^2)^{1/2}} \exp\left\{ -\frac{1}{2}\left(\frac{u^2}{\sigma^2} + \left\{ \frac{\sigma v - \tau\rho u}{\sigma\tau(1 - \rho^2)^{1/2}} \right\}^2 \right) \right\}$$

just as before. ○

Example 6.13.2: order-statistics. Let $X_1, \ldots, X_n$ be independent and identically distributed with density $f(x)$ and distribution $F(x)$.

Let $X_{(1)}$ be the smallest of the X_i, $X_{(2)}$ the next smallest, and so on to $X_{(n)}$, which is the largest. Thus we obtain the so-called *order-statistics*; obviously

$$X_{(1)} < X_{(2)} < \cdots < X_{(n)}.$$

Note that the probability of any two being equal is zero, because they are continuous. What is the joint density of $X_{(1)}, \ldots, X_{(n)}$?

The first point to make is that the transformation

$$(X_1, \ldots, X_n) \to (X_{(1)}, \ldots, X_{(n)})$$

is not one-to-one. In fact for any fixed values of the X_i, any of their $n!$ permutations give the same order-statistics.

However, for any given permutation the map is one-to-one, and J is a matrix in which each row and column contains $n - 1$ zeros and one entry of unity. Hence $|J| = 1$ and, for this permutation, by independence,

$$f(x_{(1)}, \ldots, x_{(n)}) = f(x_{(1)})f(x_{(2)}) \cdots f(x_{(n)}).$$

Since the permutations are mutually exclusive we add their probabilities, to find that the joint density of the order statistics is

$$(4) \qquad f(x_{(1)}, \ldots, (x_{(n)}) = n!f(x_{(1)}) \cdots f(x_{(n)}), \qquad x_{(1)} < \cdots < x_{(n)}.$$

From this it is routine, though dull, to find the marginal densities of any subset of the order-statistics by integration. Eventually you find that

$$(5) \qquad f_{X_{(r)}}(x) = nf(x)\binom{n - 1}{r - 1}F(x)^{r-1}\{1 - F(x)\}^{n-r}.$$

When the X_i are uniform on $(0, 1)$ this looks a little more attractive, and in this case even joint densities are not too bad:

$$(6) \qquad f_{X_{(r)}, X_{(s)}}(x, y) = \frac{n!}{(r - 1)!(s - r - 1)!(n - s)!}x^{r-1}(y - x)^{s-r-1}(1 - y)^{n-s}.$$

The change-of-variables technique is further explored in the exercises and problems. ○

Exercises for section 6.13

1. Let X and Y be continuous random variables with joint density $f(x, y)$.
   (a) Let $U = X$ and $V = XY$. Show that the joint density of U and V is

$$f(u, v) = \frac{1}{|u|} f\left(u, \frac{v}{u}\right)$$

   and hence find the density of V.
   (b) Let $W = X/Y$ and $Z = Y$. Show that the joint density of W and Z is

$$f(w, z) = |z| f(zw, z)$$

   and hence find the density of W.
   (c) When X and Y are independent standard normal random variables, show that $W = X/Y$ has a Cauchy density.
   (d) When X has density $f(x) = x \exp\left(-\frac{1}{2}x^2\right)$, $x > 0$, and Y has density $f(y) = \pi^{-1}(1 - y^2)^{-1/2}$, $|y| < 1$, and X and Y are independent, show that XY has a normal density.

2. Let X and Y be independent, having gamma distributions with parameters n and λ and m and λ respectively. Let $U = X + Y$ and $V = X/(X + Y)$.
   Find the joint density of U and V and show that they are independent.

3. Let $X_{(1)}, \ldots, X_{(n)}$ be the order-statistics of $X_1, \ldots, X_n$. Argue directly that

$$P(X_{(r)} \leq x) = \sum_{k=r}^{n} \binom{n}{k} \{F(x)\}^k \{1 - F(x)\}^{n-k}$$

   and hence obtain (5) by differentiating.

4. Let $X_1, \ldots, X_n$ be independent and uniform on $(0, 1)$. Let $X_{(r)}$ be the rth order-statistic of the X_i. Find the distribution of $X_{(r)}$, and show that

$$E(X_1^2) = 2/\{(n + 1)(n + 2)\},$$

$$f_{X_{(r)}}(x) = n\binom{n-1}{r-1} x^{r-1}(1 - x)^{n-r},$$

$$E(X_{(r)}) = \frac{r}{n+1},$$

$$\operatorname{var} X_{(r)} = \frac{r(n - r + 1)}{(n + 1)^2(n + 2)}.$$

   Note that integration is not needed for the last two. Now let $X_{(s)}$ be the sth order-statistic, $r < s$. Show that the joint density $f(x, y)$ of $X_{(r)}$ and $X_{(s)}$ is

$$f(x, y) = \frac{n!}{(r - 1)!(s - r - 1)!(n - s)!} x^{r-1}(y - x)^{s-r-1}(1 - y)^{n-s}.$$

   Would you like to find $E(X_{(r)}X_{(s)})$ from this? If not, how else would you find it? Without using integration, show that

$$\operatorname{cov}(X_{(r)}, X_{(s)}) = \frac{r(n + 1 - s)}{(n + 1)^2(n + 2)}.$$

5. Let X and Y be independent exponentially distributed random variables with parameters λ and μ respectively. Find the density of $Z = X/(X + Y)$.

6.14 REVIEW

In this chapter we have considered jointly distributed random variables. Such variables X and Y have a distribution function

$$F(x, y) = P(X \leq x, Y \leq y)$$

where

$$P(a < X \leq b, c < Y \leq d) = F(b, d) - F(a, d) - F(b, c) + F(a, c).$$

Random variables (X, Y) when discrete have a distribution $p(x, y)$ and when continuous have a density $f(x, y)$ such that:

in the discrete case

in the continuous case

$$P((X, Y) \in C) = \sum_{(x,y) \in C} p(x, y), \qquad P((X, Y) \in C) = \iint_{(x,y) \in C} f(x, y)\, dx\, dy,$$

$$f(x, y) = \frac{\partial^2 F}{\partial x \partial y}.$$

Marginals
In the discrete case

in the continuous case

$$p_X(x) = \sum_y p(x, y), \qquad f_X(x) = \int f(x, y)\, dy,$$

$$p_Y(y) = \sum_x p(x, y). \qquad f_Y(y) = \int f(x, y)\, dx.$$

Functions
In the discrete case

in the continuous case

$$P(g(X, Y) = z) = \sum_{g=z} p(x, y) \qquad P(g(X, Y) \leq z) = \iint_{g \leq z} f(x, y)\, dx\, dy$$

$$P(X + Y = z) = \sum_x p(x, z - x) \qquad f_{X+Y}(z) = \int_z f(x, z - x)\, dx$$

Independence. Random variables X and Y are independent if and only if for all x and y

$$F(x, y) = F_X(x) F_Y(y);$$

in the discrete case

in the continuous case

$$p(x, y) = p_X(x) p_Y(y) \qquad f(x, y) = f_X(x) f_Y(y).$$

Conditional. The conditional distribution of X given Y is

in the discrete case

in the continuous case

$$p_{X|Y}(x|y) = \frac{p(x, y)}{p_Y(y)} \qquad f_{X|Y}(x|y) = \frac{f(x, y)}{f_Y(y)}.$$

Expectation.   The law of the unconscious statistician states that

in the discrete case   in the continuous case

$$Eg(X, Y) = \sum_x \sum_y g(x, y)p(x, y) \qquad Eg(X, Y) = \iint g(x, y)f(x, y)\, dx\, dy.$$

In particular this yields
$$E\{g(X, Y) + h(X, Y)\} = Eg(X, Y) + Eh(X, Y).$$

Moments.   The covariance of X and Y is
$$\text{cov}(X, Y) = E\{(X - EX)(Y - EY)\} = E(XY) - EX\,EY$$
and the correlation coefficient is
$$\rho(X, Y) = \frac{\text{cov}(X, Y)}{\{\text{var } X \text{ var } Y\}^{1/2}}$$

Independence.   When X and Y are independent
$$E(XY) = EX\,EY,$$
so they are uncorrelated, and
$$\text{cov}(X, Y) = \rho(X, Y) = 0.$$

Conditioning.   For any pair of random variables,
$$E(X) = E\{E(X|Y)\}.$$

6.15 PROBLEMS

1. You flip a fair coin repeatedly; let X be the number of flips required to obtain n heads.
    (a) Show that $EX = 2n$.
    (b) Show also that $P(X < 2n) = \frac{1}{2}$.
         (Elaborate calculations are not required.)
    (c) Atropos flips a fair coin $n + 1$ times; Belladonna flips a fair coin n times. The one with the larger number of heads wins. Show that $P(A \text{ wins}) = \frac{1}{2}$.

2. Each packet of Acme Gunk is equally likely to contain one of three different types of coupon. Let the number of packets required to complete your set of the three different types be X. Find the distribution of X.

3. Your coin shows heads with probability p. You flip it repeatedly; let X be the number of flips until the first head, and Y the number until the first tail. Find $E(\min\{X, Y\})$, and $E|X - Y|$.

4. The random variable X is uniform on $[0, 1]$, and conditional on $X = x$; Y is uniform on $(0, x)$. What is the joint density of X and Y? Find
    (a) $f_Y(y)$,   (b) $f_{X|Y}(x|y)$,   (c) $E(X|Y)$.

5. Rod and Fred play the best of five sets at tennis. Either Fred wins a set with probability ϕ or Rod wins with probability ρ, independently of other sets. Let X be the number of sets Fred wins and Y the number of sets Rod wins. Find the joint distribution of X and Y, and calculate cov (X, Y) when $\rho = \psi = \frac{1}{2}$.

6. You roll n dice obtaining X sixes. The dice showing a six are rolled again yielding Y sixes. Find the joint distribution of X and Y. What is $E(X|Y)$? What is $\text{var}(X|Y)$?

7. You make n robots. Any robot is independently faulty with probability ϕ. You test all the robots; if any robot is faulty your test will detect the fault with probability δ, independently of the other tests and robots. Let X be the number of faulty robots, and Y the number detected as faulty. Show that, given Y, the expected number of faulty robots is

$$E(X|Y) = \frac{n\phi(1-\delta) + (1-\phi)Y}{1-\phi\delta}.$$

8. Two impatient, but also unpunctual, people arrange to meet at noon. Art arrives at X hours after noon and Bart at Y hours after noon, where X and Y are independently and uniformly distributed on $(0, 1)$. Each will wait at most 10 minutes before leaving, and neither will wait after 1.00 p.m. What is the probability that they do in fact meet? What is the probability that they meet given that neither has arrived by 12.30?

9. An urn contains three tickets bearing the numbers a, b, and c. Two are taken at random without replacement; let their numbers be X and Y. Write down the joint distribution of X and Y. Show that $\text{cov}(X, Y) = \frac{1}{9}(ab + bc + ca - a^2 - b^2 - c^2)$.

10. You roll two dice, showing X and Y. Let U be the minimum of X and Y. Write down the joint distribution of U and Y, and find $\text{cov}(U, X)$.

11. **Runs revisited.** A coin shows heads with probability p, or tails with probability q. You flip it repeatedly; let X be the length of the opening run until a fresh face appears, and Y the length of the second run. Find the joint distribution of X and Y, and show that

$$\text{cov}(X, Y) = -\frac{(p-q)^2}{pq}.$$

12. Let X and Y be independent standard normal random variables. Let $Z = X^2/(X^2 + Y^2)$. Show that $EZ = \frac{1}{2}$ and $\text{var } Z = \frac{1}{8}$.

13. A casino offers the following game. There are three fair coins on the table, values 5, 10, and 20 respectively. You can nominate one coin; if its value is a, then your entry fee is a. Denote the values of the remaining coins by b and c.
Now the coins are flipped; let X be the indicator of the event that yours shows a head, and Y and Z the indicators of heads for the other two. If $aX > bY + cZ$ then you win all three coins; if $aX < bY + cZ$ then you get nothing; if $aX = bY + cZ$ then your stake is returned. Which coin would you nominate?

14. Suppose that $X_1, \ldots, X_n$ are independent and that each has a distribution which is symmetric about 0. Let $S_n = \sum_{r=1}^{n} X_r$.
(a) Show that the distribution of S_n is also symmetric about zero.
(b) Is this still necessarily true if the X_r are not independent?

15. Let X and Y be independent standard normal random variables. Show that

$$E(\min\{|X|, |Y|\}) = 2(\sqrt{2} - 1)/\sqrt{\pi}$$

16. Let U and V have joint density $f(u, v)$, and set $X = UV$. Show that

$$P(UV \leqslant x) = \int_0^\infty \int_{-\infty}^{x/v} f(u, v) \, du \, dv + \int_{-\infty}^0 \int_{x/v}^\infty f(u, v) \, du \, dv,$$

and hence deduce that X has density

$$f_X(x) = \int_{-\infty}^\infty \frac{1}{|v|} f\left(\frac{x}{v}, v\right) dv.$$

17. **Buffon's needle.**   The floorboards in a large hall are planks of width 1. A needle of length $a \leqslant 1$ is dropped at random onto the floor. Let X be the distance from the centre of the needle to the nearest joint between the boards, and Θ the angle which the needle makes with the joint. Argue that X and Θ have density

$$f(x, \theta) = \pi^{-1}, \quad 0 \leqslant x \leqslant \tfrac{1}{2}; \, 0 \leqslant \theta \leqslant 2\pi.$$

Deduce that the probability that the needle lies across a joint is $2a/\pi$.

18. An urn contains j jet and k khaki balls. They are removed at random until all the balls remaining are of the same colour. Find the expected number left in the urn.

19. Show that $\operatorname{cov}(X, Y + Z) = \operatorname{cov}(X, Y) + \operatorname{cov}(X, Z)$. When Z and Y are uncorrelated show that

$$\rho(X, Y + Z) = a\rho(X, Y) + b\rho(X, Z),$$

where

$$a^2 + b^2 = 1.$$

Suppose $Y_1, \dots, Y_n$ are uncorrelated. Show that

$$\rho(X, \textstyle\sum_{r=1}^n Y_r) = \sum_{r=1}^n a_r\rho(X, Y_r),$$

where

$$\sum_{r=1}^n a_r^2 = 1$$

Can you see a link with Pythagoras' theorem?

20. You are the finance director of a company that operates an extremely volatile business. You assume that the quarterly profits in any year, X_1, X_2, X_3, X_4, are independent and identically distributed continuous random variables. What is the probability that $X_1 > X_2 > X_3 > X_4$? Would you be very concerned if this happened? Would you be thrilled if $X_1 > X_2 < X_3 < X_4$? Would you expect a bonus if profits increased monotonically for six successive quarters?

21. Let X and Y be independent and uniform on $\{1, 2, \dots, n\}$ and let $U = \min\{X, Y\}$ and $V = \max\{X, Y\}$. Show that

$$\mathrm{E}U = \frac{(n + 1)(2n + 1)}{6n}, \quad \mathrm{E}V = \frac{(4n - 1)(n + 1)}{6n},$$

$$\operatorname{cov}(U, V) = \left(\frac{n^2 - 1}{6n}\right)^2, \quad \operatorname{var} U = \operatorname{var} V = \frac{(n^2 - 1)(2n^2 + 1)}{36n^2},$$

and that

$$\rho(U, V) = \frac{n^2 - 1}{2n^2 + 1} \to \frac{1}{2} \text{ as } n \to \infty.$$

22. Let X and Y have joint density $f(x, y) = 1, 0 \leqslant x, y \leqslant 1$. Let $U = X \wedge Y$ and $V = X \vee Y$. Show that $\rho(U, V) = \tfrac{1}{2}$. Explain the connexion with the preceding problem.

23. There are 10 people in a circle, and each of them flips a fair coin. Let X be the number whose coin shows the same as both neighbours.
   (a) Find $P(X = 10)$ and $P(X = 9)$.
   (b) Show that $\mathrm{E}X = \tfrac{5}{2}$ and $\operatorname{var} X = \tfrac{25}{8}$.

24. Let X and Y be random variables, such that $Y = e^X$. If X has the $\mathrm{N}(\mu, \sigma^2)$ normal distribution, find the density function of Y. This is known as the *log normal* density, with parameters μ and σ^2. Find the mean and variance of Y.

25. You are selling your house. You receive a succession of offers $X_0, X_1, X_2, \ldots$, where we assume that the X_i are independent identically distributed random variables. Let N be the number of offers necessary until you get an offer better than the first one, X_0. That is,

$$N = \min\{n: X_n > X_0\}, \quad n \geq 1.$$

Show that

$$P(N \geq n) = P(X_0 \geq X_1, X_0 \geq X_2, \ldots, X_0 \geq X_{n-1})$$

$$\geq n^{-1}.$$

Deduce that $EN = \infty$. Why is this a poor model? (Hint: Let A_k be the event that no offer is larger than the kth. Then $P(N \geq n) = P(A_0)$ and $1 = P(\bigcup_{k=0}^{n-1} A_k)$; use Boole's inequality.)

26. Let X, Y, and Z be jointly distributed random variables such that $P(X \geq Y) = 1$.
    (a) Is it true that $P(X \geq x) \geq P(Y \geq x)$ for all x?
    (b) Is it true that $P(X \geq Z) \geq P(Y \geq Z)$?
    (c) Show that $EX \geq EY$.

27. Let X and Y have the joint distribution

$$p(x, y) = c\{(x + y - 1)(x + y)(x + y + 1)\}^{-1}, \quad x, y \geq 1.$$

Find the value of c, and EX.

28. You plant n seeds; each germinates independently with probability γ. You transplant the resulting X seedlings; each succumbs independently to wilt with probability δ. Let Y be the number of plants you raise. Find the distribution of X and Y, and calculate $\text{cov}(X, Y)$.

29. *Are you normal?*   When a medical test measuring some biophysical quantity is administered to a population, the outcomes are usually approximately normally distributed, $N(\mu, \sigma^2)$. If your measurement is further than 2σ from μ, you are diagnosed as abnormal and a candidate for medical treatment. Suppose you undergo a sequence of such tests with independent results.
    (a) What is the probability that any given test indicates that you are abnormal?
    (b) After how many tests will you expect to have at least one abnormal result?
    (c) After how many tests will the probability of your being regarded as abnormal exceed $\frac{1}{2}$?

30. Recall the gambler's ruin problem: at each flip of a fair coin you win $1 from the casino if it is heads, or lose $1 if it is tails.
    (a) Suppose initially you have k, and the casino has $(n - k)$. Let X_k be the number of flips until you or the casino has nothing. If $m_k = EX_k$, show that for $0 < k < n - 1$

$$2m_k = m_{k+1} + m_{k-1} + 2.$$

    Deduce that $m_k = k(n - k)$.
    (b) Suppose initially that you have Y where Y is binomial $B(n, p)$ and the casino has $(n - Y)$, and let the duration of the game be X. Find EX and $\text{cov}(X, Y)$. Show that X and Y are uncorrelated if $p = \frac{1}{2}$.
    (c) Suppose that initially you have Y and the casino has Z, where Y and Z are independent. Find EX and $\text{cov}(X, Y)$, where X is the duration of the game.

31. (a) Let $X_1, X_2, \ldots, X_n$ be independent and uniformly distributed on $(0, 1)$, and define $M_n = \max\{X_1, \ldots, X_n\}$. Show that as $n \to \infty$,

$$P(n(1 - M_n) > y) \to e^{-y}, \quad y > 0.$$

    (b) Let $X_1, X_2, \ldots, X_n$ be independent and identically distributed on $(0, 1)$, with density $f(x)$, where $0 < f(1) < \infty$. Let $M_n = \max\{X_1, \ldots, X_n\}$. Show that as $n \to \infty$

$$P(n(1 - M_n) > y) \to \exp\{-f(1)y\}.$$

32. Let $X_1, X_2, \ldots, X_n$ be independent with common density
$$f(x) = 6x(1-x), \quad 0 < x < 1.$$
Let $M_n = \max\{X_1, \ldots, X_n\}$. Can you find a result for M_n similar to those proved in problem 31?

33. (a) A certain university faculty comprises 52 mathematicians. Examination of their birthdates shows that three dates are each shared by two people; the remaining 46 do not share a birthday. Would you have expected more or fewer coincidences?

(b) Suppose 52 birthdays are uniformly and independently distributed over the year; ignore 29 February. Let S be the number of birthdays enjoyed alone, D the number shared by exactly two people, and M the number of birthdays shared by three or more. Show that, approximately,

$$ES = 45.4$$

$$ED = 3$$

and

$$P(M > 0) \leqslant 0.2.$$

Do you need to reassess your answer to (a)?

34. (a) You roll a conventional die repeatedly. Let D be the number of rolls until all six faces have appeared at least once. Show that $ED \simeq 14.7$.

(b) If your die is regular with 10 faces, show that the expected number of rolls for all to appear is approximately 29.3.

35. Let $X_1, \ldots, X_n$ be independent and identically distributed with mean μ and variance σ^2. Let $\overline{X} = n^{-1}\sum_{r=1}^{n}X_r$. Show that

$$\sum_{r=1}^{n}(X_r - \overline{X})^2 = \sum_{r=1}^{n}(X_r - \mu)^2 - n(\overline{X} - \mu)^2,$$

and deduce that

$$E\sum_{r=1}^{n}(X_r - \overline{X})^2 = (n-1)\sigma^2.$$

36. Let X and Y have zero mean and unit variance, with correlation coefficient ρ. Define $U = X$ and $V = Y - \rho X$. Show that U and V are uncorrelated.

37. Alf, Betty, and Cyril each have a torch; the lifetimes of the single battery in each are independent and exponentially distributed with parameters λ, μ, and ν respectively. Find the probability that they fail in the order: Alf's, Betty's, Cyril's.

38. Let X and Y be independent standard normal random variables, and let $Z = X + Y$. Find the distribution and density of Z given that $X \geqslant 0$ and $Y \geqslant 0$.
(Hint: No integration is required; use the circular symmetry of the joint density of X and Y.)
Show that $E(Z|X > 0, Y > 0) = 2\sqrt{2/\pi}$.

39. Let $X_1, \ldots, X_n$ be a collection of random variables with the property that the collection $\{X_1, \ldots, X_r\}$ is independent of X_{r+1} for $1 \leqslant r \leqslant n$. Prove that the X_i are all mutually independent.

40. Let X and Y be independent and let each have the distribution N(0, 1). Let

$$U = \frac{2XY}{(X^2 + Y^2)^{1/2}} \quad \text{and} \quad V = \frac{X^2 - Y^2}{(X^2 + Y^2)^{1/2}}.$$

show that U and V are also N(0, 1), and independent.

41. Let X, Y, Z be independent N(0, 1) random variables and define $T = X^2/(X^2 + Y^2 + Z^2)$. Show that T has a beta distribution and that T is independent of $X^2 + Y^2 + Z^2$.

42. What is the distribution of $T = X^2/(X^2 + Y^2 + Z^2)$ when the point (X, Y, Z) is picked uniformly
  (a) on the surface of a sphere?
  (b) in the interior of a sphere?

43. A coin shows a head with probability p, or a tail with probability $q = 1 - p$. You flip the coin n times; it yields X heads and R runs.
  (Remember that a run of heads is an unbroken sequence of heads either between two tails, or at the ends of the trial. A run of tails is defined likewise. Thus for the sequence $HHTH$, $X = 3$ and $R = 3$; for the sequence HHH, $X = 3$ and $R = 1$. Recall also that $ER = 1 + 2(n - p)pq$.)
  (a) Show that $\operatorname{cov}(X, R) = 2(n - 1)pq(q - p)$.
  (b) Show that $\operatorname{var} R = 4pqn(1 - 3pq) - pq\{1 + 5(p - q)^2\}$.
  (c) Show that for large n

$$\rho(X, R) \simeq \frac{q - p}{(1 - 3pq)^{1/2}}.$$

  (d) It follows from (a) that when $p = q$, X and R are uncorrelated. Are they independent in this case?

44. **Random stake.**   You make n wagers, where the probability of winning each wager is p. The stake is the same for each, but is a random variable X chosen before the wagers. Thus your winnings are $W = \sum_{r=1}^{n} XY_r$, where

$$Y_r = \begin{cases} 1 & \text{with probability } p \\ -1 & \text{with probability } 1 - p. \end{cases}$$

Find $\operatorname{cov}(X, W)$ and show that X and W are uncorrelated if the game is fair. Are they independent?

45. For any random variable X with finite fourth moment we define the *kurtosis* of X by

$$\operatorname{kur} X = E\{(X - \mu)^4\}/\{E(X - \mu)^2\}^2$$

$$= \sigma_4/\sigma^4$$

  (a) If X is N(μ, σ^2), show that kur $X = 3$.
  (b) If X is exponential with parameter λ, show that kur $X = 9$.
  (c) If $X_1, \ldots, X_n$ are independent and identically distributed and $S_n = \sum_{r=1}^{n} X_r$, show that kur $S_n = 3 + (\operatorname{kur} X_1 - 3)/n$.
  (d) If X is Poisson with parameter λ, show that kur $X = 3 + \lambda^{-1}$.
  (e) if X is binomial B(n, p), show that kur $X = 3 + (1 - 6pq)/(npq)$.

Remark.   Kurtosis is an indication of how peaked the distribution of X is above its mean; small values indicate strong peaking.

46. Let $X_1, X_2, X_3, \ldots$ be a sequence of independent identically distributed random variables, each uniform on (0, 1). Let

$$N = \min\left\{n: \sum_{r=1}^{n} X_r > x\right\}, \quad 0 < x < 1.$$

Show that, for $n \geq 2$,

$$G_n(x) = P(N > n) = \int_0^x G_{n-1}(x - u)\, du = \int_0^x G_{n-1}(y)\, dy,$$

and deduce that

$$p_N(k) = P(N = k) = \frac{x^{k-1}}{(k-1)!} - \frac{x^k}{k!}, \quad k \geqslant 1.$$

What is EN?

47. If X, Y, and Z are independent and uniformly distributed on $(0, 1)$, find $\text{var}(XY)$, and $\rho\{XY, (1 - X)Z\}$.

48. **Plates.**   Your factory produces rectangular plates of length X and width Y, where X and Y are independent; they have respective means μ_X, μ_Y and respective variances σ_X^2, σ_Y^2. Find the variance of the perimeter B and the area A of any plate, and show that A and B are not independent.

49. **Chi-squared.**   Let $X_1 \ldots, X_n$ be independent $N(0, 1)$ random variables. Show that the density of $Z = \sum_{r=1}^n X_r^2$ is

$$2^{-n/2}\left\{\Gamma\left(\tfrac{1}{2}n\right)\right\}^{-1} x^{n/2-1} e^{-x/2},$$

known as the chi-squared density, with n degrees of freedom, $\chi^2(n)$.

50. **Student's t-density.** Let X have the $\chi^2(n)$ density, and let Y be $N(0, 1)$ and independent of X. Show that the density of $Z = Y(X/n)^{-1/2}$ is

$$\frac{\Gamma\left(\tfrac{1}{2}(n + 1)\right)}{(n\pi)^{1/2}\Gamma\left(\tfrac{1}{2}n\right)}\left(1 + \frac{x^2}{n}\right)^{-(n+1)/2},$$

known as 'Student's t-density', with n degrees of freedom, $t(n)$.

7

Generating functions

7.1 PREVIEW

The most important thing about random variables is that they all have a distribution. Where they have moments, we like to know those as well. However, distributions are often cumbersome to deal with; as an example of this, recall the convolution formulas for the sums of independent random variables in section 6.6. And even simple tasks, such as finding moments, are often wearisome. We may suspect, furthermore, that there are even more tedious computations to come.

Fortunately there are miraculous devices to help us with many of these humdrum tasks; they are called generating functions. In this chapter we define the probability generating function and the moment generating function; then we explore some of their simpler properties and applications. These functions were first used by Euler and de Moivre in the 18th century; Euler used them in number theory and de Moivre actually used them to help with probability distributions. They seem to have hit on the idea independently, a classic illustration of the theory that great minds think alike.

Prerequisites.   We use no new mathematical techniques in this chapter.

7.2 INTRODUCTION

We introduce the basic idea with two examples. First, recall that the binomial distribution is

$$(1) \qquad p(r) = \binom{n}{r} p^r q^{n-r}, \quad 0 \leqslant r \leqslant n.$$

But we know, by the binomial theorem, that these numbers are the coefficients of s^r in the function

$$(2) \qquad G(s) = (q + ps)^n = \sum_{r=0}^{n} \binom{n}{r} p^r q^{n-r} s^r = \sum_{r=0}^{n} p(r) s^r.$$

Therefore the collection of probabilities in (1) is exactly equivalent to the function $G(s)$ in (2), in the sense that if we know $G(s)$ we can find the $p(r)$, and if we know the $p(r)$ we can find $G(s)$. The function $G(s)$ has effectively bundled up the $n + 1$ probabilities in (1) into a single entity, and they can then be generated from $G(s)$ whenever we so desire.

Second, we likewise know that the moments of the exponential density are

$$(3) \qquad \mu_r = \int_0^\infty x^r \lambda e^{-\lambda x} dx = \frac{r!}{\lambda^r},$$

after prolonged integration. But we also know, by Taylor's theorem, that these numbers appear in the coefficients of t^r in the expansion of the function

$$(4) \qquad M(t) = \frac{\lambda}{\lambda - t} = \sum_{r=0}^\infty \frac{t^r}{\lambda^r} = \sum_{r=0}^\infty \frac{\mu_r}{r!} t^r.$$

Therefore the collection of moments in (3) is exactly equivalent to the function $M(t)$ in (4), in the sense defined above. The function $M(t)$ has effectively bundled up the moments in (3) into a single entity, and the moments μ_r can thus be generated from $M(t)$, whenever we so desire.

With this preliminary, the following definitions are obvious.

Definition.   Let X be a non-negative integer-valued random variable with distribution $p(x)$. Then

$$(5) \qquad G(s) = p(0) + p(1)s + p(2)s^2 + \cdots$$

$$= \sum_{r=0}^\infty p(r)s^r$$

$$= \mathrm{E} s^X,$$

is the *probability generating function* of X. Sometimes we denote it by $G_X(s)$, and we may refer to it as the p.g.f. of X. $\triangle$

Definition.   Let the random variable X have moments $\mu_r = \mathrm{E} X^r$, $r \geqslant 0$. Then

$$(6) \qquad M(t) = 1 + \mathrm{E} X t + \frac{\mathrm{E} X^2}{2!} t^2 + \cdots$$

$$= \sum_{r=0}^\infty \frac{\mathrm{E} X^r}{r!} t^r,$$

is the *moment generating function* of X. If the sum in (6) converges for $|t| < d > 0$, then we can continue thus:

$$(7) \qquad M(t) = \mathrm{E} \sum_{r=0}^\infty \frac{X^r}{r!} t^r$$

$$= \mathrm{E}(e^{tX}).$$

Sometimes we denote it by $M_X(t)$, and we may refer to it as the m.g.f. of X. $\triangle$

Note that if X has a probability generating function, then (5) and (6) yield

$$(8) \qquad M_X(t) = G_X(e^t).$$

Next we remark that the series in (5) converges for $|s| \leqslant 1$, because $\sum_r p(r) = 1$. Moments are not so well behaved, which is why we need the extra condition to derive (7). We shall always assume that, for some $d > 0$, $M(t)$ does converge for $|t| < d$.

The functions $G(s)$ and $M(t)$ have the following essential properties.

First, $G(s)$ determines the probabilities $p(r)$ uniquely; that is, $G_X(s) = G_Y(s)$ if and only if $p_Y(r) = p_X(r)$.

Second, $M(t)$ determines the moments uniquely; that is, if $M_X(t) = M_Y(t)$ (and both exist for $|t| < d$), then $EX^r = EY^r$, and conversely.

However, there is more. In fact $G_X(s)$ also determines the moments of X, and $M_X(t)$ also determines the distribution of X. We emphasize these two results as follows:

Theorem.   Let $Es^X = G(s)$, and let $G^{(r)}(s)$ be the rth derivative of $G(s)$. Then

(9) $$G^{(r)}(1) = E\{X(X - 1) \cdots (X - r + 1)\}.$$

In particular

(10) $$G'(1) = G^{(1)}(1) = EX.$$

Proof.   For $r = 1$, we note that for $|s| < 1$ we can write

$$\frac{dG(s)}{ds} = \frac{d}{ds} \sum_{k=0}^{\infty} p(k)s^k$$

$$= \sum_{k=0}^{\infty} kp(k)s^{k-1}.$$

Because the sum of the series converges for $|s| < 1$, we can let $s \to 1$ to obtain (10). A similar argument proves (9). Note that (9) may be written as

$$G^{(r)}(1) = r!\,E\binom{X}{r} \qquad\qquad \square$$

and for this reason $G^{(r)}(1)$ is called the rth factorial moment.

Next we have

Theorem.   Let $E(e^{tX}) = M(t) < \infty$, for $|t| < d$, where X is continuous with density $f(x)$. Then

(11) $$f(x) \propto \int e^{-tx} M(t)\,dt.$$

This looks nice enough, but unfortunately the integrand is a function of a complex variable, and the integral is taken around a curve in the Argand plane. So we neither prove (11), nor do we ever evaluate it; it is enough to know it is there. It is called the *inversion theorem* for the m.g.f. $M(t)$.

To sum up: if we know that

$$G(s) = \sum_{k=0}^{\infty} p(k)s^k = Es^X$$

then $P(X = k) = p(k)$, and

$$G^{(r)}(1) = E\{X(X - 1) - (X - r + 1)\};$$

if we know that

$$M(t) = \int_{-\infty}^{\infty} e^{tx} f(x)\,dx = Ee^{tX} < \infty, \qquad |t| < d,$$

then X has density $f(x)$ and

$$M^{(r)}(0) = EX^r.$$

Exercises for section 7.2

1. If $Es^X = G(s)$, show that
$$\text{var } X = G^{(2)}(1) + G'(1) - \{G'(1)\}^2.$$

2. (a) Let X have the geometric distribution
$$p(k) = q^{k-1}p, \quad k \geqslant 1.$$
    Show that X has p.g.f.
$$G_X(s) = \frac{ps}{1 - qs}.$$

    (b) What is the generating function of the distribution
$$p(k) = q^k p, \quad k \geqslant 0?$$

3. Find the distributions that have the generating functions
    (a) $G(s) = \frac{1}{6}s(1 - s^7)/(1 - s)$ (b) $G(s) = q + ps$.

4. Let X have p.g.f. $G_X(s)$, and let $Y = aX + b$, where a and b are integers. Find the p.g.f. of Y, and use it to show that $EY = a\,EX + b$, and var $Y = a^2$ var X.

5. Let X have density $f(x) = 2x, 0 \leqslant x \leqslant 1$. Find the moment generating function of X.

6. Let X have distribution
$$p(0) = \lambda, \quad p(k) = (1 - \lambda)q^{k-1}p, \quad k \geqslant 1.$$
    Find the moment generating function of X, and hence find EX and var X.

7.3 EXAMPLES OF GENERATING FUNCTIONS

Before we can use generating functions we need to learn to recognize them (very much as drivers need to learn to recognize road signs before setting off). Here are some popular varieties.

Discrete uniform.   Here $p(k) = n^{-1}, 1 \leqslant k \leqslant n$, so

(1)
$$G(s) = \sum_{k=1}^{n} \frac{s^k}{n} = \frac{1}{n}\left(\frac{s - s^{n+1}}{1 - s}\right).$$

Continuous uniform.   Here $f(x) = (b - a)^{-1}, a < x < b$, so

(2)
$$M(t) = \int_a^b \frac{e^{tx}}{b - a}\,dx = \frac{e^{bt} - e^{at}}{(b - a)t}.$$

Binomial.   As we noted above, if $p(k) = \binom{n}{k}p^k q^{n-k}$, then

(3)
$$G(s) = (q + ps)^n.$$

Hence when X is binomial $B(n, p)$

$$EX = G'(1) = np,$$

and
$$\operatorname{var} X = G''(1) + G'(1) - \{G'(1)\}^2 = npq.$$

Poisson. Let X be Poisson with parameter λ; then X has p.g.f.

(4)
$$G(s) = \sum_{k=0}^{\infty} \lambda^k e^{-\lambda} \frac{s^k}{k!} = e^{\lambda(s-1)}.$$

It is very striking that $G^{(r)}(1) = \lambda^r$, so the factorial moments of the Poisson distribution are particularly simple.

Geometric. If X is geometric, then X has p.g.f.

(5)
$$G(s) = \sum_{k=1}^{\infty} q^{k-1} p s^k = \frac{ps}{1 - qs}.$$

Two-sided geometric. If X has distribution
$$P(X = x) = cq^{|x|}, \quad 0 < q < 1, \ -\infty < x < \infty$$
then X has p.g.f.

(6)
$$\mathrm{E}s^X = \sum_{-\infty}^{\infty} s^x cq^{|x|} = c\left(1 + \frac{qs}{1 - qs} + \frac{q/s}{1 - q/s}\right), \quad q < s < q^{-1}.$$

Thus probability generating functions can be defined for random variables taking negative values. We mainly use them in this case when examining the simple random walk.

Example 7.3.1. If X is exponential with parameter λ, its m.g.f. is

(7)
$$M(t) = \int_0^\infty e^{tx} \lambda e^{-\lambda x} dx = \frac{\lambda}{\lambda - t}, \quad \text{when } \lambda > t,$$

$$= \sum_{r=0}^{\infty} \frac{t^r}{\lambda^r}.$$

Hence the moments are very easily found to be
$$\mu_r = \lambda^{-r} r! \qquad \qquad \bigcirc$$

Example 7.3.2: normal density. If X has the standard normal density $\phi(x)$, then its m.g.f. is

$$\sqrt{2\pi} M(t) = \int_{-\infty}^{\infty} e^{tx} e^{-x^2/2} dx$$

$$= \int_{\infty}^{\infty} \exp\{-\tfrac{1}{2}(x - t)^2 + \tfrac{1}{2}t^2\} dx$$

$$= \exp\left(\tfrac{1}{2}t^2\right) \int_{-\infty}^{\infty} e^{-v^2/2} dv, \quad \text{setting } x - t = v.$$

Hence

$$M(t) = e^{t^2/2} = \sum_{r=0}^{\infty} \frac{t^{2t}}{r!\,2^r}$$

and so X has moments

(8) $$\mu_{2r} = \frac{(2r)!}{r!\,2^r}.$$ ○

Of course, by the law of the unconscious statistician, we do not need to find the density of $g(X)$ in order to find the m.g.f. of $g(X)$, as the following demonstrates.

Example 7.3.3: squared normal.   Let X be a standard normal random variable. Find the m.g.f. of $Y = X^2$.

Solution.   We seek

$$M_Y(t) = \mathrm{E}e^{tY} = \mathrm{E}e^{tX^2}$$

$$= \frac{1}{(2\pi)^{1/2}} \int_{-\infty}^{\infty} \exp\left(tx^2 - \tfrac{1}{2}x^2\right) dx$$

$$= \int_{-\infty}^{\infty} \frac{1}{(2\pi)^{1/2}} \frac{1}{(1-2t)^{1/2}} \exp\left(-\tfrac{1}{2}y^2\right) dy;$$

on setting

$$x = \frac{y}{(1-2t)^{1/2}}$$

we obtain

$$M_Y(t) = \frac{1}{(1-2t)^{1/2}}.$$ ○

Exercises for section 7.3

1. If X has m.g.f. $M(t)$ and $Y = a + bX$, show that Y has m.g.f.
$$M_Y(t) = e^{at} M(bt).$$

   Hence show that if Z has a normal density with mean μ and variance σ^2, then
   $$M_Z(t) = \exp\left(\mu t + \tfrac{1}{2}\sigma^2 t^2\right).$$

2. If X is uniform on $(0, a)$, show by use of the m.g.f. that
   $$\mu_r = \mathrm{E}X^r = a^r/(r+1).$$

3. Let X have the gamma density $f(x) = x^{r-1}e^{-\lambda x}\lambda^r/(r-1)!$
   Show that
   $$M_X(t) = \{\lambda/(\lambda - t)\}^r, \quad \lambda > t.$$

4. Find the mean and variance of the (a) binomial, (b) geometric, and (c) gamma distributions, by using their generating functions.

7.4 APPLICATIONS OF GENERATING FUNCTIONS

Generating functions are remarkably versatile objects, and can be used to do many things. But for us, they are mostly used for two purposes: *finding limits* and *working with sums*.

I Limits

The following results are crucial.

The continuity theorems

(i) Let $X, X_1, X_2, \ldots$ be a collection of random variables with distributions $F(x)$, $F_1(x), \ldots$ and moment generating functions $M(t), M_1(t), M_2(t), \ldots, |t| < d$. Then if

$$\lim_{n \to \infty} M_n(t) = M(t), \quad |t| < d,$$

it follows that

$$\lim_{n \to \infty} F_n(x) = F(x).$$

(ii) If $X, X_1, X_2, \ldots$ have probability generating functions $G(s), G_1(s), G_2(s), \ldots$ and

$$\lim_{n \to \infty} G_n(s) = G(s), \quad |s| \leq 1$$

then

$$\lim_{n \to \infty} P(X_n = k) = P(X = k).$$

These are not easy to prove, and we make no attempt to do so.

Results like this are known as *continuity theorems*. Of course, in (i), if all the random variables have densities, then $f_n(x) \to f(x)$, also.

Roughly speaking, such results show that we can often investigate the limits of a sequence of distributions by looking at the limit of their generating functions. This seems rather devious, but the advantages are made clear by looking at some examples, several of them familiar to us from chapter 4.

Example 7.4.1: uniform limit. Let Y_n be uniform on $\{1, 2, \ldots, n\}$. Then Y_n/n has moment generating function

$$M_n(t) = \mathrm{E} e^{tY_n/n} = n^{-1}\{e^{t/n} + e^{2t/n} + \cdots + e^{nt/n}\}$$

$$= \frac{1}{n}\left\{\frac{e^{t/n} - e^{(n+1)t/n}}{1 - e^{t/n}}\right\} = \frac{1}{n}\left\{\frac{1 - e^t}{e^{-t/n} - 1}\right\}$$

$$= \frac{e^t - 1}{t - \dfrac{t^2}{2n} + \dfrac{t^3}{6n^2} \cdots} \quad \to \quad \frac{e^t - 1}{t}, \text{ as } n \to \infty,$$

$$= \mathrm{E} e^{tU}$$

where U is uniform on $[0, 1]$, by (2) of section 7.3. Hence, as $n \to \infty$,

$$P\left(\frac{Y_n}{n} \leqslant x\right) \to x, \quad 0 \leqslant x \leqslant 1,$$

$$= P(U \leqslant x). \qquad \bigcirc$$

Example 7.4.2: geometric-exponential limit.   Let X_n be geometric with parameter $p = \lambda/n$. Then X_n/n has moment generating function

$$M_n(t) = \frac{\lambda}{n} e^{t/n} \left\{ 1 - \left(1 - \frac{\lambda}{n}\right) e^{t/n} \right\}^{-1}$$

$$= \lambda \left(1 + \frac{t}{n} + \cdots \right) \left(\lambda - t + \frac{\lambda t^2 - t^2}{2n} + \cdots \right)^{-1}$$

$$\to \frac{\lambda}{\lambda - t}, \quad \text{as } n \to \infty,$$

$$= E e^{tY},$$

where Y is exponential with parameter λ. Hence, as $n \to \infty$,

$$P\left(\frac{X_n}{n} \leqslant x\right) \to 1 - e^{-\lambda x}, \quad x \geqslant 0.$$

$$= P(Y \leqslant x). \qquad \bigcirc$$

Compare the transparent simplicity of these derivations with the tedious chore of working directly with the distributions, as we did in chapter 4. Many other limit theorems can be proved this way, but we have to move on.

II Sums of independent random variables

Here the crucial result is already well known to you: if X and Y are independent, then

$$E\{g(X)h(Y)\} = Eg(X) Eh(Y).$$

In particular, for independent X and Y, when they have probability generating functions G_X and G_Y,

$$Es^{X+Y} = Es^X Es^Y = G_X(s) G_Y(s);$$

and, in any case, when X and Y are independent,

$$Ee^{t(X+Y)} = Ee^{tX} Ee^{tY} = M_X(t) M_Y(t).$$

Obviously, similar results hold true for any number of independent random variables. Here are several examples that use this idea to help in dealing simply with sums of independent random variables.

Example 7.4.3: Bernoulli trials and the binomial distribution.   Let $(I_k; 1 \leqslant k \leqslant n)$ be a collection of independent Bernoulli trials with

$$P(X_k = 1) = p = 1 - q.$$

Then $Es^{I_k} = q + ps$, and for the sum $X = \sum_{k=1}^n I_k$ we have

$$\mathrm{E}s^X = \prod_{k=1}^{n} \mathrm{E}s^{I_k}, \quad \text{by independence}$$

$$= (q + ps)^n.$$

Therefore X is binomial with parameters n and p. We already know this, of course, but compare the brevity and elegance of this line of reasoning with more primitive methods. ○

Example 7.4.4: geometric sum.   Let X and Y be independent geometric random variables with parameter p, and let $Z = X + Y$. Now

$$G_Z(s) = \mathrm{E}s^Z = \mathrm{E}(s^{X+Y})$$

$$= \mathrm{E}s^X \mathrm{E}s^Y, \quad \text{by independence}$$

$$= \left(\frac{ps}{1 - qs}\right)^2, \quad \text{by (5) of section 7.3}$$

$$= \sum_{k=1}^{\infty} kp^2 q^{k-1} s^{k+1}, \quad \text{by expanding in series.}$$

Hence

(1) $$P(Z = k) = (k - 1)p^2 q^{k-2}.$$

By comparison with (7) of section 4.4 we see that Z has a negative binomial mass function. ○

The above example was especially simple because X and Y had the same parameter p. It is interesting to compare this with the case when they have different parameters.

Example 7.4.5: geometric sum revisited.   Let X and Y be independent geometric random variables, with parameters α and β respectively. That is to say,

$$f_X(x) = (1 - \alpha)^{x-1}\alpha, \quad x \geq 1; \quad G_X(s) = \frac{\alpha s}{1 - (1 - \alpha)s}$$

and

$$f_Y(y) = (1 - \beta)^{y-1}\beta, \quad y \geq 1; \quad G_Y(s) = \frac{\beta s}{1 - (1 - \beta)s}.$$

Hence, if $Z = X + Y$, we have

$$G_Z(s) = \mathrm{E}(s^{X+Y})$$

$$= \frac{\alpha s}{1 - (1 - \alpha)s} \left\{ \frac{\beta s}{1 - (1 - \beta)s} \right\}$$

$$= \frac{\alpha \beta s}{\alpha - \beta} \left\{ \frac{1}{1 - (1 - \beta)s} - \frac{1}{1 - (1 - \alpha)s} \right\}$$

$$= \frac{\alpha \beta}{\alpha - \beta} s \left\{ \sum_{r=0}^{\infty} \{(1 - \beta)^r - (1 - \alpha)^r\} s^r \right\}$$

By looking at the coefficient of s^z we obtain

(2) $$P(Z = z) = \frac{\alpha\beta}{\alpha - \beta}\{(1 - \beta)^{z-1} - (1 - \alpha)^{z-1}\}.$$ ○

The same idea works for continuous random variables if we use the moment generating function. Here is an illustration.

Example 7.4.6: normal sums.   Let X and Y be normal and independent, with distributions $N(\mu, \sigma^2)$ and $N(\nu, \tau^2)$ respectively. Find the distribution of $Z = X + Y$.

Solution.   We know that the $N(\mu, \sigma^2)$ distribution has moment generating function
$$M(t) = e^{\mu t + \sigma^2 t^2 / 2}.$$

Hence, remarkably easily,

$$\mathrm{E}e^{tZ} = \mathrm{E}e^{tX}\mathrm{E}e^{tY}$$

$$= e^{(\mu+\nu)t + (\sigma^2 + \tau^2)t^2/2}$$

Therefore Z is $N(\mu + \nu, \sigma^2 + \tau^2)$. Compare this with the dreary convolution integrals necessary to find $f_Z(z)$ directly.   ○

Example 7.4.7.   Let X and Y be independent, with gamma distributions having parameters n, λ and m, λ respectively. Find the distribution of $X + Y$.

Solution.   First we need $M_X(t)$, which is given by

$$\mathrm{E}e^{tX} = \int_0^\infty \frac{1}{(n - 1)!}\lambda^n x^{n-1} e^{-\lambda x + xt}\, dx$$

$$= \left(\frac{\lambda}{\lambda - t}\right)^n \int_0^\infty \frac{1}{(n-1)!}(\lambda - t)^n x^{n-1} e^{-x(\lambda - t)}\, dx$$

$$= \left(\frac{\lambda}{\lambda - t}\right)^n,$$

where the last step follows because the integrand is the gamma density with parameters n and $\lambda - t$. Hence

$$\mathrm{E}e^{t(X+Y)} = \left(\frac{\lambda}{\lambda - t}\right)^{m+n}$$

and Z is gamma with parameters $m + n$ and λ.   ○

Exercises for section 7.4

1. Let X and Y be Poisson random variables with parameters λ and μ respectively. Show that $X + Y$ has a Poisson distribution, by finding its generating function.

2. Let X_n have the binomial distribution with parameters n and $p_n = \lambda/n$. By considering the probability generating function, show that as $n \to \infty$, for fixed k,
$$P(X_n = k) \to e^{-\lambda}\lambda^k / k!.$$

3. I have a die in which the three pairs of opposite faces bear the numbers 1, 3, and 5. I roll the die twice and add the scores, then flip a fair coin twice and add the number of heads to the sum of the scores shown by the die. Find the distribution of the total.

4. ***Compounding.*** Let X have a binomial distribution with parameters n and Θ, where Θ is uniformly distributed on $(0, 1)$. Find $E(s^X|\Theta)$, and hence show that

$$Es^X = \frac{1}{n+1}\left(\frac{1-s^{n+1}}{1-s}\right).$$

Thus X is uniform on $\{0, 1, \ldots, n\}$, a result which we showed with very much more effort in chapter 6.

7.5 RANDOM SUMS AND BRANCHING PROCESSES

Generating functions are even more useful in dealing with the sum of a random number of random variables. This may sound a little arcane, but it is a very commonly arising problem, as we noted when we proved Wald's equation in chapter 6.

Example 7.5.1: insurance. Claims from an insurance company can be regarded as a collection of independent identically distributed random variables, $X_1, X_2, \ldots$. In one year there are N claims, where N is a random variable independent of the X_r. Define the total annual payout

$$(1) \qquad\qquad T = \sum_{r=0}^{N} X_r.$$

What is the distribution of T? Direct evaluation by summing probabilities looks rather foul; we prefer the following approach, which yields the generating function of T. First, suppose T is discrete; also recall that if $N = 0$ we interpret the empty sum as zero, to give $T = 0$. By conditional expectation,

$$(2) \qquad\qquad E(s^T) = E\{E(s^T|N)\}.$$

Now, conditional on $N = n$,

$$E(s^T|N = n) = E(s^{X_1+X_2+\cdots+X_n})$$

$$= \{G_X(s)\}^n, \quad \text{by independence.}$$

Hence

$$E(s^T|N) = \{G_X(s)\}^N,$$

and substituting in (2) gives

$$E(s^T) = E\{(G_X)^N\}$$

$$= G_N(G_X(s)),$$

because $E(x^N) = G_N(x)$ for any x in $[0, 1]$. We usually write this argument out even more succinctly as follows:

(3) $E(s^T) = E\{E(s^T|N)\}$

                         $= E(G_X^N),$ by independence

                         $= G_N\{G_X(s)\}.$ ○

Now we notice that the above argument works just as well for continuous random variables $X_1, X_2, \ldots,$ provided that we use the moment generating function. Thus we have in general

Random sum of random variables.   Let $(X_r; r \geqslant 1)$ be independent random variables with common moment generating function $M(t)$, and let N be non-negative, and independent of all the X_r, with probability generating function $G(s)$. Then when $T = \sum_{r=0}^{N} X_r$ we have

(4) $M_T(t) = E(e^{tT}) = E\{E(e^{tT}|N)\}$

                         $= E\{M(t)^N\}$

                         $= G(M(t)).$

Naturally this greatly simplifies the task of finding the moments of T; for example, on using (4) we have very easily, by differentiating,

(5) $ET = M_T'(0) = G'\{M(0)\}M'(0)$

                         $= EN\,EX_1,$

which we have proved already in chapter 6. For the variance we calculate

$$ET^2 = M_T''(0) = G''\{M(0)\}\{M'(0)\}^2 + G'\{M(0)\}\,M''(0)$$

$$= \{E(N^2) - EN\}\{EX_1\}^2 + EN\,E(X_1^2)$$

and hence

(6) $\operatorname{var} T = (EX_1)^2 \operatorname{var} N + EN \operatorname{var} X_1.$

Very often the distribution of T can be readily found from (3) or (4).

Example 7.5.2.   Let each X_r be exponential with parameter λ, and let N be geometric with parameter p. Then

(7) $M(t) = \dfrac{\lambda}{\lambda - t}$ and $G(s) = \dfrac{ps}{1 - qs}.$

Therefore

(8) $G(M(t)) = \dfrac{p\lambda}{\lambda - t - q\lambda} = \dfrac{p\lambda}{p\lambda - t}.$

Therefore $T = \sum_{r=1}^{N} X_r$ has an exponential distribution with parameter λp. (Remember that the sum of a *fixed* number of exponential random variables has a gamma distribution.)   ○

Probably the most famous and entertaining application of these results is to the theory of branching processes. This subject had its origins in the following basic question: why do some families die out, while others survive?

The question has been posed informally ever since people started using family names. But the first person to use generating functions in its solution was H.W. Watson in 1873, in answer to a challenge by Francis Galton. (The challenge appeared in the April 1 issue of the *Educational Times*.) For this reason it is often known as the Galton–Watson problem, though Watson's analysis was flawed. The correct solution in modern format was eventually produced by J. F. Steffensen in 1930. I. J. Bienaymé had earlier realized what the answer should be, but failed to supply any reasons.

The problem they were all interested in is as follows. A population reproduces itself in generations; the number in the nth generation is Z_n. The rules for reproduction are these.

(i)   Each member of the nth generation produces a family (maybe of size zero) in the $(n + 1)$th generation.

(ii)   Family sizes of all individuals are independent and identically distributed random variables, with distribution $(p(x);\ x \geqslant 0)$ and probability generating function $G(s)$.

With these rules, can we describe Z_n in the long run? Let $\mathrm{E}s^{Z_n} = G_n(s)$, and assume $Z_0 = 1$. Then the solution to the problem is based on the following result:

$$(9) \qquad G_n(s) = G(G_{n-1}(s)) = G_{n-1}(G(s))$$

$$= G(G(\cdots (G(s)) \cdots))$$

where the right-hand side is the n-fold iterate of the function $G(\cdot)$.

The proof of (9) relies on the following observations:

(i)   every member of the nth generation has an ancestor in the first generation;

(ii)   the rth member of the first generation has $Z_{n-1}^{(r)}$ descendants in the nth generation, where $Z_{n-1}^{(r)}$ has the same distribution as Z_{n-1}. Hence

$$(10) \qquad Z_n = Z_{n-1}^{(1)} + \cdots + Z_{n-1}^{(Z_1)}.$$

Now this is a random sum of independent random variables, so

$$(11) \qquad \mathrm{E}s^{Z_n} = G_n(s) = G_{Z_1}(G_{n-1}(s)) = G_1(G_{n-1}(s)).$$

The same argument applied to the $(n-1)$th generation shows that

$$(12) \qquad Z_n = Z_1^{(1)} + Z_1^{(2)} + \cdots + Z_1^{(Z_{n-1})},$$

which gives

$$(13) \qquad G_n(s) = G_{n-1}(G(s)).$$

Iterating either (11) or (13) gives (9). Decompositions such as (10) and (12) lie at the heart of many arguments in the theory of branching processes.

Now recall that our interest was motivated by the question, does the family of descendants of the first individual become extinct? We can examine this by considering the events

$$A_n = \{Z_n = 0\}.$$

Obviously if A_n occurs, then the descendants of the first individual have become extinct by the nth generation. Furthermore, setting $s = 0$ in (9) shows that

(14) $P\{A_n\} = G_n(0) = G(G_{n-1}(0))$

$$= G(P(A_{n-1})).$$

At this point we assume that var $Z_1 > 0$, to exclude trivial cases.

Because $G(s)$ is increasing and convex, it is now a routine task in analysis (which we omit) to prove the following.

(i)   If $EZ_1 \leq 1$ then, as $n \to \infty$, $P(A_n) \to 1$.

(ii) If $EZ_1 > 1$ then, as $n \to \infty$, $P(A_n) \to \eta$, where η is the smallest positive root of the equation $G(x) = x$.

In words, this says that if the expected family size is less than or equal to 1, then extinction is certain; but if the expected family size is greater than 1, then the chance of extinction is less than 1, and is given by $G(x) = x$. This result is attractive, and simple.

What is not so simple is evaluating $G_n(s)$ explicitly. There are few really easy non-trivial cases when we can do it; here is one.

Example 7.5.3: geometric branching.   Suppose that, in the branching process defined above, each family size has a type of geometric distribution such that

(15) $P(Z_1 = k) = q^k p, \quad k \geq 0; \ p + q = 1.$

Then $G(s) = p/(1 - qs)$, and we can show by induction that the nth iterate of $G(\cdot)$ is

(16) $$G_n(s) = \begin{cases} \dfrac{p\{q^n - p^n - qs(q^{n-1} - p^{n-1})\}}{q^{n+1} - p^{n+1} - qs(q^n - p^n)}, & p \neq q \\[2ex] \dfrac{n - (n-1)s}{n + 1 - ns}, & p = q. \end{cases}$$

Now $EZ_1 = q/p$, and from (19) we have that

$$G_n(0) = \begin{cases} \dfrac{p(q^n - p^n)}{q^{n+1} - p^{n+1}}, & p \neq q \\[2ex] \dfrac{n}{n + 1}, & p = q \end{cases}$$

$$\to \begin{cases} 1, & q \leq p \\ pq^{-1} & p < q, \end{cases}$$

as $n \to \infty$. This all agrees with what we said above, of course.   ○

Exercises for section 7.5

1. Let $T = \sum_{r=1}^{N} X_r$, where N has a Poisson distribution with parameter λ and the X_r are independent and identically distributed with distribution

$$P(X_r = k) = \frac{p^k}{-k \log(1 - p)}, \quad k \geq 1; \ 0 \leq p \leq 1.$$

Show that T has a negative-binomial distribution.

2. Carry out the proof of (16) by induction.

3. Differentiate (9) to show that $E Z_n = E Z_1 E Z_{n-1}$. Deduce that $E Z_n = (E Z_1)^n$.
Let $E Z_1 = \mu$, and var $Z_1 = \sigma^2$. Show that

$$\text{var } Z_n = \begin{cases} \mu^{n-1}\sigma^2(\mu^n - 1)/(\mu - 1), & \mu \neq 1 \\ n\sigma^2, & \mu = 1. \end{cases}$$

7.6 CENTRAL LIMIT THEOREM

Recall the normal approximation to the binomial distribution, which we sketched in section 4.10. In our new notation, we write this as follows: if S_n is binomial with mean np and variance npq then, as $n \to \infty$,

$$(1) \qquad P\left(\frac{S_n - np}{(npq)^{1/2}} \leq x\right) \to \Phi(x),$$

where $\Phi(x)$ is the standard normal distribution function. The central limit theorem extends this result to a much wider range of random variables, as follows.

Central limit theorem. Let $X_1, X_2, \ldots$ be a collection of independent identically distributed random variables with $E X_r = \mu$,

$$0 < \text{var } X_r = \sigma^2 < \infty,$$

and

$$E e^{t X_r} = M(t), \quad |t| < \delta > 0.$$

Then if we let $S_n = \sum_{r=1}^n X_r$, we have

$$(2) \qquad P\left(\frac{S_n - n\mu}{n^{1/2}\sigma} \leq x\right) \to \Phi(x).$$

First, notice that if we let each X_r be a Bernoulli trial with parameter p, then S_n is binomial $B(n, p)$. In this case (2) is exactly the same as (1).

Second, notice that the theorem considers a sum of independent random variables, and then takes a scaled limit of this sum. We have seen that moment generating functions are very adept at dealing with each of these operations separately, so we may expect them to be particularly excellent at handling them together. This is indeed the case, as we now see.

Proof. Let $Y_r = (X_r - \mu)/\sigma$, $r \geq 1$. Then $E Y_r = 0$ and var $Y_r = 1$, so that

$$(3) \qquad M_Y(t) = E e^{t Y_r}$$

$$= 1 + \frac{t^2}{2} + \frac{\text{skw } X_r}{6} t^3 + \cdots.$$

(Recall that skw $X = E(X - \mu)^3/\sigma^3$.) Next we calculate the moment generating function of the random variables in question:

(4) $$\mathrm{E}\exp\left\{t\left(\frac{S_n - n\mu}{n^{1/2}\sigma}\right)\right\} = \mathrm{E}\exp\left\{\frac{t}{n^{1/2}}\sum_{r=1}^{n}Y_r\right\}$$

$$= \left\{M_Y\left(\frac{t}{n^{1/2}}\right)\right\}^n, \quad \text{by independence}$$

$$= \left(1 + \frac{t^2}{2n} + \frac{\mathrm{skw}\,X_1}{6}\frac{t^3}{n^{3/2}} + \cdots\right)^n, \quad \text{by (3)}$$

(5) $$\to \exp\left(\frac{t^2}{2}\right), \quad \text{as } n \to \infty,$$

by the exponential limit stated in appendix 3.12.III.

Now $\exp(\frac{1}{2}t^2)$ is the moment generating function of a standard normal random variable. Therefore the continuity theorem applies, and the distribution of $(S_n - n\mu)/(n^{1/2}\sigma)$ converges to $\mathrm{N}(0, 1)$ as $n \to \infty$. And this is just (2), which we wished to prove.   □

It follows that the techniques of approximation that we used in section 4.10 can be used to deal with sums of any independent random variables with a moment generating function, not just sums of Bernoulli trials.

Exercises for section 7.6

1. A fair die is rolled 12 000 times; let X be the number of sixes. Show that
$$\mathrm{P}(1900 < X < 2200) \simeq \Phi(2\sqrt{6}) - \Phi(-\sqrt{6}).$$

2. *Rounding errors.*   Suppose that you round off 108 numbers to the nearest integer, and then add them to get the total S. Assume that the rounding errors are independent and uniform on $[-\frac{1}{2}, \frac{1}{2}]$. What is the probability that S is wrong by (a) more than 3, (b) more than 6?

3. Let X be Poisson with parameter λ, and let Y be gamma with parameters r and 1. Explain why we can say, without any elaborate calculations, that $(X - \lambda)/\sqrt{\lambda}$ is approximately normally distributed as $\lambda \to \infty$, and $(Y - r)/\sqrt{r}$ is approximately normally distributed as $r \to \infty$.

7.7 RANDOM WALKS AND OTHER DIVERSIONS

Generating functions can be applied to many new problems, and also provide new ways of doing old problems. We give a few randomly selected examples here. Many of them rely on a particular application of conditional expectation, that is, the fact that

(1) $$\mathrm{E}s^X = \mathrm{E}\{\mathrm{E}(s^X|Y)\}$$

for any discrete random variables X and Y. Similarly, you may sometimes use
$$\mathrm{E}\exp(tX) = \mathrm{E}[\mathrm{E}\{\exp(tX)|Y\}],$$

in the continuous case.

Our first example is extremely famous, and arises in a startling number of applications, in various disguises. As usual we prefer to use a standard format and nomenclature; the following is hallowed by tradition.

Example 7.7.1: simple random walk.   Starting from the origin, a particle performs a random walk on the integers, with independent and identically distributed jumps

$X_1, X_2, \ldots$ such that

(2) $\qquad\qquad P(X_1 = 1) = p \quad \text{and} \quad P(X_1 = -1) = q = 1 - p.$

Its position after n steps is $S_n = \sum_{r=1}^{n} X_r$. Let T_r be the number of jumps until the particle visits r for the first time $(r > 0)$, and let T_0 be the number of jumps until it first revisits the origin. Show that

(3) $\qquad\qquad E(s^{T_r}) = (Es^{T_1})^r, \quad r > 0.$

Hence find Es^{T_1}. Use this to deduce that

(4) $\qquad\qquad Es^{T_0} = 1 - (1 - 4pqs^2)^{1/2}.$

Solution.   For any k, let $T_{k,k+1}$ be the number of steps from T_k until T_{k+1}, that is, the number of steps to reach $k + 1$ after first having arrived at k. Then $T_{0,1} = T_1$, and the random variables $T_{k,k+1}$ are independent and identically distributed. Furthermore

(5) $\qquad\qquad T_r = T_{0,1} + T_{1,2} + \cdots + T_{r-1,r}.$

Hence

(6) $\qquad\qquad Es^{T_r} = (Es^{T_{01}})^r = (Es^{T_1})^r.$

Next we observe that

(7) $\qquad\qquad Es^{T_1} = E\{E(s^{T_1} | X_1)\}.$

Now, trivially,

$$E(s^{T_1} | X_1 = 1) = s$$

and not quite so trivially

$$E(s^{T_1} | X_1 = -1) = E(s^{1 + T_{-1,0} + T_{0,1}}) = s(Es^{T_1})^2, \quad \text{by (6)}$$

Therefore, by conditional expectation (conditioning on X_1)

(8) $\qquad\qquad Es^{T_1} = ps + qs(Es^{T_1})^2.$

Hence Es^{T_1} is a root of the quadratic $qsx^2 - x + ps = 0$. Only one of these roots is a probability generating function that converges for $|s| \leq 1$, and it is

(9) $\qquad\qquad Es^{T_1} = \dfrac{1 - (1 - 4pqs^2)^{1/2}}{2qs}.$

For the last part we note that

$$Es^{T_0} = E\{E(s^{T_0} | X_1)\}.$$

Now $E(s^{T_0} | X_1 = 1) = Es^{1 + T_{1,0}} = sE(s^{T_{0,-1}})$ and $E(s^{T_0} | X_1 = -1) = sEs^{T_{0,1}}$. From their definitions we see that $Es^{T_{0,-1}}$ is obtained from $Es^{T_{0,1}}$ by simply interchanging p and q. Then, by conditional expectation again,

(10) $\qquad\qquad Es^{T_0} = psEs^{T_{1,0}} + qsEs^{T_{0,1}}$

$$= 1 - (1 - 4pqs^2)^{1/2}. \qquad\qquad\qquad\qquad \bigcirc$$

Remark.   The name 'random walk' was invented by Karl Pearson in 1905, to describe a similar problem in two (or more) dimensions. Following the solution of that problem by Rayleigh, Pearson noted the corollary that '... in open country the most likely place to find a drunken man is somewhere near his starting point'. Since then it has also been known as the 'drunkard's walk' problem. The solution of Rayleigh's problem of random flights leads to a similar corollary for drunken birds in the open air.

Example 7.7.2: Huygens' problem.   Two coins are flipped alternately; they show heads with respective probabilities α and β. Let X be the number of flips up to and including the first head. Find $\mathrm{E}s^X$.

Solution.   The sequence must begin with one of the three mutually exclusive outcomes H or TH or TT. Now

$$\mathrm{E}(s^X|H) = s, \quad \mathrm{E}(s^X|TH) = s^2, \quad \mathrm{E}(s^X|TT) = \mathrm{E}(s^{2+X^*}),$$

where X^* has the same distribution as X. Hence

$$(11) \qquad \mathrm{E}s^X = \alpha\mathrm{E}(s^X|H) + (1-\alpha)\beta\mathrm{E}(s^X|TH)$$

$$+ (1-\alpha)(1-\beta)\mathrm{E}(s^X|TT)$$

$$= \alpha s + (1-\alpha)\beta s^2 + (1-\alpha)(1-\beta)s^2\mathrm{E}s^X.$$

Thus

$$(12) \qquad \mathrm{E}s^X = \frac{\alpha s + \beta(1-\alpha)s^2}{1 - (1-\alpha)(1-\beta)s^2}.$$

From this it is a trivial matter to calculate any desired property of X. For example, the probability that the second coin flipped first shows a head is just the sum of the coefficients of the even powers of s in $\mathrm{E}s^X$, and from (12) this is simply

$$\frac{\beta(1-\alpha)}{1-(1-\alpha)(1-\beta)}. \qquad \bigcirc$$

Example 7.7.3: Waldegrave's problem revisited.   In our final visit to this problem, we find the generating function $\mathrm{E}s^N$ of the number of rounds played in this game. In the usual notation we write $N = 1 + X$, where X is the number of flips of a coin until it first shows $n-1$ consecutive heads. Then by conditional probability and independence,

$$(13) \qquad \mathrm{E}s^X = \tfrac{1}{2}\mathrm{E}(s^X|T) + \left(\tfrac{1}{2}\right)^2\mathrm{E}(s^X|HT) + \cdots$$

$$+ \left(\tfrac{1}{2}\right)^{n-1}\mathrm{E}(s^X|H^{n-2}T) + \left(\tfrac{1}{2}\right)^{n-1}\mathrm{E}(s^X|H^{n-1})$$

$$= \left\{\tfrac{1}{2}s + \left(\tfrac{1}{2}\right)^2 s^2 + \cdots + \left(\tfrac{1}{2}\right)^{n-1}s^{n-1}\right\}\mathrm{E}s^X + \left(\tfrac{1}{2}s\right)^{n-1}.$$

Hence

$$\mathrm{E}s^X = \frac{\left(\tfrac{1}{2}s\right)^{n-1}}{1 - \left\{\dfrac{\tfrac{1}{2}s - \left(\tfrac{1}{2}s\right)^n}{1 - \tfrac{1}{2}s}\right\}}$$

and so

$$(14) \qquad \mathrm{E}s^N = s\mathrm{E}s^X = \frac{s^n\left(\tfrac{1}{2}\right)^{n-1}\left(1 - \tfrac{1}{2}s\right)}{1 - s + \left(\tfrac{1}{2}s\right)^n}. \qquad \bigcirc$$

Example 7.7.4: tail generating functions. We have seen above that it is often useful to know $P(T > n)$ for integer-valued random variables. Let

$$T(s) = \sum_{s=0}^{\infty} P(X > n)s^n.$$

Show that if $Es^X = G(s)$, then

(15)
$$T(s) = \frac{1 - G(s)}{1 - s}.$$

Solution. Let $I(X > n)$ be the indicator of the event that $X > n$. Then

$$T(s) = \sum_{n=0}^{\infty} P(X > n)s^n = \sum_{n=0}^{\infty} EI(X > n)s^n$$

$$= E \sum_{n=0}^{\infty} I(X > n)s^n$$

$$= E \sum_{n=0}^{X-1} s^n, \quad \text{because } I \text{ is an indicator}$$

$$= E \frac{1 - s^X}{1 - s}, \quad \text{summing the geometric series}$$

$$= \frac{1 - G(s)}{1 - s}. \qquad \bigcirc$$

Example 7.7.5: coupons. Suppose any packet of Acme Deathweed is equally likely to contain any one of four different types of coupon. If the number of packets you need to collect the set is T, find Es^T and $P(T = k)$.

Solution. We know that the number of packets bought between the consecutive appearances of a new type is geometric, with parameters $1, \frac{3}{4}, \frac{1}{2}, \frac{1}{4}$ respectively for each. Hence, using the geometric p.g.f.,

(16)
$$Es^T = s \left(\frac{\frac{3}{4}s}{1 - \frac{1}{4}s} \right) \left(\frac{\frac{1}{2}s}{1 - \frac{1}{2}s} \right) \left(\frac{\frac{1}{4}s}{1 - \frac{3}{4}s} \right)$$

$$= \tfrac{3}{32}s^4 \left(\frac{\frac{1}{2}}{1 - \frac{1}{4}s} - \frac{4}{1 - \frac{1}{2}s} + \frac{\frac{9}{2}}{1 - \frac{3}{4}s} \right).$$

Hence

(17)
$$P(T = k) = \tfrac{3}{32}\left\{ \tfrac{1}{2}\left(\tfrac{1}{4}\right)^{k-4} - 4\left(\tfrac{1}{2}\right)^{k-4} + \tfrac{9}{2}\left(\tfrac{3}{4}\right)^{k-4} \right\}$$

$$= 3\left\{ \left(\left(\tfrac{1}{4}\right)^{k-1}\right) - \left(\tfrac{1}{2}\right)^{k-5} + \tfrac{1}{4}\left(\tfrac{3}{4}\right)^{k-2} \right\}$$

$$\simeq \left(\tfrac{3}{4}\right)^{k-1}, \quad \text{for large } k. \qquad \bigcirc$$

Finally we note that just as pairs of random variables may have joint distributions, so too may they have joint generating functions.

Definition.   Discrete random variables X and Y have joint probability generating function

(18) $$G(s, t) = E(s^X t^Y).$$

Continuous random variables X and Y have joint moment generating function

(19) $$M(u, v) = E(e^{uX+vY}).$$ △

It is easy to see that, for example,

(20) $$E(XY) = G_{st}(1, 1) = M_{uv}(0, 0),$$

where suffices denote differentiation. Furthermore, we have a very important result, which we do not prove:

Theorem.   Random variables X and Y are independent if and only if

(21) $$M(u, v) = M_X(u)M_Y(v).$$

Example 7.7.6: de Moivre trials.   In a certain ballgame, suppose any pitch independently results in a ball, a strike, or a hit, with respective probabilities p, q, and r, where $p + q + r = 1$. Let n pitches yield X_n balls, Y_n strikes, and $n - X_n - Y_n$ hits. Then

$$E(s^{X_1} t^{Y_1}) = r + ps + qt.$$

Hence, by independence of the pitches

$$G_n(s, t) = E(s^{X_n} t^{Y_n}) = (r + ps + qs)^n.$$

We see that

(22) $$E(s^{X_n}) = G_n(s, 1) = (r + q + ps)^n$$

so the number of balls X_n is binomial, and clearly X_n and Y_n are not independent, by (21). Furthermore

(23) $$E(X_n Y_n) = \frac{\partial^2 G_n}{\partial s \partial t}(1, 1) = n(n - 1)pq.$$

Hence

$$\text{cov}(X_n, Y_n) = n(n - 1)pq - npnp = -npq.$$ ○

Exercises for section 7.7

1. **Example 7.7.1 continued: random walk**
    (a) Show that if $p > \frac{1}{2}$ then the particle is certain to visit the position $+1$ and that, in this case, the expected number of steps to do so is $(p - q)^{-1}$.
    (b) Show that if $p = \frac{1}{2}$ then the particle is certain to revisit the origin, but that in this case the expected number of steps to do so is infinite.
    (c) Find the probability generating function of S_n, the position of the walk after n steps.
    (d) If the walk is stopped after Y steps, where Y is independent of all the steps, find the generating function of the final position of the walk.

2. **Waldegrave's problem revisited.**   Four card players are bored with bridge, and play Waldegrave's game instead. North is A_0. Show that the probability that either of East or West wins is $\frac{2}{5}$.

3. **Waldegrave's problem once again.** Suppose that in each round the challenger wins with probability p; the conditions of the problem are otherwise unchanged. Show that

$$EN = 1 + p^{-1} + \cdots + p^{-(n-1)}$$

and

$$Es^N = \frac{s^n p^{n-1}(1 - ps)}{1 - s + (1 - p)p^{n-1}s^n}.$$

The bridge players in exercise 2 want the North–South pair to have the same chance of winning as the East–West pair; show that this is impossible for any choice of p such that $0 < p < 1$.

4. Suppose any pitch results in a ball, a strike, a hit, or a foul, with respective probabilities $p, q, r, 1 - p - q - r$. Then n independent pitches yield X_n balls, Y_n strikes, Z_n hits, and $n - X_n - Y_n - Z_n$ fouls. Find the joint p.g.f. of these four random variables. Now suppose the number of pitches is N, where N is Poisson with parameter λ. Show that the numbers of balls, hits, strikes, and fouls are independent Poisson random variables.

5. If $X, \ldots, X_r$ jointly have the multinomial distribution, show that

$$E(s_1^{X_1} s_2^{X} \cdots s_r^{X_r}) = (p_1 s_1 + p_2 s_2 + \cdots + p_r s_r)^n.$$

7.8 REVIEW

In this chapter we have introduced the idea of generating functions, in particular the probability generating function (p.g.f.)

$$G(s) = Es^X$$

and the moment generating function (m.g.f.)

$$M(t) = Ee^{tX}.$$

You can think of such functions as organizers which store a collection of objects that they will regurgitate on demand. Remarkably, they will often produce other information if differently stimulated: thus the p.g.f. will produce the moments, and the m.g.f. will produce the probability distribution (in most cases).

Furthermore, these functions are particularly adept at handling sequences, sums, and collections of random variables, as was exemplified in section 7.3. We applied the idea in looking at branching processes, the central limit theorem, and random walks.

7.9 APPENDIX. TABLES OF GENERATING FUNCTIONS

Table 7.1. *Discrete distributions*

Name	$p(x)$	Probability generating function
Bernoulli	$p^x(1 - p)^{1-x}; \quad x = 0, 1$	$1 - p + ps$
binomial	$\binom{n}{x} p^x(1 - p)^{n-x}; \quad 0 \leqslant x \leqslant n$	$(1 - p + ps)^n$
Poisson	$e^{-\lambda}\lambda^x/x!; \quad x \geqslant 0$	$\exp\{\lambda(s - 1)\}$
uniform	$n^{-1}; \quad 1 \leqslant x \leqslant n$	$(s^{n+1} - s)/\{n(1 - s)\}$
geometric	$p(1 - p)^{x-1}; \quad x \geqslant 1$	$ps/\{1 - (1 - p)s\}$

cont.

Table 7.1 (*cont.*)

Name	$p(x)$	Probability generating function
negative binomial	$\binom{x-1}{r-1} p^r (1-p)^{x-r}, \quad x \geqslant r$	$[ps/\{1-(1-p)s\}]^r$
hypergeometric	$\binom{b}{x}\binom{a}{n-x} / \binom{a+b}{n}$, $0 \leqslant x \leqslant b \wedge n$	coefficient of t^n in $(1+st)^b(1+t)^a / \binom{a+b}{n}$
logarithmic	$-x^{-1}p^x/\log(1-p), \quad x \geqslant 1$	$\log(1-sp)/\log(1-p)$.

Table 7.2. *Continuous distributions*

Name	$f(x)$	Moment generating function
uniform	$(b-a)^{-1}, \quad a < x < b$	$(e^{bt} - e^{at})/\{t(b-a)\}$
normal	$(2\pi)^{-1/2} \exp\{-\frac{1}{2}(x-\mu)^2/\sigma^2\}$	$e^{\mu t + \sigma^2 t^2/2}$
exponential	$\lambda e^{-\lambda x}, \quad x \geqslant 0$	$\lambda/(\lambda - t), \quad t < \lambda$
gamma	$\lambda^a x^{a-1} e^{-\lambda x}/\Gamma(a), \quad x \geqslant 0$	$\{\lambda/(\lambda - t)\}^a, \quad t < \lambda$
two-sided exponential	$\frac{1}{2}\lambda e^{-\lambda |x|}$	$\lambda^2/(\lambda^2 - t^2), \quad |t| < \lambda$
Cauchy	$\{\pi(1+x^2)\}^{-1}$	none
hyperbolic	$2(e^{\pi x} + e^{-\pi x})^{-1}$	$\sec(\frac{1}{2}t)$
doubly exponential	$\exp(-x - e^{-x})$	$\Gamma(1-t)$

7.10 PROBLEMS

1. Consider the coupon-collecting problem with three different types of coupon, and let T be the number of packets needed until you first possess all three types. Find $P(T = k)$ using a probability generating function. Show that $ET = \frac{11}{2}$ and $\operatorname{var} T = 6\frac{3}{4}$.

2. Consider Huygens' problem with three coins A, B, and C, which show heads with probability, α, β, and γ respectively. They are flipped repeatedly in the order $ABCABCAB$ .... Let X be the number of flips until the first head. Find Es^X, and hence deduce the probability that C is the first to show a head.

3. Let X_n have a negative binomial distribution with parameters n and $p = 1 - \lambda/n$. Show that X_n has probability generating function $\{ps/(1 - qs)\}^n$, and deduce that as $n \to \infty$, the distribution of $X_n - n$ converges to a Poisson distribution.

4. If X has moment generating function $M(t)$, then the function $K(t) = \log M(t)$ is called the *cumulant generating function*; if the function $K(t)$ is expanded as

$$K(t) = \sum_{r=1}^{\infty} \kappa_r t^r / r!$$

then κ_r is called the rth cumulant of X. What are the cumulants when X is (a) Poisson? (b) normal? (c) exponential?

5. You are taking a test in which the test paper contains 116 questions; you have one hour. You decide to spend no more than one minute on any one question, and the times spent on questions are independent, with density $f(x) = 6x(1-x)$, $0 \leqslant x \leqslant 1$. Show that there is a 20% chance, approximately, that you will not attempt all the questions.

6. **Gamblers and casinos**
   (a) The gambler. Let X_r be the outcome of your wagering \$1 on red at roulette, so that
   $$P(X_r = 1) = \tfrac{9}{19}, \quad P(X_r = -1) = \tfrac{10}{19},$$
   $$EX_r = -\tfrac{1}{19}, \quad \operatorname{var} X_r = 1 - \left(\tfrac{1}{19}\right)^2.$$
   Let $S_n = \sum_{r=1}^n X_r$ be the outcome of n such wagers. If $n = 40$, show that there is only one chance in a thousand that you win more than \$19 or lose more than \$23.
   (b) The casino. Let Y be the yield to the casino of 400 000 such wagers of \$1 each. Show that there is only one chance in a thousand that the casino makes a profit of less than \$19 000 or more than \$23 100.
   (Hint: $\Phi(3.3) \simeq 0.9995$.)

7. Let $X_1, X_2, \ldots$ be independent Bernoulli random variables with parameter $p = 1 - q$. Define
   $$Y_n = \left(\sum_{r=1}^n X_r - np \right) \Big/ (npq)^{1/2}.$$
   Show that
   $$E e^{tY_n} = \left[p \exp\left\{ \frac{qt}{(npq)^{1/2}} \right\} + q \exp\left\{ \frac{-pt}{(npq)^{1/2}} \right\} \right]^n$$
   and hence deduce de Moivre's central limit theorem.

8. Let X be the normal distribution $N(0, \sigma^2)$. Show that
   $$EX^{2k} = \sigma^{2k}(2k)! 2^{-k} / k!$$

9. Let X and Y be independent normal distributions $N(0, 1)$. Find $E e^{t(X^2 + Y^2)}$ and $E e^{tXY}$.

10. Let $T(s)$ be the tail generating function of X (defined in example 7.7.4). Show that
    $$EX = \lim_{s \to 1-} T(s)$$

11. Let Z_n be the size of the nth generation in an ordinary Galton–Watson branching process with $Z_0 = 1$. Let T_n be the total number of individuals who have ever lived, up to and including the nth generation. If $E s^{T_n} = H_n(s)$, show that
    $$H_n(s) = s G_1(H_{n-1}(s))$$
    where $G_1(s) = E s^{Z_1}$.

12. Let Z_n be the size of the nth generation in an ordinary Galton–Watson branching process with $EZ_1 = \mu$. Show that
    $$E(Z_m Z_n) = \mu^{n-m} E Z_m^2, \quad m \leqslant n.$$
    Hence find $\operatorname{cov}(Z_m, Z_n)$ and $\rho(Z_m, Z_n)$.

13. Find the probability of extinction of a Galton–Watson branching process when the initial population Z_0 is a random variable with probability generating function $P(s)$.

14. Let X have moments $\mu_k = EX^k$, and define

$$L(t) = \sum_{k=0}^{\infty}(EX^k)t^k = \sum_{k=0}^{\infty}\mu_k t^k.$$

Why would you not use $L(t)$ as the moment generating function of X?

15. You flip a fair coin repeatedly until the first occasion when it shows two consecutive heads; this takes X flips. Show that

$$Es^X = s^2/(4 - 2s - s^2) \quad \text{and} \quad EX = 6.$$

What is var X?

16. Let X_n have the gamma density

$$f_n(x) = e^{-\lambda x}\lambda^n x^{n-1}/(n-1)!, \quad x > 0$$

and define $Z_n = (\lambda X_n - n)/n^{1/2}$.
   (a) Find the moment generating function $M_n(t)$ of Z_n. Show that as $n \to \infty$, for fixed t
   $$M_n(t) \to \exp\left(\tfrac{1}{2}t^2\right).$$
   What do you deduce about Z_n?
   (b) Find the density of Z_n and show that as $n \to \infty$, for fixed x,
   $$f_{Z_n}(x) \to \frac{1}{\sqrt{2\pi}}\exp\left(-\tfrac{1}{2}x^2\right).$$
   Hint: Take logarithms and remember Stirling's formula.)

17. Which of the following two functions can be a moment generating function? Why? What is the corresponding distribution?
   (a) $M(t) = 1 + \sum_{r=1}^{n}p_r t^r, \quad p_r > 0;$
   (b) $M(t) = \sum_{r=1}^{n}p_r e^{a_r t}, \quad p_r > 0.$

18. You are a contestant in a quiz show, answering a series of independent questions. You answer each correctly with probability p, or wrongly with probability $q = 1 - p$. You score a point for each correct answer, and you are eliminated when you first answer two successive questions wrongly. Let D be the number of questions put to you, and S the number of points you score. Show that

$$Es^D = \frac{s^2 q^2}{1 - ps - pqs^2}.$$

What is the generating function of S? Find the joint generating function $E(s^D t^S)$ and hence find cov (D, S).

19. **Bivariate normal m.g.f.**   Let X and Y be independent standard normal random variables; let $U = X$; and $V = \rho X + (1 - \rho^2)^{1/2}Y$. Show that the joint moment generating function of U and V is

$$M(s, t) = E(e^{sU+tV})$$

$$= \exp\left\{\tfrac{1}{2}(s^2 + 2st\rho + t^2)\right\}.$$

Hence find $\rho(U, V)$.

20. Let X and Y be independent standard normal random variables and let
$$U = \mu + \sigma X \quad \text{and} \quad V = \nu + \tau\rho X + \tau(1 - \rho^2)^{1/2}Y.$$
Find the joint moment generating function of U and V.

21. Find the moment generating functions Ee^{tX} corresponding to the following density functions on $(-\infty, \infty)$:

(a) $\frac{1}{2}e^{-|x|}$;   (b) $\dfrac{1}{\cosh \pi x}$;   (c) $\exp(-x - e^{-x})$.

For (b) use the fact that $\int_0^\infty \{x^{a-1}/(1+x)\}\,dx = \pi/\sin a\pi$.)

For what values of t do they exist?

22. You flip two fair coins. Let I, J, and K be the indicators of the respective events that
    (a) the first shows a head,
    (b) the second shows a head,
    (c) they both show heads or they both show tails.
    Find the joint probability generating function
    $$G_{IJK}(x, y, z) = E(x^I y^J z^k).$$

    With obvious notation, verify that
    $$G_{IJ}(x, y) = G_I(x)G_J(y), \quad G_{JK}(y, z) = G_J(y)G_K(z), \quad G_{IK}(x, z) = G_I(x)G_K(z),$$

    but that
    $$G_{IJK}(x, y, z) \neq G_I(x)G_J(y)G_K(z).$$
    The events are pairwise independent, but not independent.

23. **Random walk in the plane.**   A particle takes a sequence of independent unit steps in the plane, starting at the origin. Each step has equal probability $\frac{1}{4}$ of being north, south, east, or west. It first reaches the line $x + y = a$ after T steps, and at the point (X, Y). Show that
    $$G_T(s) \doteq Es^T = \left[s^{-1}\left\{ 1 - (1 - s^2)^{1/2} \right\} \right]^a, \quad |s| < 1.$$
    Deduce that
    $$Es^{X-Y} = G_T\big(\tfrac{1}{2}(s + s^{-1})\big).$$

24. Two particles perform independent random walks on the vertices of a triangle; that is to say, at any step each particle moves along the clockwise edge with probability p, or the anticlockwise edge with probability $q = 1 - p$. At time $n = 0$ both are at the same vertex; let T be the number of steps until they again share a vertex. Let S be the number of steps until they first share a vertex if initially they are at different vertices. Show that
    $$Es^T = (p^2 + q^2)s + 2pqs\, Es^S,$$
    $$Es^S = pqs + (1 - pq)s\, Es^S.$$
    Hence find Es^T and show that $ET = 3$.

25. **More compounding.**   Let X have a Poisson distribution with parameter Λ, where Λ is a random variable having an exponential density with parameter μ. Find $E(s^X|\Lambda)$, and hence show that
    $$P(X = k) = \mu(\mu + 1)^{-(k+1)}, \quad 0 \leqslant k.$$

26. Show that
    $$G(x, y, z) = \tfrac{1}{8}(xyzw + xy + yz + zw + zx + yw + xz + 1)$$
    is the joint generating function of four variables that are pairwise and triple-wise independent, but which are nevertheless *not* independent.

27. Let $X \geqslant 0$ have probability generating function $G(s)$. Show that
    $$E\{(X + 1)^{-1}\} = \int_0^1 G(s)ds.$$
    Hence find $E\{(X + 1)^{-1}\}$ when X is (a) Poisson, (b) geometric, (c) binomial, (d) logarithmic.

28. Let X have moment generating function $M(t)$. Show that for $a > 0$:

(a)   if $t > 0$, $P(X \geqslant a) \leqslant e^{-at} M(t)$;

(b)   if $t < 0$, $P(X \leqslant a) \leqslant e^{-at} M(t)$.

Now let X be Poisson with parameter λ. Show that, for $b > 1$,

$$P(X \geqslant b\lambda) \leqslant \inf_{t < 0} \{e^{-b\lambda t} M(t)\} = e^{-\lambda} \left(\frac{e}{b}\right)^{\lambda b}$$

and for $b < 1$

$$P(X \leqslant \lambda b) \leqslant \inf_{t < 0} \{e^{-\lambda b t} M(t)\} = e^{-\lambda} \left(\frac{e}{b}\right)^{\lambda b}.$$

In particular, verify that

$$P(X \geqslant 2\lambda) \leqslant \left(\frac{e}{4}\right)^{\lambda} \quad \text{and} \quad P\left(X \leqslant \frac{\lambda}{2}\right) \leqslant \left(\frac{2}{e}\right)^{\lambda/2}.$$

Compare these with the bounds yielded by Chebyshov's inequality, (λ^{-1} and $4\lambda^{-1}$, respectively).

29. Three particles perform independent symmetric random walks on the vertices of a triangle; that is to say, at any step each particle moves independently to either of the other two vertices with equal probability $\frac{1}{2}$. At $n = 0$, all three are at the same vertex.

(a)   Let T be the number of steps until they all again share a vertex. (a) Find Es^T, and show that $ET = 9$.

(b)   Suppose that they all start at different vertices; let R be the time until they first share a vertex. Do you think $ER > ET$ or $ER < ET$? Find ER and test your conjecture.

(c)   Let S be the number of steps until they all again share the same vertex from which they began. Find ES.

30. ***Poisson number of de Moivre trials.***   Suppose any ball yields a wicket with probability p, or one or more runs with probability q, or neither of these with probability r, where $p + q + r = 1$. Suppose the total number of balls N has a Poisson distribution with parameter λ, independent of their outcomes. Let X be the number of wickets, and Y the number of balls from which runs are scored. Show that X and Y are independent Poisson random variables by calculating $G(s, t) = E(s^X t^Y)$.

31. ***Characteristic function.***   We have occasionally been hampered by the non-existence of a moment generating function. In such cases, in more advanced work we define the characteristic function

$$\phi(t) = E e^{itX} = E \cos(tX) + i E \sin(tX)$$

$$= \int_{-\infty}^{\infty} e^{itX} f(x) \, dx$$

where $i^2 = -1$. Show that

(a) $|\phi(t)| \leqslant 1$ for all $t \in \mathbb{R}$,   (b) $\phi(0) = 1$.

Remark.   It can be shown that the characteristic function of a Cauchy random variable X is $\phi_X(t) = e^{-|t|}$.

32. Show that if $X_1, \ldots, X_n$ are independent Cauchy random variables and $\overline{X} = n^{-1} \sum_{r=1}^{n} X_r$, then $\overline{X}$ has the same Cauchy density as the X_i.

33. **Multivariate normal.** Let $Y_1, \ldots, Y_n$ be a collection of independent $N(0, 1)$ random variables, and define $X_1, \ldots, X_n$ by

$$X_r = \sum_{k=1}^{n} a_{rk} Y_k, \quad 1 \leq r \leq n.$$

Show that X_r is normal, with mean zero and variance $\sigma_r^2 = \sum_{k=1}^{n} a_{rk}^2$. Find the mean and variance of the random variable $Z = \sum_{r=1}^{n} t_r X_r$ and hence find the joint moment generating function

$$\mathrm{E}\exp\left(\sum_{r=1}^{n} t_r X_r\right).$$

Deduce that X_j and X_k are independent if and only if $\mathrm{cov}(X_j, X_k) = 0$.

34. **Normal sample.** Let $X_1, \ldots, X_n$ be independent $N(\mu, \sigma^2)$ random variables; as usual set

$$\overline{X} = n^{-1} \sum_{r=1}^{n} X_r \quad \text{and} \quad S^2 = \sum_{r=1}^{n} (X_r - \overline{X})^2.$$

Show that $\mathrm{cov}(\overline{X}, X_r - \overline{X}) = 0$, $1 \leq r \leq n$, and deduce that $\overline{X}$ and S^2 are independent. (Hint: Consider the joint distribution of the vector $(\overline{X}, X_1 - \overline{X}, \ldots, X_n - \overline{X})$ and use the previous question.)

Hints and solutions for selected exercises and problems

Section 2.2

1. (a) All sequences of j's and k's of length m.
   (b) The non-negative integers.
   (c) New rules: $(a_1, a_2), (a_3, a_4), (a_5, a_6)$, where $a_i \leqslant 7$ $(1 \leqslant i \leqslant 6)$.
   (d) All quadruples $(\mathbf{x}_1, \mathbf{x}_2, \mathbf{x}_3, \mathbf{x}_4)$, where each $\mathbf{x}_i$ is a choice of five different elements from
   $$\Omega = (C, D, H, S) \times (A, 2, 3, 4, 5, 6, 7, 8, 9, 10, J, Q, K),$$
   and $\mathbf{x}_i \cap \mathbf{x}_j = \emptyset, i \neq j$.

Section 2.3

1. (a) $\Omega = \{i, j: 1 \leqslant i, j \leqslant 6\}$, $A = \{i, j: i + j = 3\}$.
   (b) $\Omega = \{i: 0 \leqslant i \leqslant 100\}$, $A = \{i: 0 \leqslant i \leqslant 4\}$.
   (c) $\Omega = \{B, G\} \times \{B, G\} \times \{B, G\}$, $A = \{BBB, GGG\}$.
   (d) $\Omega = $ the non-negative integers, $A = $ the integers in $[10, 15]$.
   (e) $\Omega = \{(r_1, f_1), (r_2, f_2), (r_3, f_3)\}$ where $0 \leqslant r_i, f_i \leqslant 7$,
   $A = \{r_1 = r_2 = 7\} \cup \{r_1 = r_3 = 7\} \cup \{r_2 = r_3 = 7\}$.
   (f) $\Omega = \{x, y: x, y \geqslant 0\}$, $A = \{x, y: x > y\}$.

2. Draw two Venn diagrams.

Section 2.4

1. $\frac{18}{37}$.

2. There are $\frac{1}{2} \times 52 \times 51$ pairs of cards, and 6 pairs of aces, so P(two aces) $= (13 \times 17)^{-1} = \frac{1}{221}$.

3. $\dfrac{\text{area } ABD}{\text{area } ABC} = \dfrac{\frac{1}{2} \times |BD| \times \text{height}}{\frac{1}{2} \times |BC| \times \text{height}} = \dfrac{|BD|}{|BC|}$.

4. $\dfrac{\pi(\frac{1}{2}r)^2}{\pi r^2} = \frac{1}{4}$.

5. $b/(a + b)$.

Section 2.5

1. $S \cap F = \emptyset$ and $\Omega = S \cup F$. Hence $1 = P(\Omega) = P(S) + P(F)$, by (3).

2. If $|\Omega| = n$, then there are at most 2^n different events in Ω.

336

3. $P\left(\bigcup_{r=1}^{n+1} A_r\right) = P\left(\bigcup_{r=1}^{n} A_r\right) + P(A_{n+1})$, by (3). Now induction yields (4).

4. $A = (A \setminus B) \cup (A \cap B)$. By the addition rule, $P(A) = P(A \setminus B) + P(A \cap B)$. Hence $P(A \cap B) \leqslant P(A)$. The same is true with B and A interchanged.

5. Use $1 = P(\Omega) = P(A) + P(A^c) = P(\Omega) + P(\varnothing)$.

Section 2.6

1. Obvious from $P(B) = P(A) + P(B \setminus A)$.

2. $P(A \cup B \cup C) = P(A) + P(B \cup C) - P(A \cap (B \cup C))$
   $= P(A) + P(B) + P(C) - P(B \cap C) - P((A \cap B) \cup (A \cap C))$
   and so on.

3. P(at least one double six in r throws) $= 1 - \left(\frac{35}{36}\right)^r$,
$$1 - \left(\tfrac{35}{36}\right)^{24} \simeq 0.491 < \tfrac{1}{2} < 0.506 \simeq 1 - \left(\tfrac{35}{36}\right)^{25}.$$
   So 25 is the number needed.

4. By enumeration of cases, the probabilities are $P(S_k) = a_k/216$, where, in order from a_3 to a_{18}, the a_i are
$$1, 3, 6, 10, 15, 21, 25, 27, 27, 25, 21, 15, 10, 6, 3, 1.$$

5. (a) $1 - \left(\frac{5}{6}\right)^6$.
   (b) $1 - \left(\frac{35}{36}\right)^{12} < 1 - \left(\frac{5}{6}\right)^6$.

Section 2.7

1. $0.1/0.25 = 40\%$
2. (a) We know $0 \leqslant P(A \cap B) \leqslant P(B)$. Divide by $P(B)$.
   (b) $P(\Omega|B) = P(\Omega \cap B)/P(B) = 1$.
   (c) $P(A_1 \cup A_2|B) = P(A_1 \cap B) \cup (A_2 \cap B))/P(B)$. Expand this.
   For the last part set $A_1 = A$, $A_2 = A^c$, and use (c).

3. RHS $= \dfrac{P(A \cap B \cap C)}{P(B \cap C)} \dfrac{P(B \cap C)}{P(C)} P(C) =$ LHS.

4. Use the above exercise to give
$$P(\text{all red}) = P(3 \text{ red}|2 \text{ red})P(2 \text{ red}|1 \text{ red})P(1 \text{ red})$$
$$= \tfrac{3}{13} \times \tfrac{4}{14} \times \tfrac{5}{15} = \tfrac{2}{91}.$$
   By the addition rule
$$P(\text{same colour}) = P(\text{red}) + P(\text{green}) + P(\text{blue}) = \tfrac{2}{91} + \tfrac{4}{5} \times \tfrac{1}{91} + \tfrac{4}{91}.$$

5. LHS $= P(A \cap B)/P(A \cup B)$
$$\leqslant \min\left\{\frac{P(A \cap B)}{P(A)}, \frac{P(A \cap B)}{P(B)}\right\} = \text{RHS}.$$

Section 2.8

1. (a) $\phi\beta + (1 - \phi)\sigma$;   (b) $\phi\beta/\{\phi\beta + (1 - \phi)\sigma\}$.
2. P(reject) $= 10^{-4} \times \frac{95}{100} + (1 - 10^{-3}) \times \frac{5}{100}$. Then we have (a) $10^{-4} \times \frac{95}{100}/$P(reject);
   (b) $1 - $P(reject).

3. (a) You can only say what this probability is if you assume (or know) what your friend decided to tell you in all possible circumstances. Otherwise you cannot determine this probability. Consider two cases for example.

   (i) Your friend has decided that if she has both red aces she will say 'one is the ace of diamonds'. In this case the probability that she has both red aces, if she tells you she has the ace of hearts, is zero.

   (ii) Your friend has decided that if she has one ace she will say 'I have a club', but if she has two aces, she will say 'one is the ace of hearts'. In this case the probability that she has both red aces, if she tells you she has the ace of hearts, is unity.

   (b) In this case do you know what the answers would be in all cases, because you have received a truthful answer to a fixed question. We calculate:

   $$P(\text{either is } A_H) = 1 - P(\text{neither is } A_H) = 1 - \frac{3 \times 2}{4 \times 3} = \frac{1}{2},$$

   $$P(\text{both red aces}) = \frac{2 \times 1}{4 \times 3} = \frac{1}{6},$$

   hence

   $$P(\text{both red aces}|\text{either is } A_H) = \tfrac{1}{6}/\tfrac{1}{2} = \tfrac{1}{3}.$$

4. $\sum P(A|B_i \cap C)P(B_i|C) = \sum P(A \cap B_i \cap C)/P(C) = P(A \cap C)/P(C).$

Section 2.9

1. $P(A^c \cap B^c) - P(A^c)P(B^c) = P(A \cup B)^c) - (1 - P(A))(1 - P(B)) = P(A \cap B) - P(A)P(B) = 0$ iff A and B are independent.

2. (a) 0.35,   (b) 0.2,   (c) $\tfrac{2}{9}$,   (d) 0.08.

3. $P(n - 1 \text{ tails followed by head}) = \{P(T)\}^{n-1}P(H).$

4. Let $P(G) = q, P(B) = p$, where $p + q = 1$.

   (a) $P(\text{both sexes}) = 1 - p^3 - q^3$, $P(\text{at most one girl}) = p^3 + 3p^2q$,
   $P(\text{both sexes and at most one girl}) = 3p^2q$. Then
   $$(1 - p^3 - q^3)(p^3 + 3p^2q) = 3p^2q$$
   for independence, which occurs when $p = q = \tfrac{1}{2}$ and when $pq = 0$, and not otherwise.

   (b) $P(\text{both sexes}) = 1 - p^4 - q^4$, $P(\text{at most one girl}) = p^4 + 4p^3q$,
   $P(\text{both sexes and at most one girl}) = 4p^3q$. For independence
   $$(1 - p^4 - q^4)(p^4 + 4p^3q) = 4p^3q,$$
   which occurs when $pq = 0$ and for just one other value of p, which is approximately $p = 0.4$.

5. Flip two coins, with $A \equiv$ first coin shows head, $B \equiv$ second coin shows head, $C \equiv$ both coins show heads.

6. See example 2.9.7.

Section 2.10

1. $P(A_1|A_2)P(A_1) = P(A_1 \cap A_2)$ and so on, giving the LHS by successive cancellation.

2. As figure 2.20, terminated after the sixth deuce.

3. Modified version of figure 2.19.

4. $6p^2q^2.$

Section 2.11

1. Condition on the first flip to give
$$p_n = P(\text{even number in } n) = P(\text{odd number in } n-1)P(H)$$
$$+ P(\text{even number in } n-1)P(T)$$
$$= (1 - p_{n-1})p + qp_{n-1}$$
$$= \tfrac{1}{2} + \tfrac{1}{2}(q - p)^n.$$

2. By considering the nth roll we see that $p_n = \tfrac{1}{5}(1 - p_{n-1})$. We are told $p_1 = 1$. Hence $p_n = \tfrac{1}{6} - \tfrac{1}{6}(-\tfrac{1}{5})^{n-2}$, $n \geqslant 1$.

Section 2.12

1. By definition, for a unit stake, you get $1 + \pi_a$ with probability $P(A)$. The value of this is $(1 + \pi_a)P(A)$, so the casino's take is
$$t = 1 - (1 + \pi_a)P(A).$$
From (1) we have $P(A) = (1 + \phi_a)^{-1}$, so
$$t = 1 - (1 + \pi_a)(1 + \phi_a)^{-1} = (\phi_a - \pi_a)(1 + \phi_a)^{-1}.$$

Section 2.16

1. (a) $\left(\tfrac{1}{20}\right)^2 \times \tfrac{1}{2} + 2 \times \tfrac{1}{20} \times \tfrac{1}{2} \times \tfrac{19}{20}$; (b) $\tfrac{1}{2} \times \left(\tfrac{1}{20}\right)^2$.

2. $\tfrac{32}{663}$.

3. $\tfrac{15}{442}$, $\tfrac{32}{663} \times \tfrac{15}{442}$.

4. (a) No; (b) $\tfrac{1}{4} \leqslant P(\text{rain at weekend}) \leqslant \tfrac{1}{2}$.

5. $P(3 \text{ divides PIN}) = \tfrac{2999}{8998}$, $P(7 \text{ divides PIN}) = \tfrac{1286}{8998}$, $P(21 \text{ divides PIN}) = \tfrac{169}{8990}$, so $P(\text{either one divides PIN}) = \tfrac{4116}{8998}$.

6. (a) $P(A_{12}) = 0$; (b) $P(A_3) = 60/6^3$, $P(A_6) = 100/6^4$.

7. $\tfrac{1}{4}, \tfrac{1}{3}$.

8. No. In fact 28 will do.

9. (c) (i) $\tfrac{2}{5}$, (ii) $\tfrac{2}{5}$.

10. (c) $P(T) = \tfrac{2}{5}$ (from part (c) of the previous answer). Hence (a) $\tfrac{1}{4}$; (b) $\tfrac{3}{4}$.

11. Use a Venn diagram, or check the elements;
by the first result, $A \cup B \cup C = ((A \cup B)^c \cap C^c)^c$
$$= (A^c \cap B^c \cap C^c)^c.$$

12. Use the above problem and induction.

13. $P(A \cap B) = P(A) + P(B) - P(A \cup B)$.
If $B \subseteq A$, this gives $P(A \cap B) = P(B) = \tfrac{1}{2}$.
If $A \cup B = \Omega$, this gives $P(A \cap B) = \tfrac{1}{10}$.
The bounds are as given because
$$\max\{P(A), P(B)\} \leqslant P(A \cup B) \leqslant 1.$$

14. The first inequality follows from $P(A \cap B) > P(A)P(B)$. The second follows from
$$P(A^c \cap B) = P(B) - P(A \cap B) < P(B) - P(B)P(A) = P(B)P(A^c).$$

15. $P(A \cap A) = \{P(A)\}^2$ gives $P(A) = 0$ or $P(A) = 1$.

16. (a) $1/k$;   (b) $1/k!$;   (c) k/n.

17. $A \setminus B \subseteq A$.

18. $P(\bigcup_{r=1}^{n} A_r) = P(\bigcup_{r=1}^{n-1} A_r) + P(A_n) - P(\bigcup(A_r \cap A_n))$; now use induction.

$$P(\bigcap_{r=1}^{n} A_r) = \sum P(A_r) - \sum_{i<j} P(A_i \cup A_j) + \cdots (-1)^{n+1} P(\bigcup_{r=1}^{n} A_r).$$

19. $P(A \cap B) - P(A) = P(B)P((A^c \cup B^c)^c) - (1 - P(A^c))(1 - P(B^c))$

$= 1 - P(A^c \cup B^c) - 1 + P(A^c) + P(B^c) - P(A^c)P(B^c) = $ RHS.

20. (a) $\alpha = a/(a+b)$.

(b) $P(\text{amber}) = P(\text{amber}|\text{1st amber})a/(a+b) + P(\text{amber}|\text{1st blue})b/(a+b)$.

21. (a) $P(\text{more than 3}) = 1 - P(\text{in circle radius 3}) = 1 - (9\pi/48)$.

(b) $10(\pi/48)^2 + (\pi/16)^2 = 19(\pi/48)^2$.

(c) $P(\text{total score 15}) = \{P(\text{dart scores 5})\}^3$.

22. (a) $\frac{1}{12}$;   (b) $\frac{5}{16}$;   (c) $\frac{2}{27}$.

23. $\frac{1}{3}$.

24. $P(A_1) = \frac{1}{2} = P(A_2) = P(A_3)$; $P(A_1 \cap A_2) = \frac{1}{4} = P(A_2 \cap A_3) = P(A_3 \cap A_1) = P(A_1 \cap A_2 \cap A_3)$.

25. (a) $P(\text{even}) = p_k = \frac{5}{6}p_{k-1} + \frac{1}{6}(1 - p_{k-1})$. So $p_k = \frac{1}{2}\{1 + (\frac{2}{3})^k\}$.

(b) $P(\text{divisible by 3}) = p_k$, $P(\text{one more than a multiple of 3}) = q_k$;

$$p_k = \frac{1}{6}(1 - p_{k-1} - q_{k-1}) + \frac{5}{6}p_{k-1}; \quad q_k = \frac{1}{6}p_{k-1} + \frac{5}{6}q_{k-1}.$$

Hence $p_k = \frac{1}{3}(1 + \lambda^k + \mu^k)$, where λ and μ are the roots of $12x^2 - 18x + 7 = 0$.

26. $4x^2 - 2x - 1 = 0$ has roots $(1 \pm \sqrt{5})/4$. Use $p_1 = 0$ and $p_2 = \frac{1}{4}$.

28. BC.

29. Let T denote the event that each of a series of similar answers is correct. Let S_r denote the event that you receive r similar answers. Let V denote the event that the passer-by is a tourist. Then in general we want $P(T|S_r)$, which we can rearrange as

$$P(T|S_r) = P(T \cap V|S_r), \text{ since } T \cap V^c = \varnothing$$
$$= P(T \cap V \cap S_r)/P(S_r)$$
$$= P(T \cap S_r|V P(V)/P(S_r).$$

(a) $P(T \cap S_1|V) = \frac{3}{4}$; obviously $P(S_1) = 1$. Hence $P(T|S_1) = \frac{3}{4} \times \frac{2}{3} = \frac{1}{2}$.

(b) $P(T \cap S_2|V) = (\frac{3}{4})^2$; $P(S_2) = \{(\frac{3}{4})^2 + (\frac{1}{4})^2\}\frac{2}{3} + \frac{1}{3}$. Hence $P(T|S_2) = \frac{1}{2}$.

(c) $P(T \cap S_3|V) = (\frac{3}{4})^3$; $P(S_3) = \{(\frac{3}{4})^3 + (\frac{1}{4})^3\}\frac{2}{3} + \frac{1}{3}$. Hence $P(T|S_3) = \frac{9}{20}$.

(d) Now we know the speaker is a tourist, so

$$P(\text{East is true } |EEEW) = (\tfrac{3}{4})^3 \times \tfrac{1}{4}/\{(\tfrac{3}{4})^3 \times \tfrac{1}{4} + (\tfrac{1}{4})^3 \times \tfrac{3}{4}\} = \tfrac{9}{10}.$$

(e) Thus $P(\text{East correct}|\text{fourth answer East also})$ is found to be

$$(\tfrac{3}{4})^4 \times \tfrac{2}{3}/[\{(\tfrac{3}{4})^4 + (\tfrac{1}{4})^4\}\tfrac{2}{3} + \tfrac{1}{3}] = \tfrac{9}{35}.$$

Section 3.2

1. (a) 6; (b) 0; (c) 6.

3. $6!/6^6 \simeq 0.015$.

4. For every collection of x_i such that $\sum_{i=1}^{1000} x_i = 1100$, there is a one–one map $x_i \to 7 - x_i$ to the collection of $7 - x_i$ such that $\sum(7 - x_i) = 5900$.

Section 3.3

1. (a) Use the one–one correspondence between choosing r objects from n and choosing the remaining $n - r$.
   (b) Verify trivially from (7).
   (c) Set up the same difference equation for $C(n, r)$ and $C(n, n - r)$.

2. Expand RHS to give (4).

Section 3.4

1. Each element is either in or not in any set, giving 2^n choices.

2. There are $\binom{n}{k}$ ways to choose the k brackets to supply x^k, the rest of the brackets supply y^{n-k}.

3. Set $x = y = 1$ in exercise 2.

4. The answer to the hint is $\binom{r-1}{k-1}$, as there are $r - 1$ numbers less than r, of which we choose $k - 1$. Now sum over all possibilities for the largest number selected.

5. $\binom{r+s}{r}$.

Section 3.5

1. $|\Omega| = 9 \times 10^3$. $|A| =$ number of PINs with double zero $+$ number of PINs with single zero $+$ number of PINs with no zeros $= 9 \times 8 \times 3 + 9 \times 8 \times 3 \times 3 + 9 \times \binom{8}{2} 4!/2!$.
   Hence $P(A) = 0.432$.

2. Choose three faces in $\binom{6}{3}$ ways; divide by 2! to avoid counting the pairs twice, permute these symbols in $\binom{5}{2, 2, 1}$ ways.

3. Divide the given expression by P(you have x spades)
   $= \binom{13}{x}\binom{39}{13 - x}$.

5. Choose the points to win in $\binom{n}{k}$ ways.

Section 3.6

1. Choose 5 non-adjacent objects from 48, and choose one to make a pair in 49 objects.

3. Choose k of your selection from the selected numbers and then choose $r - k$ from the $n - r$ unselected.

Section 3.9

1. The probability that a given choice of r players draw their own name and the remaining $n - r$ do not is

$$\frac{1}{n}\left(\frac{1}{n-1}\right)\cdots\left(\frac{1}{n-r+1}\right)\left\{\frac{1}{2!} - \frac{1}{3} + \cdots + \frac{(-)^{n-r}}{(n-r)!}\right\}$$

There are $\binom{n}{r}$ such choices of r players, giving the result.

Section 3.13

1. (a) $\frac{55}{96}$; (b) five will do. (P(at least two of the five share a sign) $\simeq 0.6$).

2. (b) 0.15 approximately.

3. (a) $7 \Big/ \binom{10}{4}$; (b) $1 \Big/ \binom{10}{4}$; (c) $\binom{9}{3} \Big/ \binom{10}{4} = 2/5$;

    (d) $1 - \binom{8}{4} \Big/ \binom{10}{4} = 2/3$; (e) $5 \Big/ \binom{10}{4} = 1/42$.

4. (a) $\frac{1}{2}(\frac{5}{6})^5$; (b) $1 - 6!/6^6 = 319/324$; (c) $\left\{\binom{6}{4}5^2 + \binom{6}{5}5 + 1\right\} \Big/ 6^6 = 406/6^6$;

    (d) $\dfrac{406}{6^6} \Big/ \dfrac{319}{324}$.

5. Choose the ranks of the pairs in $\binom{13}{2}$ ways, the suits of the pairs in $\binom{4}{2}\binom{4}{2}$ ways, and the other card in 44 ways. Then

$$44\binom{13}{2}\binom{4}{2}\binom{4}{2} \Big/ \binom{52}{5} \simeq 0.48.$$

6. If the first thing is not chosen there are $(n - 1)^{\underline{r}}$ permutations of r things from the remaining $n - 1$. The other term arises if the first thing is chosen. Then use the addition rule. Alternatively, substitute in the formula.

7. Use van der Monde, example 3.4.2.

8. Consider the boxes that include the first colour, and those that do not.

9. When you make the nth cut, the previous $n - 1$ cuts divide it into n segments at most. So the largest number of extra bits produced by this cut is n. Hence $R_n = R_{n-1} + n$. Now verify the given solution.

10. (a) $2(n - k - 1)/n!$; (b) $2/(n - 1)!$

11. $(n - k)!/n!$.

12. Choose the start of the run in 10 ways and the suits in 4^5 ways; exclude the straight flush.

13. $\binom{32}{13} \Big/ \binom{52}{13}$. This gives fair odds of 1800: 1, approximately, so by exercise 1 of section 2.12 the Earl's percentage take was around 44%. A nice little earner, provided he insisted on a final shuffle himself.

14. (a) This is (7) of section 3.8.

(b) Let A be the event that the wasp has visited g_{k-1} when its last flight is to g_k from g_{k+1}. And let B be the event that the wasp has visited g_{k+1} is when its last flight is to g_k from g_{k-1}. It must do one or the other, and $A \cap B = \emptyset$.

By the first part, $P(L_k|A) = P(L_k|B) = n^{-1}$. Hence, using the partition rule,

$$P(L_k) = P(L_k|A)P(A) + P(L_k|B)P(B) = n^{-1}\{P(A) + P(B)\} = n^{-1}.$$

15. (a) $\binom{6}{4}\binom{43}{2} \Big/ \binom{49}{6}$;   (b) $\binom{25}{6} \Big/ \binom{49}{6}$.

(c) Let A_r be the event that the number r has failed to turn up. Then

$$P\left(\bigcup_{r=1}^{49} A_r\right) = \sum P(A_r) - \sum P(A_s \cap A_s) + \cdots$$

$$= 49P(A_1) - \binom{49}{2}P(A_1 \cap A_2) + \cdots + \binom{49}{43}P(A_1 \cap \cdots \cap A_{43}).$$

For any set of k numbers, $P(A_1 \cap \cdots \cap A_k) = \binom{49-k}{6} \Big/ \binom{49}{6}$ and so

$$P\left(\bigcup_{r=1}^{49} A_r\right) = \left\{49\binom{48}{6} - \binom{49}{2}\binom{47}{6} + \cdots + \binom{49}{43}\right\} \Big/ \binom{49}{6}$$

Section 4.2

1. (a) $1 - (1 - p)^n$;   (b) $3p^2(1-p) + p^3$.

2. (a) $1 - q^4 - 4q^3(1 - q)$;   (b) $1 - q^3 - 3q^2(1 - q)$.
(a) $\geqslant$ (b); the moral seems obvious.

3. Imagine that you are 'in gaol', and look at Figure 4.1.

Section 4.3

1. $P(A_n) = \left(\frac{5}{6}\right)^{n-1} \times \frac{1}{6}$, $P(A_n \cap E) = \left(\frac{5}{6}\right)^{2m-1} \times \frac{1}{6}(n = 2m)$, $P(E) = \frac{5}{11}$
and $P(A_n|E) = \left(\frac{5}{6}\right)^{2m} \times \frac{11}{25}(n = 2m)$. Yes, but not *the* geometric distribution.

2. From example 4.3.4, $P(A_n|D_n) = a_n/\lambda(n) = p/(p + q) = P(A_n)$.

Section 4.4

2. $\binom{n}{k}^2 p^{2k}q^{2n-2k} \Big/ \left\{\binom{n}{k+1}p^{k+1}q^{n-k-1}\binom{n}{k-1}p^{k-1}q^{n-k+1}\right\}$
$= (k+1)(n-k+1)/\{(n-k)k\} \geqslant 1$.

3. The correspondence rule in action.

Section 4.5

1. Consider the ratio

$$\binom{a}{r}\binom{N+1-a}{n-r}\Big/\binom{N}{n} : \binom{a}{r}\binom{N+1-a}{n-r}\Big/\binom{N+1}{n},$$

which reduces to $an : r(N+1)$. This gives increasing terms up to the integer nearest to an/r; thereafter the terms decrease.

2. $\left\{\binom{93}{10} + 7\binom{93}{9}\right\}\Big/\binom{100}{10}$.

3. $\dfrac{r}{r+1}\left(\dfrac{n-r}{n-r+1}\right)\left(\dfrac{a-r}{a-r+1}\right)\left(\dfrac{N-a-n+r}{N-a-n+r+1}\right) \le 1$.

Section 4.6

1. $\mu = \sum_{x=1}^{n} x n^{-1} = n^{-1} \times \frac{1}{2}n(n+1)$.

   $\sigma^2 = \sum_{x=1}^{n} n^{-1}x^2 - \mu^2 = \frac{1}{6}(n+1)(2n+1) - \frac{1}{4}(n+1)^2$.   (See subsection 3.12.I.)

3. $\sigma^2 = \sum_{x=1}^{\infty} x^2 q^{x-1} p - \mu^2$. Use the negative binomial theorem from subsection 3.12.III to sum the series: $\sum_{x=1}^{\infty} \frac{1}{2}x(x+1)q^{x-1} = (1-q)^{-3}$.

4. $\sum k \lambda^k e^{-\lambda}/k! = \lambda \sum \lambda^{k-1} e^{-\lambda}/(k-1)!$

Section 4.8

1. $n = 200$, $p = \frac{1}{40}$, $\lambda = np = 5$. So

   (a) $1 - P(\text{less than } 4) = 1 - \sum_{r=0}^{3} 5^r e^{-5}/r! \simeq 0.74$,

   (b) $P(\text{none}) = e^{-5} \simeq 0.0067$.

2. $n = 404$; $p = 10^{-2}$; $\lambda = np = 4.04$,

   $$P(\text{bump at least one}) = e^{-4.04}\left\{1 + 4.04 + \tfrac{1}{2}(4.04)^2 + \tfrac{1}{6}(4.04)^3\right\} \simeq 0.43.$$

3. $k = [\lambda]$ if λ is not an integer. If λ is an integer then $p(\lambda - 1) = p(\lambda) = \lambda^\lambda e^{-\lambda}/\lambda!$.

Section 4.9

1. $T(x) = 1 - (1-x)^2$; $t(x) = 2(1-x)$.

Section 4.10

1. Exact binomial, $p(12) = 0.028$, $p(16) = 0.0018$;
   normal approximation, $p(12) \simeq 0.027$, $p(16) \simeq 0.0022$.

2. Show that the mode m is $[np]$ and then use Stirling's formula.

3. $\mu = np = 800$, $\sigma = (npq)^{1/2} = 20$; the probability is $1 - \Phi(4) \simeq \frac{1}{4}\phi(4) \simeq 0.00003$, which is extremely small. But if observed it would suggest the new one is better.

Section 4.11

1. By symmetry we might just as well pick a point in the semicircle $0 \leqslant y \leqslant (1 - x^2)^{1/2}$. Now use example 4.11.4.

2. By (5), $\frac{1}{2} \times 3 \times 3a + \frac{1}{2} \times 2 \times 3a = 1$; thus $a = \frac{2}{15}$.

$$P(|X| > 1) = P(-3 \leqslant X \leqslant -1) + P(1 \leqslant X \leqslant 2)$$
$$= \frac{1}{2} \times 2 \times 2a + \frac{1}{2} \times 1 \times \frac{3}{2}a = \frac{11}{30}.$$

Section 4.12

1. $p(x, y) = \frac{1}{6}$, $x \neq y$; $P(X + Y = 3) = \frac{1}{3}$; $P(X = x) = \frac{1}{3} = P(Y = y)$; so each is uniform on $\{1, 2, 3\}$ with mean 2 and variance $\frac{2}{3}$.

2. $p(x, y) = \frac{1}{6}\binom{x}{y}2^{-x}, \quad 0 \leqslant y \leqslant x \leqslant 6.$

$$\bar{y} = \sum_{x,y} \frac{y}{6}\binom{x}{y}2^{-x} = \sum_{x=1}^{6} \frac{1}{6} \times \frac{x}{2} = \frac{7}{4}.$$

Section 4.16

1. X is binomial $B(n, 6^{-2})$ with mean $6^{-2}n$ and variance $6^{-2}n(1 - 6^{-2})$.

2. (a) $c_1 = 2/\{n(n + 1)\}$;   (b) $c_2 = 1$.

3. $\sum x^2 \lambda^2 e^{-\lambda}/x! = \sum x(x - 1)\lambda^x e^{-\lambda}/x! + \sum x\lambda^x e^{\lambda}/x! = \lambda^2 + \lambda.$

4. When $[np] = k$.

5. $x^8 \geqslant (x^2 - 1)^4$; the distribution is geometric.

6. $P(X = x) = p\{(q + r)^{x-1} - r^{x-1}\} + q\{(p + r)^{x-1} - r^{x-1}\}, \quad x \leqslant 2.$

$$P(Y = y) = \binom{y - 1}{j - 1}p^j \sum_{i=k}^{y-j}\binom{y - j}{i}q^i r^{y-j-i}$$

$$+ \binom{y - 1}{k - 1}q^k \sum_{i=j}^{y-k}\binom{y - k}{i}p^i r^{y-k-i}, \quad y \geqslant j = k.$$

7. (a) $pq/(1 - p^2)$;   (b) $p^{r-1}q/(1 - p^r)$;   (c) $p^r(1 - p^s)/(1 - p^{r+s})$.

8. $P(X = 2n) = \frac{1}{2}(1 - \alpha)^{n-1}(1 - \beta)^{n-1}\{(1 - \alpha)\beta + (1 - \beta)\alpha\}$;

$P(X = 2n - 1) = \frac{1}{2}(1 - \alpha)^{n-1}(1 - \beta)^{n-1}(\alpha + \beta), \quad n \geqslant 1.$

$P(E) = \alpha + \beta - 2\alpha\beta/\{2(\alpha + \beta - \alpha\beta)\}$. Not in general, but B is independent of E and $\{X = 2n\}$ when $\alpha = \beta$.

9. $p(n + k) = \binom{n + k - 1}{k - 1}p^k q^n$

$$= \frac{(n + k - 1)(n + k - 2)\ldots(k + 1)k}{n!}\left(1 - \frac{\lambda}{k}\right)^k \left(\frac{\lambda}{k}\right)^n$$

$$= \frac{\lambda^n}{n!}\left(1 + \frac{n - 1}{k}\right)\cdots\left(1 + \frac{1}{k}\right)\left(1 - \frac{\lambda}{k}\right)^k \to \frac{\lambda^n}{n!}e^{-\lambda}.$$

$p(n + k)$ is the probability that in repeated Bernoulli trials the $(n + k)$th trial is the kth success; this is the probability of exactly n failures before the kth success. The result shows that as failures become rarer, in the long run they have a Poisson distribution. This is consistent with everything in section 4.8.

10. P(kth recapture is mth tagged)
    = P(kth is tagged) × P(1st $k - 1$ recaptures include exactly $m - 1$ tagged).

    Now P(kth is tagged) is t/n, and second term is hypergeometric. Hence the result.

11. (a) The total number of possible sequences is $\binom{h+t}{h}$. The number of ways of having x runs of heads is the number of ways of dividing the h heads into x groups, which we may do with $x - 1$ bars placed in any of the $h - 1$ gaps, that is in $\binom{h-1}{x-1}$ ways. The t tails must then be distributed with at least one in these $x - 1$ positions, and any number (including zero) at each end. Adding 2 (notional) tails to go at the ends shows that this is the same as the number of ways of dividing $t + 2$ tails into $x + 1$ groups, none of which is empty. This may be done in $\binom{t+2-1}{x+1-1}$ ways, by problem 17 of section 3.13, and the result follows.

12. $t(x) = 1 - |x|, \quad |x| \leqslant 1.$
    $$T(x) = \begin{cases} \frac{1}{2}(1+x)^2, & -1 \leqslant x \leqslant 0 \\ 1 - \frac{1}{2}(1-x)^2, & 0 \leqslant x \leqslant 1. \end{cases}$$

13. $p(x, w) = \frac{1}{36}, \quad x + 1 \leqslant w \leqslant x + 6, \quad 1 \leqslant x \leqslant 6.$

14. (a) Choose the numbers less than x in $\binom{x-1}{5}$ ways.

    (b) $\binom{49-x}{5} \Big/ \binom{49}{6}, \quad 1 \leqslant x \leqslant 44.$

15. 7×10^{-4} approximately.

16. $P(X = k) = (1 - pt)^{k-1} pt.$

17. $p(\hat{k})$ in (2) of section 4.5, where in the general case $\hat{k}$ is the integer part of
    $$\frac{(m+1)r - (w+1)}{m+1+w+1}$$

    What are the special cases?
    The ratio $p_t(k)/p_{t-1}(k)$ is
    $$\frac{(t-m)(t-r)}{t(t-m-r+k)}.$$
    So $p(k)$ is largest for fixed m, r, and k when $t = [mr/k]$.

19. (a) $\binom{x-y-1}{4} \Big/ \binom{49}{6}, \quad 1 \leqslant y < x - 4 \leqslant 45.$

    (b) $\binom{49-z-1}{6-z} \Big/ \binom{49}{6}.$

20. $P(X = 0) = n^{-1}, \quad P(X = x) = 2n^{-2}(n - x), \quad 1 \leqslant x \leqslant n - 1.$
    $$\text{mean} = \sum_{x=1}^{n-1} xp(x) = \sum_{x=1}^{n-1} \frac{2}{n^2}(nx - x^2) = \frac{n^2 - 1}{3n}.$$

21. $P(X = n) = (n - 1)p^2 q^{n-2} + (n - 1)q^2 p^{n-2}, \quad n \geqslant 4.$

mean $= \sum_{n=4}^{\infty} n(n-1)(q^{n-2}p^2 + p^{n-2}q^2)$. Now use the negative binomial theorem from 3.12.III with $n = 3$ to obtain the result.

23. (b) Deaths per day may be taken to be Poisson with parameter 2. P(5 or more in one day) $= 1 - 7e^{-2} \simeq 0.054$, just over 5%. This is not so unlikely. You would expect at least one such day each month. However, as deaths per annum are approximately normal, we calculate, using part (a),

$$P(X > 850) = P\left(\frac{x - 730}{\sqrt{730}} > \frac{120}{\sqrt{730}}\right) \simeq 1 - \Phi\left(\frac{120}{\sqrt{730}}\right) \simeq 3 \times 10^{-5}.$$

So 2000 really was an extremely bad year, compared with the previous two decades.

24. (a) 2; (b) 3, using problem 6; (c) $\frac{11}{2}$, using problem 21.

Section 5.2

1. Yes, because for any $\omega \in \Omega$, $X(\omega)$ is fixed uniquely; so $X(\omega) - X(\omega) = 0$, etc.

2. X and Y must be defined on the same sample space Ω. Then $X(\omega) + Y(\omega)$ is a real-valued function on Ω, which is a random variable if the outcomes ω such that $X(\omega) + Y(\omega) \leq z$ are an event. Likewise for $X - Y$, XY. These conditions are always satisfied if Ω is countable, so that all its subsets are events.

3. $W \in \{-n, -n+1, \ldots, n-1, n\}$, the elements denoting your position in metres east of the start. You can write $W = X - Y$.

4. Pick a point at random on a dartboard, and let X be the score.

Section 5.3

1. $F(x) = \begin{cases} \frac{1}{2}n^{-2}(n+x)(n+x+1), & -n \leq x \leq 0 \\ 1 - \frac{1}{2}n^{-2}(n-x)(n-x-1), & 0 \leq x \leq n. \end{cases}$

2. If $P(X = x) = p(x)$, $x \in D$, then we let $\Omega = D$ and define $X(\omega) = \omega$, together with $P(A) = \sum_{x \in A} p(x)$ for any event in Ω.

3. If $b > 0$, $\quad p_Y(y) = p_X\left(\frac{y-a}{b}\right)$, $\quad F_Y(y) = F_X\left(\frac{y-a}{b}\right)$.

If $b = 0$ then $P(Y = a) = 1$. If $b < 0$ then $p_Y(y)$ is as above, but

$$F_Y(y) = P\left(X \geq \frac{y-a}{b}\right) = \sum_{x \geq (y-a)/b} p_X(x).$$

Section 5.4

1. $c = 2a^{-2}$; $F(x) = x^2/a^2$.

2. $\lambda f_1 + (1 - \lambda)f_2 \geq 0$ and $\int\{\lambda_1 + (1 - \lambda)f_2\} \, dx = \lambda + 1 - \lambda = 1$.

(b) Not in general. For example, if $f_1 = f_2 = \frac{1}{2}$, $0 \leq x \leq 2$, then $f_2 f_2 = \frac{1}{4}$, which is not a density. But consider $f_1 = f_2 = 1$, $0 \leq x \leq 1$, when $f_1 f_2$ is a density.

3. Check that (20) and (21) hold. Yes; likewise.

Section 5.5

1. From example 5.5.6,
$$f_Y(y) = \frac{1}{2\sqrt{y}}\{\phi(\sqrt{y}) + \phi(-\sqrt{y})\} = \frac{1}{\sqrt{2\pi y}}\exp(-\tfrac{1}{2}y), \ y \geqslant 0.$$

2. Uniform on the integers $\{0, 1, \ldots, m-1\}$; $p_Y(k) = m^{-1}$.

3. $\tfrac{1}{3}y^{-2/3}f(y^{1/3})$.

4. $f_Y(y) = f(y)$, by symmetry about $y = \tfrac{1}{2}$.

Section 5.6

1. $\int_0^1 2x^2\,dx = \tfrac{2}{3}$.

2. $\dfrac{2}{n(n+1)}\sum_{x=1}^{n}x^2 = \dfrac{2}{n(n+1)} \times \tfrac{1}{6} \times n(n+1)(2n+1) = \tfrac{1}{3}(2n+1)$.

3. $EX = \displaystyle\int_0^\infty \{\lambda^r x^r e^{-\lambda x}/(r-1)!\}\,dx = \dfrac{r}{\lambda}\int\{\lambda^{r+1}x^r e^{-\lambda x}/r!\}\,dx = \dfrac{r}{\lambda}$, because the integrand is a density with integral unity.

Section 5.7

1. $B(n, p)$ has mean np and variance npq. Hence $EX = n/2$, $EY = 2n/4$; $\mathrm{var}\,X = n/4$, $\mathrm{var}\,Y = 6n/16$. Use example 5.7.4 to give $EZ = n$, $\mathrm{var}\,Z = 2n$.

2. Let I be the indicator of the event $h(X) \geqslant a$. Then $aI \leqslant h(X)$ always. Now take the expected value of each side, and use the fact that $EI = P(I = 1) = P(h(X) \geqslant a)$.

3. $P(X = n) = 2^{-n}$, $EX = 2$, $\mathrm{var}\,X = 2$.

    (a) $P(|X - 2| \geqslant 2) \leqslant E(|X - 2|^2)/4 = \tfrac{1}{4}\mathrm{var}\,X = \tfrac{1}{2}$.
    Actually $P(|X - 2| \geqslant 2) = P(X \geqslant 4) = 2^{-4}$.

    (b) $P(X \geqslant 4) \leqslant E|X|/4 = 2/4$. Actually $P(X \geqslant 4) = 2^{-4}$.

Section 5.8

1. (a) Condition on the first point, and then on the second, to get (with an obvious notation) first
    $EX = \rho E(X|R) + \phi E(X|F) + 1$ and second $E(X|R) = 1 + \phi E(X|F)$, $E(X|F) = 1 + \rho E(X|R)$. Hence $E(X) = (2 + \rho\phi)/(1 - \rho\phi)$.

    (b) $EY = (2 + EY)2\rho\phi + 2(\rho^2 + \phi^2)$. So $EY = 2/(1 - 2\rho\phi)$.

    (c) $E(X|L) = EX$ and $E(Y|L) = EY$.

2. Condition on the first point. Check that τ_k satisfies the recurrence, together with $\tau_0 = \tau_n = 0$.

Section 5.9

1. Given that $X \geqslant t$, we have shown that $X = t + Y$, where Y is exponential. Hence $E(X|X \geqslant t) = t + \lambda^{-1}$, and $\mathrm{var}\,(X|X \geqslant t) = \mathrm{var}\,Y = \lambda^{-2}$.

2. $E(X|B)P(B) + E(X|B^c)P(B^c)$

$$= \int [xf(x|B)P(B) + xf(x)|B^c)P(B^c)] \, dx$$

$$= \int \left[x\frac{d}{dx}F(x|B)P(B) + x\frac{d}{dx}F(x|B^c)P(B^c) \right] dx$$

$$= \int \left[x\frac{d}{dx}P(\{X \leqslant x\} \cap B) + x\frac{d}{dx}P(\{X \leqslant x\} \cap B^c) \right] dx, \quad \text{by (1) of section 5.9.}$$

$$= \int x\frac{d}{dx}F(x) \, dx = EX.$$

3. Immediate from (8).

Section 5.12

1. (a) $P(X \geqslant x) = \{(6 - x + 1)/6\}^5, 1 \leqslant x \leqslant 6$. So

$$p(x) = P(X = x) = P(X \geqslant x) - P(X \geqslant x + 1)$$

$$= \left(\frac{6 - x + 1}{6}\right)^5 - \left(\frac{6 - x}{6}\right)^5.$$

$$EX = \sum_{x=1}^{6} P(X \geqslant x) = \left(\frac{6}{6}\right)^5 + \left(\frac{5}{6}\right)^5 + \cdots + \left(\frac{1}{6}\right)^5 = \frac{4062}{2592} \simeq 1.57.$$

(b) $P(Y \geqslant y) = 1 - P(Y < y) = 1 - \{(y - 1)/6\}^5$. So $p(y) = P(Y = y) = (y/6)^5 - \{(y - 1)/6\}^5$.

By symmetry Y has the same distribution as $7 - X$, so $EY = 7 - EX \simeq 5.43$. Or do the sum.

2. (a) $\frac{1}{2}$, by symmetry. (b) Let A_r be the event that the first two dice sum to r, and B_r the event that the other two sum to r. Then we know

$$P(A_r) = P(B_r) = 6^{-2} \min\{r - 1, 13 - r\}, \quad 2 \leqslant r \leqslant 12.$$

Then $P(\text{sum of 4 dice} = 14) = \sum_{r=2}^{12} P(A_r)P(B_{14-r}) = \sum_{r=2}^{12} \{P(A_r)\}^2 = 6^{-4}(1^2 + 2^2 + 3^2 + 4^2 + 5^2 + 6^2 + 5^2 + 4^2 + 3^2 + 2^2 + 1^2) = 6^{-4} \times 146$. Hence by symmetry

$$P(\text{sum of 4 dice} \geqslant 14) = \frac{1}{2} + \frac{1}{2} \times \frac{146}{6^4}.$$

3. (a) $c = d - 1$, $d > 1$; (b) $EX = c/(d - 2)$, $d > 2$, $EX = \infty$, $1 < d \leqslant 2$; (c) $\text{var } X = c/(d - 3) - \{c/(d - 2)\}^2, d > 3, \text{var } X = \infty, 2 < d \leqslant 3$; undefined for $1 < d \leqslant 2$.

4. (a) $P(\sin X > \frac{1}{2}) = P(\{\frac{1}{6}\pi < X < \frac{5}{6}\pi\} \cup \{2\pi + \frac{1}{6}\pi < X < 2\pi + \frac{5}{6}\pi\} \cup + \cdots)$

$$= \sum_{n=0}^{\infty} P(2n\pi + \frac{1}{6}\pi < X < 2n\pi + \frac{5}{6}\pi)$$

$$= \sum_{n=0}^{\infty} \{F_X(2n\pi + \frac{5}{6}\pi) - F_X(2n\pi + \frac{1}{6}\pi)\}$$

$$= \sum_{n=0}^{\infty} [\exp\{-\lambda(2n\pi + \frac{1}{6}\pi)\} - \exp\{-\lambda(2n\pi + \frac{5}{6}\pi)\}]$$

$$= \frac{\exp(-\frac{1}{6}\lambda\pi) - \exp(-\frac{5}{6}\lambda\pi)}{1 - \exp(-2\lambda\pi)}.$$

(b) $n!\lambda^{-n}$.

5. Trivial if $EX^2 = \infty$. If $EX^2 < \infty$ then $(X - EX)^2 \geq 0$, so by the dominance inequality, example 5.7.1, $E\{(X - EX)^2\} = EX^2 - (EX)^2 \geq 0$.

6. (a) $c = 6$, $F(x) = 3x^2 - 2x^3$.   (b) It cannot.
   (c) $\exp(-x^2 + 4x) = \exp[-\frac{1}{2}\{\sqrt{2}(x - 2)\}^2 + 4]$; compare with $N(2, \frac{1}{2})$ to give $c = e^{-4}\pi^{-1/2}$, $F(x) = \Phi\{\sqrt{2}(x - 2)\}$.
   (d) $c = 1$, $F(x) = e^x/(1 + e^x)$.

7. (a) $f(x) = 2xe^{-x^2}$;   (b) $f(x) = x^{-2}e^{-1/x}$;
   (c) $f(x) = 2/(e^x + e^{-x})^2$;   (d) It cannot (consider $F'(1)$).

8. $P(X = 2n + 1 | X \text{ odd}) = \dfrac{\lambda^{2n+1}e^{-\lambda}}{(2n + 1)!}\dfrac{2}{1 - e^{-2\lambda}}$,

   $E(X | X \text{ odd}) = \lambda\dfrac{1 + e^{-2\lambda}}{1 - e^{-2\lambda}} = \lambda \coth \lambda$.

9. $\int_{-a}^{a} xe^x \, dx = 2a \cosh a - 2 \sinh a$;
   $\int_{-a}^{a} x^2 e^x \, dx = 2a^2 \sinh a - 4a \cosh a + 4 \sinh a$.

10. If the board has radius a, $f_Y(y) = \dfrac{2}{\pi a^2}(a^2 - y^2)^{1/2}$,

    $F_Y(y) = \dfrac{1}{2} + \dfrac{1}{\pi} \sin^{-1}\dfrac{y}{a} + \dfrac{y}{\pi a^2}(a^2 - y^2)^{1/2}$, $|y| \leq a$;

    $f_R(r) = \dfrac{2r}{a^2}$, $F_R(r) = \dfrac{r^2}{a^2}$, $0 \leq r \leq a$; $ER = \frac{2}{3}a$.

11. $p_X(x) = \frac{1}{2} \times \frac{2}{3} \times \frac{3}{4} \times \cdots \times \left(\dfrac{x - 1}{x}\right)\left(\dfrac{1}{x + 1}\right)$, $1 \leq x \leq 9$,

    $p_X(10) = \frac{1}{10}$; $X = 1 + \frac{1}{2} + \cdots + \frac{1}{9} + \frac{1}{10}$.

    $p_Y(y) = \{y(y + 1)\}^{-1}$, $y \geq 1$; $EY = \infty$.

12. Let C have radius a. Now $\Theta = \widehat{BOP}$ is uniform on $(0, 2\pi)$, where O is the centre of the circle.

    $X = 2a \sin \dfrac{\Theta}{2}$, hence

    $F_X(x) = P(2a \sin(\Theta/2) \leq x) = (2/\pi)\sin^{-1}(x/2a)$, $0 \leq x \leq 2a$,

    $f_X(x) = (2/\pi)(4a^2 - x^2)^{-1/2}$

    $EX = \dfrac{a}{\pi}\int_0^{2\pi} \sin \dfrac{\theta}{2} \, d\theta = \dfrac{2a}{\pi}$.

13. (a) $F(x) = (2/\pi) \sin^{-1}(x/V^2)$, $0 \leq x \leq V^2$, whence
         $f(x) = (2/\pi)(V^4 - x^2)^{-1/2}$.
    (b) $f(x) = \frac{1}{2}c\sqrt{x}(\sin 2\Theta)^{-3/2} \exp\{-x/(\sin 2\Theta)^2\}$, $c = 4/\sqrt{\pi}$.

14. $f(w) = \frac{1}{3}cw^{-2/3}f(cw^{1/3})$, $c = \{6/(\pi\rho g)\}^{1/3}$.

15. $EX = 2$; $EY = 3$.

16. $F_Y(y) = P(\tan \Theta \leq x) = (1/\pi)\tan^{-1}x$.

17. $a = (\sqrt{2\pi}\sigma^3)^{-1}$; $f(x) = \left(\dfrac{x}{\pi m^3 \sigma^3}\right)^{1/2} \exp\left(-\dfrac{x}{m\sigma^2}\right)$.

18. First note that $x\{1 - F(x)\} \leq \int_x^\infty y\{1 - F(y)\} \, dy \to 0$ as $x \to \infty$, since $EX < \infty$. Now write $EX = \int_0^a xf(x) \, dx = -x[1 - F(x)]_0^a + \int_0^a \{1 - F(x)\} \, dx$; let $a \to \infty$.

19. Use conditioning on the first roll.
    $EX_n = p(1 + E(X_n | 1 \text{ shown})) + q(1 + E(X_n | 2 \text{ shown})) + r(1 + E(X_n | 3 \text{ shown}))$, whence
    $EX_n - pEX_{n-1} - qEX_{n-2} - rEX_{n-3} = 1$.

21. (a) $c = a$, $EX = a\Gamma\left(\dfrac{a+1}{a}\right)$, where $\Gamma(a)$ is the gamma function

$$\Gamma(a) = \int_0^\infty x^{a-1}e^{-x}\,dx; \ a > 0,$$

$$= (a-1)! \text{ for integer } a.$$

(b) $P(X > s + t\,|\,X > t)/P(X > s) = a^{-1}\exp\{s^a + t^a - (s + t)^a\}$.
If $a > 1$ then $s^a + t^a - (s + t)^a < 0$, as required.
If $a < 1$ then $s^a + t^a - (s + t)^a > 0$, and

$$P(X > s + t\,|\,X > t) > P(X > s).$$

22. Let $B(a, b) = \int_0^1 x^{a-1}(1 - x)^{b-1}\,dx$. Integrate by parts to find $(a + b - 1)B(a, b) = (b - 1)B(a, b - 1) = (a - 1)B(a - 1, b)$ and iterate to get $B(a, b) = \{(a - 1)!(b - 1)!\}/(a + b - 1)!$. Then use $EX = B(a + 1, b)$, $EX^2 = B(a + 2, b)$.

24. After r readings a character is erroneous with probability $p(1 - \delta)^r$; there are fn characters. So the number of errors X is binomial $B(fn, p(1 - \delta)^r)$. $P(X = 0) = \{1 - p(1 - \delta)^r\}^{fn}$, which exceeds $\frac{1}{2}$ for the given values if $(1 - 2^{-8-r})^{2^{17}} > \frac{1}{2}$.

25. As in the solution for problem 18 above, note that $x^2\{1 - F(x)\} \leqslant \int_x^\infty y^2\{1 - F(y)\}\,dy \to 0$ as $x \to \infty$, since $EX^2 < \infty$. Now write

$$EX^2 = \int_0^a x^2 f(x)\,dx = -x^2[1 - F(x)]_0^a + \int_0^a \{1 - F(x)\}2x\,dx,$$

and let $a \to \infty$.

Section 6.2

1. $p(1, 6) \to 1$, and $p(x, y) \to 0$ otherwise.

2. (a) $\dfrac{n - 1}{2n}$, (b) $\dfrac{1}{n}$.

3. If the parties are P, Q, and R, then if voters' preferences are distributed like this, it follows that $\frac{2}{3}$ of them prefer P to Q, $\frac{2}{3}$ prefer Q to R, and $\frac{2}{3}$ prefer R to P. So whoever is elected, $\frac{2}{3}$ of the voters preferred some other party.

Section 6.3

1. $F(x, y) = \begin{cases} 1 - e^{-x} - xe^{-y}, & 0 \leqslant x \leqslant y < \infty \\ 1 - e^{-y} - ye^{-y}, & 0 \leqslant y \leqslant x < \infty; \end{cases}$

$F_X(x) = 1 - e^{-x}$, $F_Y(y) = 1 - e^{-y} - ye^{-y}$

$f_X(x) = e^{-x}$, $f_Y(y) = ye^{-y}$.

2. X and Y are not jointly continuous, so there is no contradiction. (Their joint distribution is said to be *singular*.)

3. $\dfrac{\partial^2 F}{\partial x \partial y} < 0$, so this cannot be a joint distribution.

Section 6.4

1. $\int_0^1\int_0^1\int_0^{xy} dz\,dx\,dy = \frac{1}{4}$.

2. There is a unique $\theta \in (0,1)$ such that $\theta^2 + \theta - 1 = 0$.
$$p_X(x) = \theta^{2x+1}; \quad p_Y(y) = (1-\theta^y)\theta^y, \; y \geq 1.$$

3. $F(u,v) = P(U < u < V < v) + F(u,u) = 2u(v-u) + F(u,u)$. Now differentiate.

4. Follow example 6.4.7.

Section 6.5

1. $P(W \leq w)$ is the volume of the pyramid $x \geq 0$, $y \geq 0$, $z \geq 0$, $x+y+z \leq w$. The volume of such a pyramid is $\frac{1}{3} \times$ area of base $\times$ height $= \frac{1}{3} \times \frac{1}{2}w^2 \times w = \frac{1}{6}w^3$. Hence, differentiating, $f_W(w) = \frac{1}{2}w^2$, $0 \leq w \leq 1$. Consideration of other pyramids in the cube $0 \leq x, y, z \leq 1$, yields the rest of $f_W(w)$.

2. From the solution to example 6.5.2(ii), either by symmetry or by using similar arguments
$$p_Y(y) = \binom{49-y}{5} \Big/ \binom{49}{6}, \quad 1 \leq y \leq 44.$$

Section 6.6

1. $P(A_n) = P(\text{at least } n \text{ sixes})$
$$= 1 - P(\text{less than } n \text{ sixes in } n \text{ rolls})$$
$$= 1 - \sum_{r=0}^{n-1} \binom{6n}{r}\left(\frac{5}{6}\right)^{6n-r}\left(\frac{1}{6}\right)^r \geq P(A_{n+1}), \text{ as we showed in (11). The inequality follows.}$$
Alternatively, if you have a computer big enough for symbolic algebra, it will rewrite the expression for $P(A_n)$ in a form which is monotone in n.

2. $f_Z(z) = e^{-z/2} - e^{-z}$.

3. The number of flips of an unfair coin until the nth head has the negative binomial distribution. The waiting times between heads are geometric. (Or use induction.)

4. Recall from section 4.8 that the number of meteorites up to time t is Poisson, and the gaps between meteorites are exponential. Or verify the induction, using
$$\int_0^z f(z-x)\lambda e^{-\lambda x}\,dx = \int_0^z f(x)\lambda e^{-\lambda(z-x)}\,dx$$
$$= \int_0^z \frac{\lambda^n x^{n-1} e^{-\lambda x}}{(n-1)!}\lambda e^{-\lambda(z-x)}\,dx = \frac{\lambda^{n+1}e^{-\lambda z}}{(n-1)!}\int_0^z x^{n-1}\,dx.$$

Section 6.7

1. Let I_r be the indicator of the event that your n coupons include the rth type. Find $E\sum_r I_r$.

2. Show that $\prod_r(1-I_r) \geq 1 - \sum_r I_r$. (Induction is easy.)

3. $E(S_r/S_k) = r/k$, for $r \leq k$. The cobalt balls divide the dun balls into $c+1$ groups, with the same expectation. Hence $EX = d/(c+1)$, since the sum of the groups is d.

Section 6.8

1. (a) $E\{(X+Y)(X-Y)\} = EX^2 - EY^2 = 0$.
   (b) $E\{X/(X+Y)\} = E\{Y/(X+Y)\} = \frac{1}{2}$, if X and Y have the same distribution.

2. $E(XY) = 0$ by symmetry, so $\text{cov}(X, Y) = 0$, but
$$P(X = 0)P(Y = 0) = \frac{(2n+1)^2}{\{2n(n+1)+1\}^2} \neq \frac{1}{2n(n+1)+1}$$
$$= P(X = Y = 0).$$

3. Use the fact that $\text{skw } Y = (p - q)/\sqrt{pq}$ if Y is Bernoulli.

4. $0 \leqslant E\{(sX - tY)^2\} = s^2EX^2 - 2stEXEY + t^2EY^2$. This is a quadratic with at most one real root, and the inequality is the condition for this to hold. For the last part consider $\{s(X - EX) - t(Y - EY)\}^2$.

5. $E\overline{X} = \mu$, $\text{var } \overline{X} = \sigma^2/n$, and so
$$\text{cov}(\overline{X}, X_r - \overline{X}) = E(X_r\overline{X}) - E(\overline{X}^2) - E\overline{X}(EX_r - E\overline{X})$$
$$= n^{-1}(n-1)\mu^2 + n^{-1}E(X_r^2) - n^{-1}\sigma^2 - \mu^2 = 0.$$

Section 6.9

1. Discrete case: $p(x|y) = p_X(x) \Leftrightarrow p(x, y) = p_X(x)p_Y(y)$. By the above, when X and Y are independent $E(X|Y) = EX$.

2. $p(v|u) = \begin{cases} 2(13 - 2u)^{-1}, & u < v \\ (13 - 2u)^{-1}, & u = v. \end{cases}$
   $E(V|U = u) = \sum_{v=u}^{6} vp(v|u) = \frac{42 - u^2}{13 - 2u}$
   $E(UV) = \frac{49}{4}$; $\text{cov}(U, V) = \frac{49}{4} - \frac{91}{36} \times \frac{161}{36} = \left(\frac{5}{6}\right)^2\left(\frac{7}{6}\right)^2$.

3. $E\left(\sum_1^N X_r\right)^2 = E\left(\sum_1^N X_r^2 + 2\sum_{r<s} X_r X_s\right)$
$$= ENE(X_1^2) + \{E(N^2) - EN\}(EX_1)^2.$$
   Now subtract $(E\sum_1^N X_r)^2$.

Section 6.10

1. (a) $f(u, v) = 2f(u)f(v)$, $0 \leqslant u < v$; $f_V(v) = \int_0^v 2f(u) \, duf(v)$;
   $f(u|v) = f(u, v)/f_V(v) = f(u)/\int_0^v f(u) \, du$.
   $E(U|V = v) = \int_0^v uf(u) \, du/\int_0^v f(u) \, du$.
   (b) From above.

2. $Z = X + Y$ has density $\lambda^2 z e^{-\lambda z}$, and X and Z have joint density $\lambda^2 e^{-\lambda z}$, $0 \leqslant x < z < \infty$. This follows either from example 6.10.2 or directly from
$$f(x, z) = \frac{\partial^2}{\partial x \partial z}P(X \leqslant x, Z \leqslant z) = \frac{\partial^2}{\partial x \partial z}\int_0^x \int_0^{z-x} \lambda^2 e^{-\lambda y} \, dy \, e^{-\lambda x} \, dx,$$
hence $f(x|z) = f(x, z)/f_Z(z) = z^{-1}$, $0 \leqslant x \leqslant z$, which is uniform.

3. (a) Given ρ, $P(X = k|R = \rho) = \binom{n}{k}\rho^k(1 - \rho)^{n-k}$. Hence

$$P(X = k) = \int_0^1 P(X = k|R = \rho)\,d\rho.$$

Now recall the beta distribution (see problem 22 in section 5.12) to find that

$$\int_0^1 \rho^k(1 - \rho)^{n-k}\,d\rho = k!(n - k)!/(n + 1)!$$

(b) $\dfrac{P(X = k|R = \rho)f_R(\rho)}{P(X = k)} = (n+1)\binom{n}{k}\rho^k(1 - \rho)^{n-k}$, which is a beta density,

$\beta(k + 1,\ n + 1 - k)$.

Section 6.11

1. $\operatorname{var}(Y|X) = E(Y^2|X) - \{\psi(X)\}^2$, and $\operatorname{var}\{\psi(X)\} = E\{\psi(X)^2\} - (EY)^2$.

2. We need to use the fact that

$$E[\{X - E(X|Y)\}\{E(X|Y) - g(Y)\}] = E([E\{X - E(X|Y)|Y\}]\{E(X|Y) - g(Y)\}) = 0$$

So $E(X - g)^2 = E(X - \psi + \psi - g)^2 = E(X - \psi)^2 + E(g - \psi)^2 + 2E\{(X - \psi)(g - \psi)\}$
$= E(X - \psi)^2 + E(g - \psi)^2 \geqslant E(X - \psi)^2$.

3. $p_{X|Y}(0|0) = \dfrac{a}{a + c}$, $\quad p_{X|Y}(1|0) = \dfrac{c}{a + c}$,

$p_{X|Y}(0|1) = \dfrac{b}{b + d}$, $\quad p_{X|Y}(1|1) = \dfrac{d}{b + d}$.

So $E(X|Y = 0) = \dfrac{c}{a + c}$, $\quad E(X|Y = 1) = \dfrac{d}{b + d}$, and

$E(X|Y) = \dfrac{c(1 - Y)}{a + c} + \dfrac{dY}{b + d}$, as required.

Section 6.12

1. By definition $V = v + (\tau\rho/\sigma)(U - \mu) + \tau(1 - \rho^2)^{1/2}Y$. Hence the conditional density of V, given $U = u$, is normal with mean $E(V|U) = v + (\tau\rho/\sigma)(U - \mu)$ and variance $\operatorname{var}(V|U) = \tau^2(1 - \rho^2)$.

2. (a) Using (2) and (3) with $\sigma = \tau = 1$,

$$P(U > 0,\ V > 0) = P(X > 0,\ \rho X + (1 - \rho^2)^{1/2}Y > 0).$$

In polar coordinates the region $(x > 0,\ \rho x + (1 - \rho^2)^{1/2}y > 0)$ is the region

$$\left(r > 0,\ -\frac{\rho}{(1 - \rho^2)^{1/2}} < \tan\theta < \infty\right) \equiv (r > 0,\ -\rho < \sin\theta < 1).$$

Hence

$$P(U > 0,\ V > 0) = \int_0^\infty re^{-r^2/2}\,dr\int_{-(\sin^{-1}\rho)}^{\pi/2}\frac{1}{2\pi}\,d\theta$$

$$= \frac{1}{2\pi}\left(\frac{\pi}{2} + \sin^{-1}\rho\right)$$

(b) $P(0 < U < V) = P(0 < V < U) = \frac{1}{2}P(0 < U,\ 0 < V)$, by symmetry.

(c) $\max(U, V) = \max\{X,\ \rho X + (1 - \rho^2)^{1/2}Y\}$. The line

$$y = \frac{1-\rho}{(1-\rho^2)^{1/2}} x$$

divides the plane into two regions above the line

$$x < \rho x + (1-\rho^2)^{1/2} y.$$

Below the line the inequality is reversed. In polars this line is given by

$$\tan \psi = \left(\frac{1-\rho}{1+\rho}\right)^{1/2}.$$

Note that

$$\sin \psi = \left(\frac{1-\rho}{2}\right)^{1/2}, \quad \cos \psi = \left(\frac{1+\rho}{2}\right)^{1/2}.$$

Hence

$$E\{\max(U, V)\} = \int_0^\infty \frac{re^{-r^2/2}}{2\pi} \left[\int_\psi^{\pi+\psi} \{\rho r \cos \theta + (1-\rho^2) r \sin \theta\} d\theta + \int_{\psi-\pi}^\psi r \cos \theta d\theta\right] dr$$

$$= \sqrt{\frac{\pi}{2}} \times \frac{1}{2\pi} [\rho\{\sin(\pi + \psi) - \sin \psi\} - (1-\rho^2)\{\cos(\pi + \psi) - \cos \psi\}$$

$$+ \{\sin \psi - \sin(\psi - \pi)\}]$$

$$= \frac{1}{\sqrt{2\pi}} \{\sqrt{2}(1-\rho)^{1/2}\}.$$

3. $\theta = \frac{1}{2} \cot^{-1}\left(\frac{\sigma^2 - \tau^2}{2\rho\sigma\tau}\right)$. To see this, recall that U and V are independent if and only if they are uncorrelated, which is to say that $0 = E(UV) = (EY^2 - EX^2)\frac{1}{2} \sin 2\theta + EXY \cos 2\theta$.

Section 6.13

1. (a) The inverse is $x = u$, $y = v/u$, so $|J| = \begin{vmatrix} 1 & 0 \\ -v/u^2 & 1/u \end{vmatrix} = |u|^{-1}$.

    The density of V is the marginal $\int_{-\infty}^\infty \frac{1}{|u|} f\left(u, \frac{v}{u}\right) du$.

    (b) In this case $|J| = |z|$.

    (c) Use (b), or use the circular symmetry and problem 16 of section 5.12.

    (d) Using (a) we have

    $$f_V(v) = \int f(u, v) du = \int \frac{1}{|u|} f_X(u) f_Y\left(\frac{v}{u}\right) du$$

    $$= \int_z^\infty \pi^{-1} e^{-u^2/2} (u^2 - v^2)^{-1/2} u \, du = \int_0^\infty \pi^{-1} e^{-v^2/2} e^{-y^2/2} dy$$

2. $u = uv$, $y = u - uv$, $|J| = \begin{vmatrix} v & u \\ 1-v & -u \end{vmatrix} = |u|$.

    $$f(u, v) = f_X(uv) f_Y(u - uv)|u|$$

    $$= u \frac{\lambda^n}{(n-1)!} (uv)^{n-1} e^{-\lambda uv} \frac{\lambda^m}{(m-1)!} \{u(1-v)\}^{m-1} e^{-\lambda(u-uv)}$$

    $$= \frac{\lambda^{n+m}}{(n-1)!(m-1)!} (1-v)^{m-1} v^{n-1} u^{m+n-1} e^{-\lambda u}, \quad 0 \le v \le, u \ge 0.$$

    As this factorizes, U and V are independent.

3. Condition on the exact number of $X_{(i)}$ less than x.

4. Use exercise 3, and differentiate to get the density. Now remember the beta density (find it in section 5.10).

5. Set $W = X + Y$. Then $|J| = |w|$, and

$$f_Z(z) = \int f(w, z)\, dw = \int_0^\infty \lambda\mu e^{-\lambda wz} e^{-\mu w + \mu wz} w\, dw$$

$$= \frac{\lambda\mu}{\{\lambda z + \mu(1 - z)\}^2}, \quad 0 \leqslant z \leqslant 1.$$

Section 6.15

1. (a) Expected number to get one head is 2.
   (b) Exploit the symmetry of the binomial $B\left(2n - 1, \frac{1}{2}\right)$ distribution.

2. $X = 1 + S + T$, where S is geometric with parameter $\frac{2}{3}$, and T is geometric (and independent of S) with parameter $\frac{1}{3}$. By the convolution rule,

$$p_X(x) = P(S + T = x - 1) = \sum_{r=1}^{x-2} P(S = r)P(T = x - r - 1)$$

$$= \sum_{r=1}^{x-2} \left(\frac{1}{3}\right)^{r-1} \times \frac{2}{3} \times \left(\frac{2}{3}\right)^{x-r-2} \times \frac{1}{3} = \left(\frac{2}{3}\right)^{x-1}\left\{\frac{1}{2} - \left(\frac{1}{2}\right)^{x-1}\right\}.$$

3. For any numbers x and y, by inspection we have

$$\min\{x, y\} + \frac{1}{2}|x - y| = \frac{1}{2}(x + y).$$

Hence $E \min\{X, Y\} + \frac{1}{2}E|X - Y| = \frac{1}{2}EX + \frac{1}{2}EY$. But $\min\{X, Y\} = 1$, $EX = p^{-1}$; $EY = q^{-1}$. Hence

$$E|X - Y| = \frac{1}{p} + \frac{1}{q} - 2 = \frac{1}{pq} - 2 = \frac{p}{q} + \frac{q}{p}.$$

4. $f(x, y) = x^{-1}, 0 \leqslant y \leqslant x \leqslant 1$.
   (a) $f_Y(y) = -\log y, 0 < y \leqslant 1$.
   (b) $f_{X|Y}(x|y) = -x^{-1}/\log y$.
   (c) $E(X|Y) = -(1 - Y)/\log Y$.

5. $\mathrm{cov}(X, Y) = \frac{27}{8} - \left(\frac{33}{16}\right)^2$.

6. $p(x, y) = \dfrac{n!}{(n - x)!(x - y)!\, y!} \left(\frac{5}{6}\right)^{n-x}\left(\frac{5}{36}\right)^{x-y}\left(\frac{1}{36}\right)^y.$

   From this, or by direct argument, $p(x|y) = \dbinom{n - y}{x - y}\left(\frac{30}{35}\right)^{n-x}\left(\frac{5}{35}\right)^{x-y}$, which is to say that, given Y, $X - Y$ is binomial $B\left(n - Y, \frac{5}{35}\right)$. Hence $E(X - Y|Y) = (n - Y)\frac{5}{35}$ and $\mathrm{var}(X - Y|Y) = (n - Y)\frac{30}{35} \times \frac{5}{35}$. Therefore $E(X|Y) = \frac{1}{7}(n + 6Y)$, and $\mathrm{var}(X|Y) = \frac{6}{49}(n - Y)$.

7. This is essentially the same as problem 6:

$$P(\text{faulty}|\text{not detected}) = \frac{P(\text{faulty} \cap \text{not detected})}{P(\text{not detected as faulty})} = \frac{\phi(1 - \delta)}{1 - \phi\delta}.$$

   Hence, given Y, X is then binomial $B\left\{b - Y, \dfrac{\phi(1 - \delta)}{1 - \phi\delta}\right\}$.

8. $P(\text{meet}) = \frac{11}{36}$; $P(\text{meet}|\text{after } 12.30) = \frac{5}{9}$.

9. $p(a, b) = \frac{1}{6}$, etc.; $E(XY) = \frac{1}{3}(ab + bc + ca)$, $EX = \frac{1}{3}(a + b + c)$.

10. $p(u, y) = \begin{cases} \frac{1}{36}, & 1 \leqslant u < y \leqslant 6 \\ \frac{1}{36}(6 - u + 1), & u = y, \end{cases}$

    $\mathrm{cov}(U, X) = -\frac{35}{24}.$

11. $p(x, y) = p^{x+1}q^y + q^{x+1}p^y; \quad x, y \geqslant 1.$

$p_X(x) = qp^x + pq^x; \quad p_Y(y) = p^2 q^{y-1} + q^2 p^{y-1}.$

$E(XY) = \dfrac{1}{q} + \dfrac{1}{p}; \quad EX = \dfrac{q}{p} + \dfrac{p}{q}; \quad EY = 2.$

12. $EZ = \frac{1}{2}$ by symmetry. Now if we transform to polar coordinates $R^2 = X^2 + Y^2$, $\Theta = \tan^{-1}(Y/X)$, then R and Θ have joint density $re^{-r^2/2}/(2\pi)$, which is to say that Θ is uniform on $(0, 2\pi)$. Thus

$$EZ^2 = E\left\{\left(\frac{X^2}{X^2 + Y^2}\right)^2\right\} = E(\cos^4 \Theta) = \frac{1}{2\pi}\int_0^{2\pi} \cos^4 \theta \, d\theta = \frac{3}{8}.$$

Hence var $Z = \frac{1}{8}$.

13. Let U, V, and W denote your net winnings from the coins with value 5, 10, or 20 respectively. Then, by considering cases,

$$P(U = 30) = \tfrac{1}{8}, \quad P(U = -5) = \tfrac{6}{8},$$
$$P(V = 25) = \tfrac{1}{4}, \quad P(V = -10) = \tfrac{5}{8},$$
$$P(W = 15) = \tfrac{1}{2}, \quad P(W = -20) = \tfrac{3}{8};$$

$P(U = 0) = P(V = 0) = P(W = 0) = \frac{1}{8}$. Hence $E(U) = E(V) = E(W) = 0$, so the expected net gain is nil for any nomination. However, $4 \operatorname{var} U = 525$, $4 \operatorname{var} V = 875$, and $4 \operatorname{var} W = 950$, so you will choose a if you are risk-averse, but c if you are risk-friendly.

14. (a) The random variables $-X_1, \ldots, -X_n$ are also independent and have the same joint distribution as $X_1, \ldots, X_n$. Thus $-\sum_1^n X_i$ has the same distribution as $\sum_1^n X_i$, as required.

(b) No. For example:

	1	$\frac{7}{36}$	0	$\frac{5}{36}$
Y	0	$\frac{2}{36}$	$\frac{4}{36}$	$\frac{6}{36}$
	-1	$\frac{3}{36}$	$\frac{8}{36}$	$\frac{1}{36}$
		-1	0	1
			X	

Here $p_Y(j) = p_X(j) = \frac{1}{3}$, for all j, and each is symmetric about zero, but

$$P(X + Y = -2) = \tfrac{3}{36} \neq \tfrac{5}{36} = P(X + Y = 2)$$

15. If S is the sector where $0 \leqslant \theta \leqslant \pi/4$,

$$E|X \wedge Y| = 8\iint_S \frac{y}{2\pi} \exp\{-\tfrac{1}{2}(x^2 + y^2)\} \, dx \, dy$$

$$= \int_0^\infty r^2 \exp(-\tfrac{1}{2}r^2) \, dr \int_0^{\pi/4} \frac{4}{\pi} \sin\theta \, d\theta$$

$$= 2(\sqrt{2} - 1)/\sqrt{\pi}.$$

17. The needle intersects a joint if $\left|\frac{1}{2}a\sin\Theta\right| > x$, so

$$P(\text{intersection}) = 2\int_0^\pi \int_0^{a\sin\theta/2} \frac{1}{\pi}\, dx\, d\theta = 2\int_0^\pi \frac{a}{2\pi}\sin\theta\, d\theta = \frac{2a}{\pi}.$$

18. P(last ball jet) $= j/(j+k)$; P(last ball khaki) $= k/(j+k)$. If the last is jet, then the remaining $j-1$ jet balls are divided into $k+1$ groups $S_1, \ldots, S_{k+1}$, where $ES_1 = \cdots = ES_{k+1}$, and $S_1 + \cdots + S_{k+1} = j-1$. Hence $ES_{k+1} = (j-1)/(k+1)$, and so

$$E(\text{number left}|\text{last is jet}) = \frac{j-1}{k+1} + 1 = \left(\frac{j+k}{k+1}\right).$$

Hence $E(\text{number left}) = \dfrac{j+k}{k+1}\left(\dfrac{j}{k+j}\right) + \dfrac{j+k}{j+1}\left(\dfrac{k}{k+j}\right) = \dfrac{j}{k+1} + \dfrac{k}{j+1}.$

19. $\text{var}(Y+Z) = \text{var }Y + \text{var }Z$ when uncorrelated, so

$$a^2 = \frac{\text{var }X}{\text{var }X + \text{var }Y}, \qquad b^2 = \frac{\text{var }Y}{\text{var }X + \text{var }Y}$$

20. Since the X_i are continuous, $P(X_i = X_j) = 0$ for $i \neq j$. Then, by symmetry, all 4! possible orderings of X_1, X_2, X_3, X_4 are equally likely. So $P(X_1 > X_2 > X_3 > X_4) = \frac{1}{24}$. You expect one such year in any 24 years, so your concern would depend on just how many times this had happened before. Since $P(X_1 > X_2 < X_3 < X_4) = \frac{3}{16}$, you should not get too excited; you expect this around once in five years.

21. (a) $EU = \displaystyle\sum_{k=0}^n P(U > k) = \sum_{k=1}^n \left(\frac{n-k}{n}\right)^2 = \sum_{k=0}^n \frac{k^2}{n^2}.$

(b) $EU + EV = EX + EY = n+1.$

(c) $E(UV) = \displaystyle\sum_{v=1}^n v\left\{\frac{2}{n^2}\sum_{u=1}^{v-1}u + \frac{v}{n^2}\right\} = \frac{(n+1)^2}{4}.$

(d) $EU^2 = \displaystyle\sum_{k=0}^n k^2 P(U=k) = \frac{1}{n^2}\sum_{k=0}^n k^2\{2(n-k)+1\}$

$$= \frac{n+1}{6n}(n^2+n+1).$$

So $\text{var }U = \dfrac{(n^2-1)(2n^2+1)}{36n^2}.$

(e) $\text{var }V = \text{var }U$, by symmetry.

22. $f(u, v) = 2, \quad 0 \leqslant u < v \leqslant 1.$
$f_U(u) = 2(1-u); \ f_V(v) = v; \quad 0 \leqslant u, v \leqslant 1.$
$EU = \frac{1}{3}, \ EU^2 = \frac{1}{6}, \ \text{var }U = \text{var }V = \frac{1}{18}, \ E(UV) = \frac{1}{4}$, etc. As $n \to \infty$ the discrete uniform distribution converges to the continuous.

23. (a) $P(X = 10) = 2^{-10} + 2^{-10} = 2^{-9}; \quad P(X = 9) = 0.$

(b) Let I_j be the indicator of the event A_j that the jth coin shows the same as its neighbours; then

$$EX = E\textstyle\sum_{j=1}^{10} I_j = \sum_{j=1}^{10} P(A_j) = \sum_{j=1}^{10} \frac{1}{4} = \frac{5}{2}.$$

Now calculate $\text{var }I_j = \frac{3}{16}$, and

$$\text{cov}(I_j, I_k) = \begin{cases} \frac{1}{16}, & |j-k| = 1 \\ 0 & \text{otherwise.} \end{cases}$$

Then

$$\text{var }X = \textstyle\sum \text{var }I_j + \sum \text{cov}(I_j, I_k)$$

$$= \tfrac{30}{16} + \tfrac{20}{16}.$$

24. (a) $\dfrac{1}{\sigma y}\phi\left(\dfrac{\log y - \mu}{\sigma}\right)$, where ϕ is standard normal density.

$EY = Ee^X = e^{\mu + \sigma^2/2}$; $EY^2 = Ee^{2X} = e^{2\mu + 2\sigma^2}$

(b) $\log Z$ has mean $\mu - 2\nu$ and variance $\sigma^2 - 2\sigma\tau\rho + 4\tau^2$.

25. $1 = P\left(\bigcup_{k=0}^{n-1} A_k\right) \leqslant \sum_{k=1}^{n-1} P(A_k) = nP(A_0)$, because $P(A_j) = P(A_k)$ for all j, k, by symmetry. Early bidders try a low offer, later bidders know they have to do better. The model works if you are opening simultaneous sealed bids.

26. (a) Yes. Let A be the event where $X(\omega) \geqslant x$, and B the event where $Y(\omega) \geqslant x$. Since $X(\omega) \geqslant Y(\omega)$ with probability 1, it follows that $P(B) \leqslant P(A)$.

(b) Yes, because wherever $Y(\omega) \geqslant Z(\omega)$, $X(\omega) \geqslant Y(\omega) \geqslant Z(\omega)$.

27. Use $p(x, y) = \dfrac{c}{2(x + y - 1)(x + y)} - \dfrac{c}{2(x + y)(x + y + 1)}$ and successive cancellation to find $c = 2$, and $p_X(x) = \{x(x + 1)\}^{-1}$. Hence EX is not finite.

28. X is binomial $B(n, \gamma)$; Y is binomial $B\{n, \gamma(1 - \delta)\}$, X and Y are jointly trinomial; $E(XY|X) = X^2(1 - \delta)$, so $E(XY) = \{n^2\gamma^2 + n\gamma(1 - \gamma)\}(1 - \delta)$, and hence $\text{cov}(X, Y) = n\gamma(1 - \gamma)(1 - \delta)$.

29. (a) $\Phi(-2) + 1 - \Phi(2) = 0.046$; (b) after 22 tests; (c) after 12 tests.

30. (a) Remember $m_0 = m_n = 0$.

(b) Conditional on Y, by (a) we have $E(X|Y) = Y(n - Y)$, and $E(XY|Y) = Y^2(n - Y)$. Hence, after some algebra,

$$EX = nEY - EY^2 = n(n - 1)pq,$$

$$E(XY) = nEY^2 - EY^3 = n^2(n - 1)p^2 + n^2 p - n(n - 1)(n - 2)p^3 - np - 3n(n - 1)p^2.$$

So $\text{cov}(X, Y) = n^2 qp(q - p) - npq(q - p)$
$$= n(n - 1)pq(q - p) = (q - p)EX,$$

which is zero when $p = \frac{1}{2} = q$.

(c) $EX = EYEZ$; $\text{cov}(X, Y) = EZ \text{ var } Y$.

31. (a) $P(n(1 - M_n) > y) = P(M_n < 1 - y/n) = (1 - y/n)^n$

(b) Using Taylor's theorem,

$$P\left(M_n < 1 - \frac{y}{n}\right) = \left\{F_X\left(1 - \frac{y}{n}\right)\right\}^n \simeq \left\{1 - \frac{y}{n}f_X(1) + O\left(\frac{1}{n^2}\right)\right\}^n$$
$$\to \exp\{-f(1)y\}.$$

32. One possibility follows from the fact that $F_X(1 - x) = 1 - 6x^2 + o(x^2)$. Therefore

$$P(\sqrt{n}(1 - M_n) > y) = \left\{F_X\left(1 - \frac{y}{\sqrt{n}}\right)\right\}^n$$
$$= \left\{1 - \frac{6y^2}{n} + o(n^{-1})\right\}^n \to \exp(-6y^2).$$

33. Let I_j be the indicator of the event that the jth person does not share a birthdate. Then $P(I_j = 1) = \left(\frac{364}{365}\right)^{51}$. Thus $ES = E\sum_1^{52} I_j = 52\left(\frac{364}{365}\right)^{51} \simeq 45.4$. Likewise, if K_j is the indicator of the event that the jth person shares a birthdate with exactly one other, then

$$2ED = E\sum_1^{52} K_j = 52 \times \frac{51}{365}\left(\frac{364}{365}\right)^{50} \simeq 6.$$

Finally, since there are 52 individuals, $S + 2D + 3M \leqslant 52$, and taking expectations shows

$EM \leqslant 0.2$; now use $P(M > 0) \leqslant EM$, which is obvious (and in any case follows from Markov's inequality).

34. These are just special cases of the coupon collector's problem. Use your calculator.

38. $\{2\Phi(z/\sqrt{2}) - 1\}^2 = F_Z(z)$. Differentiate for density.
$$E(Z \mid X > 0, \ Y > 0) = E(X \mid X > 0, \ Y > 0) + E(Y \mid X > 0, \ Y > 0)$$
$$= 2E(X \mid X > 0).$$

43. Use indicators: I_k indicates a head on the kth flip; J_k indicates that the $(k-1)$th and kth flips are different. Then $X = \sum_1^n I_k$, $R = 1 + \sum_2^n J_k$, and so $E(XR) = E\{\sum_{k=1}^n I_k(1 + \sum_{k=2}^n J_k)\}$. Now calculate $EI_k = p$, $EJ_k = 2pq$, $E(I_k J_k) = qp$, $E(I_k J_{k+1}) = pq$, and so on.

Section 7.2

1. From (9), $G^{(2)}(1) = E\{X(X-1)\} = EX^2 - EX = \operatorname{var} X + (EX)^2 - EX$. Now use $G'(1) = EX$.

2. (a) $\sum_1^\infty q^{k-1} ps^k = ps \sum_1^\infty (ps)^{k-1}$;   (b) $\dfrac{p}{1 - qs}$.

3. (a) $\frac{1}{6} s\left(\dfrac{1 - s^7}{1 - s}\right) = \sum_{r=1}^6 \frac{1}{6} s^r$, so $p(r) = \frac{1}{6}$; this is a die.

    (b) $p(0) = q$, $p(1) = p$. This is an indicator, or a Bernoulli trial.

4. $G_Y(s) = Es^Y = Es^{aX+b} = s^b E\{(s^a)^X\} = s^b G_X(s^a)$.
$EY = G'_Y(1) = \left[bs^{b-1} G_X(s) + as^b G'_X(s^a)\right]_{s=1} = b + aEX$.

5. $2(1 - e^t + te^t)t^{-2}$.

6. $\lambda + (1 - \lambda)pe^t(1 - qe^t)^{-1}$;
$EX = (1 - \lambda)p^{-1}$, $\operatorname{var} X = (1 - \lambda)(\lambda + q)p^{-2}$.

Section 7.3

1. $Ee^{tY} = Ee^{t(a+bX)} = e^{at}M_X(bt)$. We know that if X is $N(0, 1)$, then $\mu + \sigma X$ is $N(\mu, \sigma^2)$. Hence $M_Z(t) = M_X(\sigma t)e^{\mu t} = e^{\mu t}\exp\{\frac{1}{2}(\sigma t)^2\}$.

2. $M_X(t) = \dfrac{e^{at} - 1}{at} = \displaystyle\sum_{r=0}^\infty \dfrac{(at)^r}{(r+1)!} = \sum_{r=0}^\infty \dfrac{EX^r}{r!} t^r$.

3. $M_X(t) = \displaystyle\int_0^\infty \{e^{tx - \lambda x} x^{r-1} \lambda^r / (r-1)!\}\, dx$. Now set
$$y = (\lambda - t)x/\lambda.$$

4. (a) $G'(1) = [np(q + ps)^{n-1}]_{s=1} = np$, $G^{(2)}(1) = n^2 p^2 - np^2$.

    (b) $G'(s) = \dfrac{p}{1 - qs} + \dfrac{pqs}{(1 - qs)^2}$, $G^{(2)}(s) = \dfrac{2pq}{(1 - qs)^3}$,   so
$$\operatorname{var} X = \frac{2q}{p^2} + \frac{1}{p} - \frac{1}{p^2} = \frac{q}{p^2}.$$

    (c) $M'(t) = r\lambda^r(\lambda - t)^{-r-1}$, $M^{(2)}(t) = r(r+1)\lambda^r(\lambda - t)^{-r-2}$,   so
$$\operatorname{var} X = \frac{r(r+1)}{\lambda^2} - \left(\frac{r}{\lambda}\right)^2.$$

Section 7.4

1. $\exp\{(\lambda + \mu)(s - 1)\}$.

2. $(q_n + p_n s)^n = \{1 + (s - 1)(\lambda/n)\}^n \to \exp\{\lambda(s - 1)\}$.

3. $\frac{1}{3}(s + s^3 + s^5)\frac{1}{2}(1 + s) = \frac{1}{6}(s + s^2 + s^3 + s^4 + s^5 + s^6)$, so the distribution is the same as that of the sum of two conventional fair dice, namely triangular.

Section 7.5

1. $\mathrm{E}s^X = -\sum \dfrac{(sp)^k}{k\log(1 - p)} = \dfrac{\log(1 - sp)}{\log(1 - p)}$,   so

$$\mathrm{E}s^T = G_N(G_X(s)) = e^{-\lambda}\exp\left\{\frac{\lambda}{\log(1 - p)}\log(1 - sp)\right\}$$

$$= \left(\frac{1 - p}{1 - ps}\right)^{-\lambda/\log(1-p)}.$$

This is a negative binomial p.g.f., since $(1 - x)^{-\nu}$ is expanded in series by the negative binomial theorem.

Section 7.6

1. $n\mu = 2000$, $\sqrt{n}\sigma = 100/\sqrt{6}$. So

$$P(1900 < X < 2200) = P\left(\frac{1900 - 2000}{100/\sqrt{6}} < \frac{X - 2000}{100/\sqrt{6}} < \frac{2200 - 2000}{100/\sqrt{6}}\right).$$

Now use the central limit theorem.

2. $n\mu = 0$; $\sqrt{n}\sigma = 3$. Hence
$P(-3 < \text{error} < 3) \simeq \Phi(1) - \Phi(-1)$, $P(-6 < \text{error} < 6) \simeq \Phi(2) - \Phi(-2)$.
So (a) $2\{(1 - \Phi(1)\} \simeq 0.32$, (b) $2\{(1 - \Phi(2)\} \simeq 0.04$.

3. If λ is an integer then X has the same distribution as the sum of λ independent Poisson random variables, each with parameter 1. The central limit theorem applies to this sum. Y has the same distribution as the sum of r independent exponential random variables, each with parameter 1.

Section 7.7

1. (a) $[\mathrm{E}s^{T_1}]_{s=1} = \dfrac{1 - (1 - 4pq)^{1/2}}{2q} = \dfrac{1 - (1 - 2q)}{2q} = 1$.

   We have taken as the positive root $(1 - 4pq)^{1/2} = 1 - 2q$, because $q < \frac{1}{2}$. Differentiating yields $\mathrm{E}T_1$;   otherwise,   you   can   write $\mathrm{E}T_1 = \mathrm{E}\{\mathrm{E}(T_1|X_1)\} = p + q\{\mathrm{E}(T_1 \mid X_1 = -1) + 1\}$
   $= p + q\{2\,\mathrm{E}T_1 + 1\}$, so $\mathrm{E}T_1 = (1 - 2q)^{-1} = (p - q)^{-1}$.
   (b) $[\mathrm{E}s^{T_1}]_{s=1} = 1$, but the derivative at $s = 1$ is infinite.
   (c) $\mathrm{E}s^{X_1} = ps + qs^{-1}$, so $\mathrm{E}s^{S_n} = (ps + qs^{-1})^n$.
   (d) $G_Y(ps + qs^{-1})$.

2. From (14), with $n = 3$, $P(N \text{ is even}) = \frac{1}{2}\{G_N(1) + G_N(-1)\} = \frac{1}{2}\{1 - \frac{1}{5}\}$.

3. For equal chances, they need p such that $G_N(-1) = 0$, which is impossible.

4. $\mathrm{E}(s^{X_n}t^{Y_n}u^{Z_n}v^{n-X_n-Y_n-Z_n}) = \{ps + qt + ru + (1 - p - q - r)v\}^n$.

Section 7.10

1. Like example 5, except that $\mathrm{E}s^T = \dfrac{\frac{2}{3}s}{1-\frac{1}{3}s} \times \dfrac{\frac{1}{3}s}{1-\frac{2}{3}s}$. Now use partial fractions.

2. $\mathrm{E}s^X = \dfrac{\alpha s + (1-\alpha)\beta s^2 + (1-\alpha)(1-\beta)\gamma s^3}{1-(1-\alpha)(1-\beta)(1-\gamma)s^3}$,

    and we need the sum of the coefficients of powers of s^3, which is
    $$\frac{\gamma(1-\alpha)(1-\beta)}{1-(1-\alpha)(1-\beta)(1-\gamma)}.$$

3. $\mathrm{E}s^{X_n - n} = s^{-n}\mathrm{E}s^{X_n} = \dfrac{p^n}{(1-qs)^n} = \dfrac{(1-\lambda/n)^n}{(1-\lambda s/n)^n} \to e^{-\lambda + \lambda s}$

4. (a) $\kappa(t) = \lambda(e^t - 1)$,   so $\kappa_r = \lambda$.
    (b) $\kappa(t) = \mu t + \frac{1}{2}\sigma^2 t^2$,   so $\kappa_1 = \mu$, $\kappa_2 = \sigma^2$, $\kappa_r = 0, r \geqslant 3$.
    (c) $\kappa(t) = -\log(1 - t/\lambda)$;   so $\kappa_r = (r-1)!\lambda^{-r}$.

5. P(attempt all questions)
    $$= \mathrm{P}\left(\sum_1^{116} X_r \leqslant 60\right) = \left\{\left(\sum_1^{116} X_r - 58\right) \middle/ \left(\frac{116}{20}\right)^{1/2} \leqslant 2 \middle/ \left(\frac{116}{20}\right)^{1/2}\right\}$$
    $$\simeq \Phi(0.83) \simeq 0.8.$$

6. Central limit theorem again.

7. $\left\{p\left(1 + \dfrac{qt}{(npq)^{1/2}} + \dfrac{q^2 t^2}{2npq} + \cdots\right) + q\left(1 - \dfrac{pt}{(npq)^{1/2}} + \dfrac{p^2 t^2}{2npq} + \cdots\right)\right\}^n$
    $= (1 + \frac{1}{2}n^{-1}t^2 + \cdots)^n \to e^{t^2/2}.$

8. $\exp\left(\frac{1}{2}\sigma^2 t^2\right) = 1 + \frac{1}{2}\sigma^2 t^2 + \cdots + \left(\frac{1}{2}\sigma^2 t^2\right)^k/k! + \cdots$

9. (a) $\mathrm{E}e^{t(X^2+Y^2)} = (\mathrm{E}e^{tX^2})^2 = (1-2t)^{-1}$,   by example 7.3.
    (b) $\mathrm{E}\{\mathrm{E}(e^{tXY}|Y)\} = \mathrm{E}(e^{1/2 t^2 Y^2}) = \dfrac{1}{\sqrt{2\pi}}\displaystyle\int e^{t^2 y^2/2 - y^2/2}\,dy = \dfrac{1}{(1-t^2)^{1/2}}$

10. $\displaystyle\lim_{s\uparrow 1}\frac{1-G(s)}{1-s} = G'(1)$, by l'Hôpital's rule.

11. With an obvious notation, $T_n = 1 + \sum_{r=0}^{Z_1} T_{n-1}^{(r)}$.

12. $\mathrm{E}\{\mathrm{E}(Z_m Z_n | Z_m)\} = \mathrm{E}(Z_m^2 \mu^{n-m}) = \mu^n\,\mathrm{var}\,Z_m + \mu^{n+m}$. Hence $\mathrm{cov}(Z_m, Z_n) = \mu^{n-m}\,\mathrm{var}\,Z_m$,
    and
    $$\rho(Z_m, Z_n) = \mu^{n-m}\left(\frac{\mathrm{var}\,Z_m}{\mathrm{var}\,Z_n}\right)^{1/2} = \begin{cases} \left(\dfrac{\mu^n}{\mu^m} \times \dfrac{\mu^m - 1}{\mu^n - 1}\right)^{1/2}, & \mu \neq 1 \\ \left(\dfrac{m}{n}\right)^{1/2}, & \mu = 1. \end{cases}$$

13. $\mathrm{P}(\eta)$, where η is the extinction probability.

14. If $Z = X + Y$, then $L_Z(t) = \mathrm{E}\{1 - t(X+Y)\}^{-1}$, which is not useful. And $L(t)$ often fails to exist, even when $M(t)$ exists; for example, if X is exponential with parameter 1, $\mu_r = r!$, $L(t) = \sum t^r r!$.

15. With an obvious notation

$$G = Es^X = \tfrac{1}{2}E(s^X|T) + \tfrac{1}{4}E(s^X|HT) + \tfrac{1}{4}E(s^X|HH)$$

$$= \tfrac{1}{2}sG + \tfrac{1}{4}s^2 G + \tfrac{1}{4}s^2.$$

Now either $EX = G'(1) = 6$ or $EX = \tfrac{1}{2}(1 + EX) + \tfrac{1}{4}(2 + EX) + \tfrac{1}{2}$, whence $EX = 6$. Likewise var $X = 22$.

16. (a) $P(Z_n \leqslant x) \to \Phi(x)$.

(b) Z_n has density $\dfrac{\sqrt{n}}{\lambda} f_n\!\left(\dfrac{\sqrt{n}z + n}{\lambda}\right) = g_n(z)$, say. Then

$$\log g_n(z) = \log\left\{\frac{n^{n-1/2}}{(n-1)!}\right\} - n - z\sqrt{n} + (n-1)\log\left(1 + \frac{z}{\sqrt{n}}\right)$$

$$= \log\left\{\frac{n^{n-1/2}e^{-n}}{(n-1)!}\right\} - \tfrac{1}{2}z^2 + O\!\left(\frac{z}{\sqrt{n}}\right)$$

$$\to -\log(\sqrt{2\pi}) - \tfrac{1}{2}z^2 \text{ as } n \to \infty.$$

17. (a) No, because $EX^{2n} = 0$, which entails $X = 0$.

(b) Yes, provided that $\Sigma p_r = 1$. $P(X = a_r) = p_r$.

18. With an obvious notation,

$$G = E(s^D t^S) = pE(s^D t^S|C) + pqE(s^D t^S|WC) + q^2 E(s^D t^S|WW)$$

$$= pstG + pqs^2 tG + q^2 s^2.$$

Hence

$$G = \frac{q^2 s^2}{1 - pst - pqs^2 t}.$$

Then $G_D(s) = G(s, 1)$ and $G_S(t) = G(1, t)$. Some plodding gives $\text{cov}(D, S) = p(p - q)q^{-4}$.

21. (a) $\dfrac{1}{1 - t^2}, |t| < 1;$ (b) $\dfrac{1}{\cos \frac{1}{2}t}, |t| < \pi.$

(c) Set $e^{-x} = y$ in $M(t) = \int_{-\infty}^{\infty} e^{tx} e^{-x} e^{-e^{-x}} dx$, to obtain $M(t) = \int_0^\infty y^{-t} e^{-y} dy = \Gamma(1 - t)$, where the gamma function is defined by $\Gamma(x) = \int_0^\infty e^{-y} y^{x-1} dy$, for $x > 0$.

22. $G(x, y, z) = \tfrac{1}{4}(xyz + x + y + z)$. Hence

$$G(x, y) = \tfrac{1}{4}(xy + y + x + 1) = \tfrac{1}{2}(1 + x)\tfrac{1}{2}(1 + y) = G(x)G(y)$$

and so on.

23. (a) Let (X_n, Y_n) be the position of the walk after n steps, and let $U_n = X_n + Y_n$. By inspection, U_n performs a simple random walk with $p = q = \tfrac{1}{2}$, so by example 7.7.1 the first result follows.

(b) Let $V_n = X_n - Y_n$. It is easy to show that V_n performs a simple symmetric random walk that is independent of U_n, and hence also independent of T. The result follows from exercise 1(d) at the end of section 7.7.

24. Condition on the first step. This leads to

$$Es^T = (p^2 + q^2)s + \frac{2p^2 q^2 s^2}{1 - (1 - pq)s}.$$

Differentiate the equations, or argue directly, to get $ET = 1 + 2pq ES$ and $ES = 1 + (1 - pq)ES$.

25. $E(s^X|\Lambda) = e^{\Lambda(s-1)}$; $Ee^{t\Lambda} = \mu/(\mu - t)$. Hence

$$Es^X = E\{E(s^X|\Lambda)\} = \frac{\mu}{\mu - (s-1)} = \frac{\mu}{\mu+1} \bigg/ \left(1 - \frac{s}{\mu+1}\right),$$

which is a geometric p.g.f.

26. $G(x, y, z) = \frac{1}{8}(xyz + xy + yz + xz + x + y + z + 1)$

$\qquad = \frac{1}{2}(x+1)\frac{1}{2}(y+1)\frac{1}{2}(z+1) = G(x)G(y)G(z).$

But $G(x, y, z, w) \neq G(x)G(y)G(z)G(w)$.

27. $\int_0^1 G(s)\,ds = \int_0^1 Es^X\,ds = E\int_0^1 s^X\,ds = E\{[s^{X+1}(X+1)^{-1}]_0^1\} = E\{(X+1)^{-1}\}.$

(a) $(1 - e^{-\lambda})/\lambda$;   (b) $-(p/q^2)(q + \log p)$;   (c) $(1 - q^{n+1})/\{(n+1)p\}$;

(d) $-\{1 + (q/p)\log q\}$.

29. (a) There are three configurations of the particles:

A = all at one vertex, B = at two vertices, C = at three vertices.

Let α be the expected time to return to A starting from A, β the expected time to enter A from B, and γ the expected time to enter A from C. Then looking at the first step from each configuration gives $\alpha = 1 + \frac{3}{4}\beta$, $\beta = 1 + \frac{5}{8}\beta + \frac{1}{4}\gamma$, $\gamma = 1 + \frac{1}{4}\gamma + \frac{3}{4}\beta$. Solving these gives $\alpha = 9 = ES$.

(b) In solving the above we found $\gamma = 12 = ER$.

(c) In this case we also identify three configurations:

A = none at original vertex, B = one at original vertex, C = two at original vertex.

Let α be the time to enter the original state from A, and so on. Then $ES = 1 + \alpha$ and looking at the first steps from A, B, and C gives

$$\alpha = 1 + \frac{1}{8}\alpha + \frac{3}{8}\gamma + \frac{3}{8}\beta,$$

$$\beta = 1 + \frac{1}{2}\beta + \frac{1}{4}\alpha + \frac{1}{4}\gamma,$$

$$\gamma = 1 + \frac{1}{2}\beta + \frac{1}{2}\alpha.$$

Then $\alpha = 26$ and $ES = 27$.

Index

Remember to look at the contents for larger topics. Abbreviations used in this index: m.g.f. = moment generating function; p.g.f. = probability generating function.